Collins

AF838851

Cambridge IGCSE™

Combined Science

TEACHER'S GUIDE

Malcolm Bradley, Carol Davenport, Susan Gardner,
Sam Goodman, Sarah Jinks, Sue Kearsey, Jeremy Pollard,
Mike Smith and Chris Sunley

William Collins' dream of knowledge for all began with the publication of his first book in 1819.

A self-educated mill worker, he not only enriched millions of lives, but also founded a flourishing publishing house. Today, staying true to this spirit, Collins books are packed with inspiration, innovation and practical expertise. They place you at the centre of a world of possibility and give you exactly what you need to explore it.

Collins. Freedom to teach

Published by Collins
An imprint of HarperCollins*Publishers*
The News Building, 1 London Bridge Street, London SE1 9GF, UK

HarperCollins*Publishers*
Macken House, 39/40 Mayor Street Upper, Dublin 1, D01 C9W8, Ireland

Browse the complete Collins catalogue at
collins.co.uk

10 9 8 7 6 5 4 3

ISBN 978-0-00-854591-8

British Library Cataloguing-in-Publication Data
A catalogue record for this publication is available from the British Library.

Authors: Malcolm Bradley, Carol Davenport, Susan Gardner, Sam Goodman, Sarah Jinks, Sue Kearsey, Jeremy Pollard, Mike Smith, Chris Sunley
Publisher: Elaine Higgleton
Product manager: Joanna Ramsay
Content editor: Mollie Schofield
Development editor: Gillian Lindsey and Just Content Ltd
Proof reader and answer checker: Just Content Ltd
Cover designer: Gordon MacGilp
Cover illustration: Ann Paganuzzi
Typesetter: Hugh Hillyard-Parker
Production controller: Lyndsey Rogers

Printed and bound in the UK by Ashford Colour Press Ltd

FSC www.fsc.org
MIX
Paper | Supporting responsible forestry
FSC™ C007454

This book contains FSC™ certified paper and other controlled sources to ensure responsible forest management.

For more information visit: www.harpercollins.co.uk/green

This text has not been through the Cambridge International endorsement process.

The publishers gratefully acknowledge the permission granted to reproduce the copyright material in this book. Every effort has been made to trace copyright holders and to obtain their permission for the use of copyright material. The publishers will gladly receive any information enabling them to rectify any error or omission at the first opportunity.

Biology Contents

Supplement content is labelled for students studying the extended syllabus (consisting of all core and supplement content). Content that is not labelled as supplement is core material and for all students.

Chemistry Contents

Physics Contents

Introduction

Welcome to the Collins Cambridge IGCSE™ Combined Science Teacher's Guide, which has been written by experienced authors and teachers to help you deliver an effective and successful Cambridge IGCSE Combined Science course.

Overview

This **printed book** follows the Student Book topics and includes:

- **learning episode plans** with resource lists, learning objectives, and detailed guidance
- **practical activities** including teacher demonstrations and student experiments
- **technician's notes** so **practicals and demonstrations** can be safely planned and executed
- **schemes of work** that show at a glance all the topics linked with their learning outcomes and Cambridge syllabus references

Downloadable files, available at www.collins.co.uk/IGCSEscienceresources:

Everything in the printed book is available in PDF and in MS Word format so that you can tailor lessons and activities. It also includes:

- **worksheets** for students with instructions to follow and questions to answer
- **answers** to all the questions in the Student's Book
- printable **end of topic checklists** that students can use to map their own progress

Learning episode plans

The Teacher's Guide has been matched section by section to the Student Book to allow easy cross referencing between the books. Each section is introduced with a brief introduction, links to other topics and an overview of the learning episodes in an easy-to-navigate table. Learning episode plans give guidance on delivering learning episodes that can be combined for longer lessons to give you complete flexibility and control. Your school may wish to join an organisation such as CLEAPSS for health and safety support www.cleapss.org.uk.

Teachers should make sure that they do not contravene any school, education authority or government regulations. Responsibility for safety matters rests with schools. In addition, the following general advice should be given to students:

'*Be careful with chemicals. Never ingest them and always wash your hands after handling them.*'

Learning episode plans cover the following:

- **learning objectives** from the syllabus linked to the activities and the Student Book
- **learning aims** which reflect the intended outcomes of the consolidation and summary sessions
- **common misconceptions** held by students, which may need addressing
- **resources** you will need for activities
- a detailed pick-up-and-teach approach for **differentiated**, varied tasks
- **demonstration** and **practical activity** technician's notes so you can easily check that you have
- everything you need
- **career links** are included and can be used to enrich teaching by providing context and links to careers in different fields of biology, chemistry and physics. These highlight scientific careers in the areas of these sciences being studied. Careers in many other fields use the knowledge and skills gained studying science.

Worksheets

Worksheets provide engaging activities for students. Those that deal with practical work give step-by step guidance for students to ensure that they carry out their investigations safely and successfully.

All worksheets are available in editable format at *www.collins.co.uk/IGCSEscienceresources*.

Technician's notes

Detailed **technician's notes** are provided for every learning episode, with lists of resources needed for practical activities and/or demonstrations, set-up instructions and safety notes.

For up-to-date safety information, which you will need to check before you set up any practical activity or demonstration, please refer to your employer's guidelines or visit: www.cleapss.org.uk

Technician's notes appear within the learning episode plans for your reference, and as separate printable files to download.

Answers

Answers to all questions provided in the Student's Book are available at www.collins.co.uk/IGCSEscienceresources for easy printing.

Note: practice questions, sample answers and comments have been written by the authors. The marks awarded for these questions indicate the level of detail required in the answers. In examinations, the way marks are awarded may be different. References to assessment and/or assessment preparation are the publisher's interpretation of the syllabus requirements and may not fully reflect the approach of Cambridge Assessment International Education.

Checklists

Editable checklists that students can use to record their progress are provided to download. These have the same content as those at the end of each topic in the Student's Book.

Schemes of work

The detailed **schemes of work** provide a comprehensive overview of the Cambridge IGCSE Combined Science course, matching each learning episode to the learning objectives in the specification.

Science in context boxes

Please note that the **science in context** boxes in the Student Book put the ideas that students are learning into real-life context. It is not necessary for students to learn the content of these boxes as it is beyond the requirements of the syllabus. However, they do provide interesting examples of scientific applications that are designed to enhance their understanding. Some science in context boxes contain one or more questions which provide the opportunity to explore the content more deeply.

B1 Characteristics of living organisms

Contents

Teacher Guide	Syllabus reference
B1 Characteristics of living organisms	B1.1 Characteristics of living organisms

Introduction

This section revises and extends knowledge from previous work on the seven characteristics that are displayed by living organisms. It forms the basis for many other topics covering the separate characteristics in more detail later in the course.

Links to other topics

Topics	Essential background knowledge	Useful links
B2 Cells		B2a Cell structure and size of specimens
B6 Plant nutrition		B6a Photosynthesis
B7 Human nutrition		B7a Diet
B12 Respiration		B12a Respiration
B14 Reproduction		B14a Sexual reproduction in plants B14b Sexual reproduction in humans

Topic overview

B1a	**Characteristics of living organisms** A sorting activity will help students to revise their knowledge of the characteristics of living organisms from previous work.
B1b	**Consolidation and summary** This learning episode quickly recaps the ideas encountered in this section and allows time for students to answer the End of topic questions in the Student Book.

Career Links

Although there aren't any specific careers which study this, all biological careers, such as **botanists**, **zoologists** and **microbiologists**, will draw on this information and it is the basis of study for any biological course.

Starting points

The Student Book section opener puts the ideas in the section into context and sets the scene. It also allows students to acknowledge and value their prior learning, and provides a benchmark against which future learning can be compared.

The questions provide a structure for introducing the section and can be used in a number of different ways:

- You could ask students to consider the questions as an introductory homework task.

- You could put students into groups to share their own ideas and understanding and then to report back to the whole class.
- Students could be given access to the Internet, preferably with a tight timescale, to find out the information required.

You could then use a spider chart or other form of wall chart to summarise everybody's ideas.

Recording these initial ideas allows you to retain them for reference as the individual topics are developed. In this way, your students' progress in learning can be readily acknowledged.

Learning episode B1a Characteristics of living organisms

Learning objectives

- Describe the characteristics of living organisms by defining: movement as an action by an organism or part of an organism causing a change of position or place.
- Describe respiration as the chemical reactions in cells that break down nutrient molecules and release energy for metabolism.
- Describe sensitivity as the ability to detect and respond to changes in the internal or external environment.
- Describe growth as a permanent increase in size and dry mass.
- Describe reproduction as the processes that make more of the same kind of organism.
- Describe excretion as the removal of the waste products of metabolism and substances in excess of requirements.
- Describe nutrition as the taking in of materials for energy, growth, and development.

Common misconceptions

Students are likely to have several misconceptions that will be dealt with in more detail in later topics. For example, they may think that: plants do not respire and only carry out photosynthesis; excretion and egestion mean the same thing; breathing and respiration share the same meaning. Briefly correct any misconceptions where necessary, but return to them later to make sure they are sorted properly when the section is covered in detail.

Resources

Student Book Characteristics of living organisms

Worksheet B1a.1 Characteristics of living organisms

Approach

1. Recap knowledge from previous work

Introduce the topic by giving students two minutes to write down the characteristics shown by all living organisms. Then ask them to compare what they have written with another student to identify the similarities and differences in their answers.

Take examples from around the class to collate a class list of characteristics. Encourage discussion within the class for each new suggestion. Add any suggestions to the list that the students agree on.

2. Characteristics of living organisms

Give students Worksheet B1a.1 to work on either individually, or in pairs, to help them clarify the seven key characteristics of living organisms. They could use the cards to create their own list of characteristics, or you could give them the characteristics and ask them to identify which is shown in each example. Students could add one more plant and one more animal example of their own for each characteristic.

3. Create a mnemonic (optional)

Ask students to work on their own or in pairs to create a mnemonic (word made up of a series of letters to help you to remember something) for the initials of the seven characteristics, in whichever order they like.

4. Consolidation: Create a crossword

Ask students to create the clues for a crossword that includes all seven characteristics of living organisms.

Learning episode B1b Consolidation and summary

Learning aims

- Review the learning points of the topic summarised in the End of topic checklist
- Test understanding of the topic content by answering the End of topic questions

Resources

Student Book End of topic checklist and End of topic questions

Approach

Introduce the learning episode

Ask students to work with a partner to make a list of key words from this topic. They could then work together to produce a spider diagram showing how the different concepts are linked. They could compare their list with the list of key terms given at the end of the topic in the Student Book. Discuss the checklist at the end of the topic and use questioning to see how much of the content they are comfortable with.

Students could make flashcards of the key content and then use the flashcards to quiz each other on the information.

Develop the learning episode

Ask students to work individually through the End of topic questions in the Student Book without looking at the text. As they work, walk around the classroom observing their answers and asking questions as necessary to find out which questions are causing difficulties.

Finish the learning episode

After a set period, ask the students to stop working and discuss any areas of difficulty you observed as you walked round the class. Students should complete any unanswered questions for homework, but you should stress that they should try to answer the questions without looking at the text, so that they can see how much they have remembered.

B2 Cells

Contents

Teacher Guide	Syllabus reference
B2 Cells	B2.1 Cell structure B2.2 Size of specimens

Introduction

In this section, students will learn to recognise the structures found in cells, and relate the structures that they see in plant and animal cells to their functions. In addition, students will learn how to calculate the magnification of an object under a microscope.

This section revises another fundamental building block of knowledge that students should refer to in later work in other sections. The specialised cells mentioned in the Supplement section: root hair cells in B3.3: *Active transport* (Supplement) and B8.2: *Water uptake*, palisade mesophyll cells in B6.2, neurones in B13.1 and B13.3 (Supplement), red blood cells in B9.4, and sperm and egg cells in B15.4. White blood cells will also be referred to in B10.1 *Diseases and immunity.* It revisits and extends knowledge from earlier work on the organisation of cells into tissues, organs and organ systems.

Links to other topics

Topics	Essential background knowledge	Useful links
B3 Movement into and out of cells		B3a Diffusion B3b Osmosis B3c Active transport
B6 Plant nutrition		B6a Photosynthesis B6b Leaf structure
B8 Transport in plants		B8b Water uptake
B9 Transport in animals		B9b Blood vessels and blood
B10 Disease and immunity		B10a Disease, defence and hygiene
B12 Respiration		B12a Respiration
B14 Reproduction		B14a Sexual reproduction in plants B14b Sexual reproduction in humans

Topic overview

B2a	**Cell structure and size of specimens** This learning episode covers section B2.1: *Cell structure* and section B2.2: *Size of specimens* from the syllabus. The learning episode gives students the opportunity to develop their microscope skills while studying the similarities and differences between animal cells and plant cells. They also find out about the structure of bacterial cells. Students discover a range of cells, tissues, organs and organ systems and how they work together to carry out their functions in organisms. Supplement Students will study three examples of specialised cells and will convert measurements between millimetres and micrometres.

B2b	**Consolidation and summary** This learning episode quickly recaps on the ideas encountered in this section and allows time for students to answer the End of topic questions in the Student Book.

Careers links

These are some scientific careers that focus on this area of biology but careers in many other fields use the knowledge and skills gained studying science. **Histopathologists** diagnose and study diseases in tissues and organs, using microscopes to look at cell samples taken from patients. **Vascular scientists** and technicians work with doctors to assist with the diagnosis of disorders of the circulatory system.

Starting points

The Student Book section opener puts the ideas in the section into context and sets the scene. It also allows students to acknowledge and value their prior learning, and provides a benchmark against which future learning can be compared.

The questions provide a structure for introducing the section and can be used in a number of different ways:

- You could ask students to consider the questions as an introductory homework task.
- You could put students into groups to share their own ideas and understanding and then to report back to the whole class.
- Students could be given access to the Internet, preferably with a tight timescale, to find out the information required.

You could then use a spider chart or other form of wall chart to summarise everybody's ideas.

Recording these initial ideas allows you to retain them for reference as the individual topics are developed. In this way, your students' progress in learning can be readily acknowledged.

Learning episode B2a Cell structure and size of specimens

Learning objectives

- Describe and compare the structure of a plant cell with an animal cell, limited to: cell wall, cell membrane, nucleus, cytoplasm, chloroplasts, ribosomes, mitochondria, vacuoles.
- Describe the structure of a bacterial cell, limited to: cell wall, cell membrane, cytoplasm, ribosomes, circular DNA, plasmids.
- Identify the cell structures listed above in diagrams and images of plant, animal and bacterial cells.
- Describe the functions of the structures listed above in plant and animal cells.
- Supplement State that specialised cells have specific functions, limited to: root hair cells – absorption; palisade mesophyll cells – photosynthesis; red blood cells – transport of oxygen.
- Describe the meaning of the terms: cell, tissue, organ, organ system, and organism, as illustrated by examples given in the syllabus.
- State and use the formula: magnification = image size ÷ actual size.
- Calculate magnification and size of biological specimens using millimetres as units.
- Supplement Convert measurements between millimetres (mm) and micrometres (μm).

Common misconceptions

Some students may find it hard to grasp the tiny scale of cells and their components. It may help if they look at a piece of small newsprint or a miniature letter drawn with a sharp pencil under the microscope before viewing cells.

Resources

Student Book Cell structure, Size of specimens

Worksheet B2a.1 Using a light microscope

Worksheet B2a.2 Preparing a microscope slide

Worksheet B2a.3 Organs of the human body

Resources for class practicals (see Technician's notes)

Approach

1. Introducing cells

Ask students to remember what they can about cells, and then ask one student to come to the board and sketch the outline of one cell. Ask other students to suggest other features that could be included, but encourage discussion first on whether the suggested feature is found in *all* cells and is therefore worthy of being included in the sketch. All features drawn should be labelled.

Make sure that the sketch includes the cell membrane, cytoplasm and nucleus containing DNA. Students should draw a fully labelled sketch of a cell with these features in their workbooks.

Take the opportunity to briefly revise SI units of length (based on the metre), and ask which unit might be most appropriate for measuring the length of a cell. Students should be able to respond with *millimetre*. Make sure they understand the relationship between millimetres and metres.

Supplement Ask students which unit would be more useful for measuring structures inside cells, such as the nucleus or even smaller structures. Introduce the *micrometre* and its symbol, μm, and its relationship to millimetres and metres (1000 micrometres = 1 millimetre, 1 micrometre = 1×10^{-6} m).

2. Using a light microscope

If your students are not already familiar with using a light microscope, this activity is a good opportunity to introduce it. Worksheet B2a.1 provides basic instructions on viewing a prepared slide through a light microscope. It might help to demonstrate how to set up and focus, first using the low-power objective lens and then the high-power objective lens, to avoid the risk of crashing high-power objectives into slides (see Technician's notes). Students may need reminding of this several times in order to develop good technique.

SAFETY INFORMATION: Remind students always to focus first using the low-power objective, starting with it near to the slide, then moving it upwards to avoid crashing the lens into the slide. Also remind them to use the fine-focusing knob only while working with the high-power objective lens. If using natural light to see the specimen, students should not focus the sun's rays directly through the objective and eyepiece.

Any prepared slide of plant or animal tissue where it is possible to see a simple internal structure in cells will be suitable. Ideally you should show a labelled diagram or photograph on the board, to help students interpret what they see. Do not show tissues with complex structures, such as muscle cells, as this may be confusing.

Students should be expected to draw clearly one or two labelled cells from each slide. Remind them to use a sharp pencil for their drawings, and to draw only what they see even if it is not as clear and obvious as any drawing they have seen elsewhere. Students should not use shading in their drawings.

Step 11 on the worksheet involves students calculating the magnification of their drawings. Some students may need help with this. Use the text and worked example in the section 'Size of specimens' in the Student Book to introduce the formula for magnification: magnification = image size ÷ actual size. Students could also view a transparent plastic ruler or a simple graticule under the microscope to calculate the field of view, and use that to estimate the size of the cell they have drawn.

Supplement Expect students to use units of micrometres in their answers.

3. Preparing a microscope slide

Worksheet B2a.2 provides instructions for making stained slides of plant cells (onion) and animal cells (human cheek cells). You may need to demonstrate how to place a coverslip on a slide to avoid trapping air bubbles (see Technician's notes).

Students often over- or under-stain material, so they may need more than one attempt at making a slide of each kind. If time is limited, have slides that you prepared earlier ready for display to the class in case they are unsuccessful.

This practical could be extended to look at cells from other sources, such as banana parenchyma or moss leaf. In each case, students should apply their skills and understanding of cell structure to interpret what they see under the microscope.

SAFETY INFORMATION: Before working with their own cheek cells, remind students of safety protocols and make sure that pots of disinfectant are available for discarded cotton buds, slides and coverslips. Demonstrate to students how to use a scalpel/knife safely (see Technician's notes). Check your school guidelines as using cheek cells may be prohibited. Animal liver cells could be used as an alternative or follow local advice.

While working with the high-power objective lens, use the fine-focusing knob only. If using natural light to see the specimen, remind students to make sure they do not focus the sun's rays directly through the objective and eyepiece. Handle coverslips with care – they are fragile and very sharp if broken. Students should only handle slides of their own cheek cells, and should place slides into disinfectant when they have finished. Remind students to take care when using a scalpel or knife.

4. Functions of animal and plant cell structures

Use the 'Cell structure – Animal and plant cells' and 'Functions of cell structures' sections of the Student Book to introduce animal and plant cell structures and functions. Students should make notes and use these to test each other on the functions of the cell wall, cell membrane, nucleus, cytoplasm, chloroplasts, ribosomes, mitochondria and vacuole.

5. Bacterial cell structures

Use the 'Bacterial cells' section of the Student Book to introduce bacterial cell structures. Students should make notes and use these to test each other on which cell structures are present in all of bacterial cells, animal cells and pants cells, and which are only present in bacterial cells, or only present in plant cells.

6. Cell organisation

On the board, sketch a table like the one shown below. Make sure it has at least four rows.

Organ system	Organ	Tissues	Cells

Ask students to name organ systems found in the human body, and add each name to a different row in the first column. They should then suggest the names of any organs in each system, tissues in each organ and cells in each tissue.

Alternatively, students could copy the table outline and complete it as far as possible individually or in pairs. Take examples from around the class and check that there are no errors. Keep the table for the end of this learning episode, so that students can add further examples from what they have learned.

7. Organs of the human body

Worksheet B2a.3 provides a human body outline on which students can draw the positions of some organs. They should include annotated labels to explain what each organ does.

Supplement **8. Cell fact cards**

Ask students to carry out research on the cell (or cells) of their choice to produce a 'cell fact file'. They should choose any cell found in animals or plants, such as root hair cells, palisade mesophyll cells, or red blood cells. They should draw a large, labelled diagram describing the cell structure and explaining how this is related to its function.

They can present their research in a poster or using presentation software.

9. Consolidation: chain reaction quiz

Give students a few minutes to write down as many quick quiz questions on cell structures and their functions as they can. They should also jot down the answers. Ask one student to read out a question, and choose another to answer it. If they answer correctly, they then read out one of their own questions. If they do not answer correctly, choose another student to answer. Continue round the class until most cell structures and functions have been covered.

Technician's notes

Be sure to check the latest safety notes on these resources before proceeding.

The following resources are needed for the class practical, B2a.1, per student or pair:

light microscope with low-power (e.g., ×4) and high-power (e.g., ×10) objective lenses
prepared slides of plant and/or animal tissue in which cell structures are easily visible (suitable examples are palisade cells in a leaf and squamous epithelial cells (cheek cells)) Check your school guidelines as using cheek cells may be prohibited. Animal liver cells could be used as an alternative or follow local advice.

The following resources are needed for the class practical, B2a.2, per student or pair:

eye protection
light microscope with low-power (e.g., ×4) and high-power (e.g., ×10) objective lenses, microscope slides and coverslips
mounted needle or blunt seeker
forceps
pipette
methylene blue stain
iodine solution – iodine in potassium iodide solution
onion – fewer onions are needed if a layer is peeled off for each student or pair
sharp knife or scalpel and cutting board
clean cotton buds, at least 1 per student; paper towel
pot of disinfectant, e.g., 1% Virkon
transparent plastic ruler or simple graticule
water

SAFETY INFORMATION: Cotton buds should be kept in disinfectant for long enough for them to be sterilised after students have used them. Please refer to the manufacturer's instructions. Scalpel blades break if they are used to cut whole onions, causing a risk of injury. Knives are more appropriate for cutting up onions, or students could be supplied with sections of onions from which to prepare slides.

Check your school guidelines as using cheek cells may be prohibited. Animal liver cells could be used as an alternative or follow local advice.

Learning episode B2b Consolidation and summary

Learning aims

- Review the learning points of the topic summarised in the End of topic checklist
- Test understanding of the topic content by answering the End of topic questions

Resources

Student Book End of topic checklist and End of topic questions

Approach

Introduce the learning episode

Ask students to work with a partner to make a list of key words from this topic. They could then work together to produce a spider diagram showing how the different concepts are linked. They could compare their list with the list of key terms given at the end of the topic in the Student Book. Discuss the checklist at the end of the topic and use questioning to see how much of the content they are comfortable with.

Students could make flashcards of the key content and then use the flashcards to quiz each other on the information.

Develop the learning episode

Ask students to work individually through the End of topic questions in the Student Book without looking at the text. As they work, walk around the classroom observing their answers and asking questions as necessary to find out which questions are causing difficulties.

Finish the learning episode

After a set period, ask the students to stop working and discuss any areas of difficulty you observed as you walked round the class. Students should complete any unanswered questions for homework, but you should stress that they should try to answer the questions without looking at the text, so that they can see how much they have remembered.

B3 Movement into and out of cells

Contents

Teacher Guide	Syllabus reference
B3 Movement into and out of cells	B3.1 Diffusion B3.2 Osmosis B3.3 Active transport

Introduction

This section covers the topics of diffusion, osmosis and active transport to help students understand how substances can move into and out of cells. Students will need to apply this knowledge in later topics in relation to processes such as gas exchange and transport. For students following the supplementary syllabus there is a strong connection between the processes learned in this section and the uptake of water in root hair cells.

Note that a number of practical activities are suggested for the activities on diffusion and osmosis. You may wish to select what can be fitted into your lessons – this need only be one or two practical activities.

Links to other topics

Topics	**Essential background knowledge**	**Useful links**
B1 Characteristics of living organisms	B1a Characteristics of living organisms	
B2 Cells	B2a Cell structure and size of specimens	
B4 Biological molecules		B4b Food tests
B6 Plant nutrition		B6a Photosynthesis B6b Leaf structure
B8 Transport in plants		B8b Water uptake B8c Transpiration
B9 Transport in animals		B9b Blood vessels and blood
B11 Gas exchange in humans		B11a Gas exchange in humans
B12 Respiration		B12a Respiration

Topic overview

B3a	**Diffusion** This learning episode introduces students to the concept of diffusion in general, and later to diffusion across membranes. Supplement Students will investigate factors that affect the rate of diffusion.

B3b	**Osmosis** This learning episode provides students with a range of opportunities to investigate osmosis across partially permeable membranes. Supplement Students are given the opportunity to link their learning on osmosis to water potential, and are expected to interpret the effect of solutions on living cells in terms of water potential and turgor pressure.
B3c	Supplement **Active transport** This short learning episode helps students to distinguish active transport from the passive processes of diffusion and osmosis. Students are given the opportunity to discuss the process of active transport in particular situations.
B3d	**Consolidation and summary** This learning episode quickly recaps on the ideas encountered in this section. Students answer the End of topic questions in the Student Book.

Careers links

These are some scientific careers that focus on this area of biology but careers in many other fields use the knowledge and skills gained studying science. Dialysis therapy requires collaboration between many different people. For example, **biomedical engineers** design the dialysis machines and software to run them, **clinical engineers** manage the day-to-day operation of the machines, and **renal nurses** look after patients undergoing dialysis.

Starting points

The Student Book section opener puts the ideas in the section into context and sets the scene. It also allows students to acknowledge and value their prior learning, and provides a benchmark against which future learning can be compared.

The questions provide a structure for introducing the section and can be used in a number of different ways:

- You could ask students to consider the questions as an introductory homework task.
- You could put students into groups to share their own ideas and understanding and then to report back to the whole class.
- Students could be given access to the Internet, preferably with a tight timescale, to find out the information required.

You could then use a spider chart or other form of wall chart to summarise everybody's ideas.

Recording these initial ideas allows you to retain them for reference as the individual topics are developed. In this way, your students' progress in learning can be readily acknowledged.

Learning episode B3a Diffusion

Learning objectives

- Describe diffusion as the net movement of particles from a region of their higher concentration to a region of their lower concentration (i.e., down a concentration gradient), as a result of their random movement.
- State that some substances move into and out of cells by diffusion through the cell membrane.
- Describe the importance of diffusion of gases and solutes in living organisms.
- Supplement Investigate the factors that influence diffusion, limited to: surface area, temperature, concentration gradient, and distance

Common misconceptions

Many students confuse the term *diffusion* with simple movement and think that once diffusion has stopped, the particles have stopped moving too. This can lead to difficulties in understanding the process. It is worth reinforcing the point that diffusion is *net movement*, by repeating this definition every time you use the word 'diffusion' early on in this work.

A video clip or some kinaesthetic work (with the students acting as particles) can also help, as suggested below.

Resources

Student Book Diffusion

Worksheet B3a.1 Diffusion across visking tubing (dialysis tubing)

Worksheet B3a.2 Temperature and the rate of diffusion

Worksheet B3a.3 Surface area and the rate of diffusion

Resources for class practicals (see Technician's notes)

Approach

The practical work on diffusion in Worksheets B3a.1 and B3a.2 uses identification tests for glucose and starch. These are not covered by the syllabus until Section B4 in relation to food tests. If you wish to use Worksheets B3a.1 and B3a.2 to focus on the general factors that affect diffusion, you will need to briefly explain the tests for glucose and starch beforehand. Details of these are given on Worksheet B4b.1.

1. Introduction to diffusion: demonstration

Place a crystal of potassium manganate(VII) (potassium permanganate) into a beaker of water and ask students to watch and describe what happens. Or put a slice of coloured root vegetable that will leach colour easily, such as beetroot, into a beaker of water and ask students to watch and describe what happens (see Technician's notes).

SAFETY INFORMATION: Potassium manganate(VII) is harmful and should be handled with tongs or disposable gloves. Dispose of the solution carefully to avoid splashes, as the colour stains permanently.

Alternatively, spray a small amount of strong perfume at one side of the room and ask students to put their hand up when they smell the perfume.

Ask them to think at the level of particles and to try and imagine what is happening in the beaker or room. If possible, show them a video of Brownian motion. You will need to explain that the dots they see are not molecules, but larger particles that are being constantly jiggled by the movement of molecules that are too small to see.

Although not required, you could explain that particles are always in a state of movement as a result of the energy they have – the greater the temperature, the more energy they have and so the more they move (this should reinforce learning in physics).

Ask students to consider the effect of distance on diffusion by asking how quickly particles reach a specified distance. Extrapolate this to the importance of distance for diffusion of substances within an organism. Students should appreciate that the shorter the diffusion distance, the more rapidly particles will reach their destination.

(Note that the examples given may also involve mass flow of gas or solution, e.g., by convection, more than diffusion, because diffusion is only effective quickly over a very short distance. However, the examples do help students to visualise that particles are always in constant movement, and so helps with understanding of diffusion.)

2. Moving as particles

You can model the effect of the colour, or perfume, spreading using students as particles. Choose a small number of sensible students who are wearing similar coloured tops (or provide hats or something to distinguish them from other students). Within a fairly limited area, ask students to walk in straight lines, and when they come up against another student or a boundary edge, to change direction like billiard balls. After a few minutes, ask your key group of students to raise their hands, so that everyone can see how they have become spread out.

Give students a few minutes to work in pairs or small groups to find an explanation for the movement of colour through the solution, or perfume, across the room. Then ask for a selection of answers from different groups and discuss any differences. The key point is that the movement out from the area of high concentration is a result of the random movement of the particles.

It is also worth asking whether any particles of colour/perfume are moving back towards the area of high concentration. As a result of random motion, some will be. However, there will be more particles moving from the area of higher concentration to the area of lower concentration, and this is what we notice.

Then ask, when the colour is evenly distributed in the beaker, or the perfume in the air within the room, what happens to the particles. It is essential students understand that the movement continues but that, as there is no difference in concentration between areas, we cannot perceive any change.

Introduce the idea of *net movement*, that is the sum of movements in different directions, and go on to use this to define the term *diffusion*.

3. Diffusion and cell membranes

In biology we are most interested in the effect of diffusion as a mechanism for getting substances into and out of cells. The practical work for this is covered in Worksheet B3a.1 in relation to diffusion in general.

Experiments using visking tubing (dialysis tubing) are not required by the syllabus, but can be used as a model for a cell membrane, and the practical described in the worksheet acts as a good model for understanding that some molecules (e.g., glucose) can cross cell membranes easily, and therefore can be redistributed in the system as a result of diffusion, whereas others (e.g., starch) cannot. Students often find visking tubing difficult to use. If you prefer, you could carry out the practical as a demonstration, asking students to predict what they think will happen and to explain their predictions using what they know about diffusion and partially permeable membranes (see Technician's notes).

SAFETY INFORMATION: Demonstrate how to connect the visking tubing (dialysis tubing) to the cut-off syringe. Students should wear eye protection when using iodine solution and Benedict's solution. Remind students to take care when heating solutions in the water bath for the glucose test (see Technician's notes).

4. Supplement Factors affecting the rate of diffusion

Students need to understand the main factors that affect the rate of diffusion of substances. They will need to apply this information later to their understanding of processes such as gas exchange.

Distance: The effect of distance on diffusion was covered in the introductory task. This effect will be covered again in relation to the diffusion of gases between cells, the blood and the lungs in Topic B9b: *Blood vessels and blood*, and Topic B11a: *Gas exchange system.*

Temperature*:* Worksheet B3a.2 gives students the opportunity to plan an experiment into the effect of temperature on the rate of diffusion using a visking tubing (dialysis tubing) system. Discuss the prediction with them, and if they seem unsure, remind them about the effect of increasing temperature on the energy of particles. If there is time, and their plan is acceptable, students could carry out their plan.

Encourage students to carry out two or more repeats themselves. Please note that if students do carry out two repeats at three different temperatures, several hours will be needed for the practical. If there is a limited amount of time available for the practical, students could pool their results, instead of doing the repeats. If you decide to do this, draw students' attention to the possibility of bringing in uncertainties by sharing results with neighbours.

Surface area*:* Worksheet B3a.3 gives students the opportunity to investigate the rate of diffusion into visking tubing (dialysis tubing) of different lengths, to explore the effect of surface area on the rate of diffusion. Technically, what is happening in this experiment is osmosis, as the substance that crosses into the tubing through the membrane is water. As osmosis is not covered until the next Learning episode, it may be best to ignore this point for now. However, the model is useful as it can be applied to any substance that diffuses through a partially permeable membrane.

Students should find that the tube of longest length absorbs most water because it has the largest surface area over which diffusion can take place. More able students could calculate the surface area of each piece of tubing and use this in their analysis, but using length as a proxy for surface area is sufficient as long as students are aware of its relationship to surface area.

For both these worksheets, check that students have drawn up a suitable table for their results before they carry out the tests.

Concentration gradient: For the effect of concentration gradient on the rate of diffusion, ask students to draw a cartoon, or develop a model that helps to explain why the rate of diffusion is greater when the concentration gradient is greater. This will provide an opportunity to reinforce what they have learned about diffusion generally. Take time to check students' understanding of the basic principles at this point, as this will underpin a lot of later work.

SAFETY INFORMATION: Students should wear eye protection when using iodine solution and Benedict's solution. Remind students to take care when heating solutions in a hot water bath.

Technician's notes

Be sure to check the latest safety notes on these resources before proceeding.

The following resources are needed for the demonstration in the introduction:

250 cm^3 beaker of water
crystal of potassium manganate(VII) *or* slice of a root vegetable that leaches colour easily, such as beetroot

The following resources are needed for the class practical, B3a.1, per group:

5 cm^3 10% glucose solution (100 g dm^{-3}) and 5 cm^3 10% starch solution
15 cm length of visking tubing (dialysis tubing) and elastic band, to attach visking tubing to syringe end
sawn-off end of 10 cm^3 plastic syringe (the plunger end)
100 cm^3 beaker
tap water
iodine solution in dropper bottle
spotting tile
Benedict's solution + water bath at 80 °C and heatproof tongs
boiling tube
teat pipette
eye protection

In this practical, students investigate the diffusion of substances across a partially permeable membrane using visking tubing (dialysis tubing). The visking tubing could be prepared ahead of the lesson, as students may have difficulty with it.

Students should find that glucose diffuses through the tubing into the water, but not starch. This is because the glucose molecules are small enough to pass through the holes in the membrane, and are therefore free to diffuse from the region of higher concentration (inside the tubing) to the area of lower concentration (water outside the tubing). The starch molecules are too large to move through the holes in the membrane, so are not subject to diffusion.

Note that dry visking tubing (dialysis tubing) needs to be soaked for a few minutes in water before use. This can be done ahead of the lesson.

The following resources are needed for the class practical B3a.2, per group:

5 cm^3 10% glucose solution (100 g dm^{-3})
15 cm length of visking tubing (dialysis tubing)
sawn-off end of 10 cm^3 plastic syringe (the plunger end)
elastic band, to attach visking tubing to syringe end
100 cm^3 small beaker
tap water
Benedict's solution + water bath at 80 °C and heatproof tongs
boiling tube
teat pipette
eye protection

This practical can be used as a planning exercise or, if there is time, students may be allowed to carry out their plans after approval by you. The method is based on the one used on Worksheet B31a. If students have not had the chance to carry that out, they should be given the worksheet to help them develop their planned method.

If students carry out their investigation, they should find that the higher the temperature, the faster the rate of diffusion (the sooner that glucose is detected in the water). This is because at a higher temperature a molecule has more kinetic energy and so moves faster, increasing the chances of it reaching the membrane and passing through one of the holes to the other side.

The equipment list details what students will need. They may suggest other equipment in their plans. Plans should be checked for the equipment suggested, and alternatives offered if some suggestions are not available.

The following resources are needed for the class practical B3a.3, per group:

3 × 12cm lengths of visking tubing (dialysis tubing)
15 ml sugar solution, 5% (use sucrose, not glucose and use fresh solution)
ruler
3 paper clips
beaker containing about 150 ml of distilled water
dropping pipette
tongs
access to an electronic balance
clock or watch
3 sheets of paper towel, for removing excess water
eye protection

This practical investigates the rate of diffusion of water (osmosis, but see point in notes above) using visking tubing (dialysis tubing) to represent cells of different surface area.

Visking tubing should be wet to soften it. Students may find it difficult to tie a knot in the visking tubing. The tubing could be provided with a knot already tied at one end.

Learning episode B3b Osmosis

Learning objectives

- State that water diffuses through partially permeable membranes by osmosis.
- State that water moves into and out of cells by osmosis through the cell membrane.
- Investigate and describe the effects on plant tissues of immersing them in solutions of different concentrations.
- Supplement Describe osmosis as the net movement of water molecules from a region of higher water potential (dilute solution) to a region of lower water potential (concentrated solution), through a partially permeable membrane.
- Supplement Explain the effects on plant cells of immersing them in solutions of different concentrations by using the terms: turgid, turgor pressure, plasmolysis, flaccid.

Common misconceptions

As with diffusion, the important concept is *net movement*, not simply the movement of the water particles.

This section can be very confusing because we usually talk in terms of the concentration of *solute* molecules in a solution, not the concentration of *water* molecules. If you talk in terms of the concentration of water molecules, then it should be clear to students that osmosis is just diffusion in terms of water molecules. This is particularly important in living systems, as water is essential for life – hence it is given a special name.

Expect students to be precise in their definition of which concentration they are talking about, water or solute, in their discussions and writing.

Resources

Student Book Osmosis

Worksheet B3b.1 Osmosis and visking tubing (dialysis tubing)

Worksheet B3b.2 Measuring osmosis by change in mass

Worksheet B3b.3 Osmosis and change in shape

Supplement Worksheet B3b.4 Water potential

Resources for class practicals (see Technician's notes)

Approach

1. Introduction

Show students two potted plants, one that has been well-watered and one that is wilting. (Alternatively show pictures of a well-watered plant and a wilted plant.) Give students a short while to suggest why the plants are different, and to suggest how the wilted plant could recover. Take examples from around the class to get answers related to watering. Demonstrate recovery of the wilted plant by adding water and comparing the difference at the end of the lesson (see Technician's notes).

2. Osmosis

Prepare two eggs before the lesson by placing the unboiled eggs in vinegar (or other dilute acid) for 3–4 days to dissolve the shell. Then place each egg in a conical flask and cover one egg with a concentrated sugar solution and the other with water. After 24 hours, this should result in one shrunken egg that can be removed easily from the flask, and one large 'full' egg that cannot be extracted easily from the flask.

Show these to students and explain that the egg is like a large cell, with the membrane surrounding the egg acting as a cell membrane. Ask them to suggest how you produced the difference between the eggs.

Alternatively, show them some photos of plasmolysed (shrunken) and turgid (full) plant cells and ask them to describe and explain any differences in the movement of substances in the cells.

If necessary, remind students that water molecules are small enough to pass through cell membranes. Introduce the term *osmosis* as the special case of diffusion that considers the movement of water molecules.

The investigation in Worksheet B3b.1 could be used as a demonstration, or class practical, to help students understand the effect of partially permeable membranes on osmosis and help them to visualise what effect osmosis can have on living cells (see Technician's notes). Visking tubing is no longer included within the syllabus, but it is still a useful way to demonstrate osmosis.

SAFETY INFORMATION: Demonstrate how to connect the visking tubing (dialysis tubing) to the glass tubing (see Technician's notes).

Supplement **3. Defining osmosis**

Using the definition of diffusion that they learned earlier, ask students to produce a definition for the term *osmosis*. This should raise the opportunity to discuss the need to distinguish between the concentration of solute molecules and concentration of water molecules in a solution when talking or writing about osmosis.

4. Factors affecting osmosis

Worksheets B3b.2 and B3b.3 provide two opportunities for students to investigate the effect of concentration of solution on osmosis in living systems. Worksheet B3b.2 uses a straightforward method, and uses change in mass to measure the gain or loss of water by cells. Worksheet B3b.3 is more challenging and uses the change in curvature of pieces of plant stem as a result of differential absorption of water by inner and outer cell layers. It also gives students the opportunity to produce a dilution sequence of solutes for the investigation.

For Worksheet B3b.3, the stems need to be left for at least two hours, so it may be more appropriate to leave them until the next lesson.

If necessary, discuss with students how they are going to record their results in a table and what is the most suitable graph for them to draw. Worksheet B3b.2 also asks students to consider what errors and uncertainties there may be in their results.

Supplement Give students opportunities to apply their knowledge of the effect of temperature and surface area on diffusion to their understanding of osmosis. This could be done by simple questioning, or by asking students to draw, or describe, what happens when cells are placed in solutions of different temperature, or the problems for larger organisms that need to take in water.

SAFETY INFORMATION: Students should take care with sharp knives. Students should wash their hands after handling plant material.

5 Consolidation

Return to the potted plants shown in the introduction and ask students to describe the change in the wilted plant. (Over a period of about 1 hour, it should show significant recovery.) Ask students to sketch a cell from the plant at the start of the lesson and the same cell at the end of the lesson. They should annotate their sketches to show what has changed.

Supplement Expect students to include the terms *turgor pressure*, *water potential*, *plasmolysis* or *flaccid* in their annotations.

Technician's notes

Be sure to check the latest safety notes on these resources before proceeding.

The following resources are needed for the demonstrations in tasks 1 and 2:

two similar potted plants, one well-watered and the other unwatered for long enough to be wilted but not dead (soft-leaved and soft-stemmed plants are best, e.g., *Impatiens* [busy Lizzie])

two unboiled eggs placed in vinegar (or other dilute acid) at least 4 days before the lesson and left for 3–4 days to dissolve the shell, then 24 hours before the lesson, place each egg in a conical flask and cover one egg with a concentrated sugar solution and the other with water alternatively, photomicrographs of plasmolysed and turgid plant cells

The following resources are needed for the class practical B3b.1, per group:

plastic/glass tube about 10 cm long, fairly wide bore so that the solution can be pipetted into it
15 cm length of visking tubing (dialysis tubing)
10% starch solution or other solution of substance that will not diffuse through the visking tubing
water
100 cm^3 beaker or boiling tube
clamp + stand
wire or elastic band to attach tubing to plastic/glass tube
teat pipette
marker pen

This investigation tests understanding of osmosis. It can be carried out as a class demonstration, or as a practical by students working in groups (see Technician's notes). The set-up for the apparatus is shown on the worksheet. The key point is to get students to predict what will happen to the level of liquid in the plastic/glass tube after a few hours, and to explain their prediction.

Plastic tubing should be used, if possible, as there is less danger of it shattering when students attach the visking tubing (dialysis tubing) to it.

The equipment will need to be set up and left for a few hours, or preferably overnight, to show a significant change in liquid level.

After this time, the level of liquid should have risen in the glass tube. This is because the concentration of water molecules in the starch solution inside the visking tubing is lower than in the water in the beaker, and the starch molecules do not diffuse through the membrane. The net movement of water molecules (osmosis) will be into the visking tubing, increasing the volume of the solution, so liquid level rises.

The following resources are needed for the class practical B3b.2, per group:

5 boiling tubes + rack, 100 cm^3 beaker
10 cm^3 coloured fruit juice or fruit cordial
distilled water
glass rod
paper towels
fresh root vegetable, for example a potato (see note below)
cork borer
sharp knife
weighing balance (must be capable of weighing to 0.001 g)
10 cm^3 measuring cylinder
marker pen

This investigation uses change in mass of tissue samples to measure the effect on osmosis of solutions of different concentration. Any suitable plant tissue can be used, but root vegetables are particularly useful due to their volume. Provide whole vegetables, from which students can cut their own samples. Alternatively take cores from the vegetables as close to the lesson as possible and store them in water.

Any suitable solution can be used, but choose one that is cheap and safe to handle. The worksheet suggests the use of a coloured fruit cordial, such as blackcurrant, as students will be able to see differences in concentration and are less likely to muddle tubes.

The worksheet expects students to carry out the dilutions themselves. Alternatively, these could be prepared ahead of the lesson to save time.

The discs will need to be left in solution for at least two hours to get significant changes in mass. Some students may need help with the calculations in the first step of the analysis on the worksheet.

Students should find that discs gain mass or lose mass (water) depending on the extent to which the concentration of cell cytoplasm differs from the solution in which they are placed. Mass loss is a result of the movement of water out of the cell when placed in more concentrated solutions, and mass gain is the result of the movement of water into the cell when placed in more dilute solutions.

The following resources are needed for the class practical B3b.3, per group:

3 boiling tubes and rack
10 cm^3 0.5 mol dm^{-3} sodium chloride solution
distilled water
glass rod
3 Petri dishes
fresh hollow plant stalks, e.g., dandelion, rhubarb
sharp knife or scalpel
marker pen

This investigation links to the Developing Practical Skills box in the Student Book. It is similar to the previous investigation, but uses a different method of changing stem curvature to show the effects of osmosis. It also gives students an opportunity to create and calculate a dilution sequence.

If dandelions are not available, any hollow-stalked plant with soft stalks may be suitable. It is advisable to test other plants with a trial run to make sure that they produce measurable differences in the solution concentrations offered.

The plant stalks must be fresh, or stored in water or a dilute salt solution until use to prevent them drying out.

Students should find that the stalks change in curvature as a result of osmosis, and because the outer cells of the stalk cannot absorb water whereas the inner cells can. Fig. B3.10 in the Student Book shows results from an experiment like this.

Stalks in solutions that are more dilute (have a higher concentration of water molecules) than the cytoplasm in the cells will curve with the outer layer on the inside of the curve and inner layer outside. Stalks in solutions that are more concentrated (have a lower concentration of water molecules) than the cytoplasm in the cells will curve with the inner layer on the inside of the curve and outer layer outside.

The greater the difference from the concentration of cell cytoplasm, the greater the curvature (although the pressure caused by the cell wall will prevent the curvature in distilled water going beyond a certain point).

Supplement Learning episode B3c Active transport

Learning objectives

- Supplement Describe active transport as the movement of particles through a cell membrane from a region of lower concentration to a region of higher concentration (i.e., against a concentration gradient), using energy from respiration.
- Supplement Explain the importance of active transport as a process for movement of molecules or ions across membranes, including ion uptake by root hairs.

Common misconceptions

This is an opportunity to reinforce the fact that diffusion and osmosis are passive processes that result from the random movement of particles, whereas active transport is the deliberate selection of certain molecules by a cell for transport across the cell membrane.

Resources

Student Book Active transport

Approach

Supplement 1. Introduction to active transport

Show students a plant in a pot and explain that you water the plant every one or two days, but occasionally you add some plant 'food' to the water. Ask them to suggest what would happen if you did not water the plant. Use the discussion to briefly revise what students have learned about osmosis.

Then ask what would happen if the plant did not receive any plant 'food' occasionally. (Make sure students do not confuse 'plant food', which provides the mineral nutrients that plants need to convert carbohydrates from photosynthesis into other chemicals needed for cell growth and function, with 'animal food', which supplies both the energy and nutrient needs of an animal.)

Tell students that the concentration of solutes inside the root cells is greater than the concentration of solutes in the soil water. Ask them to suggest what problems this causes for plants. They should remember from their work on diffusion that the concentration gradient is in the wrong direction for root cells to absorb what they need from soil water by diffusion.

Introduce the idea that moving molecules from an area of lower concentration to an area of higher concentration needs energy, like pushing something uphill. The energy for this work comes from respiration (which is covered in more detail in Section 12: *Respiration*).

Supplement 2. Why do organisms need active transport?

Ask students to read through the text on 'Active transport' in the Student Book and make notes to explain why certain substances move across cell membranes as a result of active transport, and why cells need special protein carrier molecules in their membranes for this purpose.

Supplement 3. Consolidation of work on movement of substances

Give each student three small pieces of paper and ask them to write *diffusion* on one and *osmosis* on another and *active transport* on the third. Then ask a series of questions to which one or more of these words is the answer. Ask students to hold up the appropriate answer(s) for each question. For example:

- Which of the processes involves the random movement of molecules? [diffusion and osmosis]
- Which process only involves water molecules? [osmosis]
- Which of the processes involves molecules other than water? [diffusion and active transport]
- Which process requires energy to happen? [active transport]

This will help you to identify any weaknesses in learning by individual students.

Learning episode B3d Consolidation and summary

Learning aims

- Review the learning points of the topic summarised in the End of topic checklist
- Test understanding of the topic content by answering the End of topic questions

Resources

Student Book End of topic checklist and End of topic questions

Approach

Introduce the learning episode

Ask students to work with a partner to make a list of key words from this topic. They could then work together to produce a spider diagram showing how the different concepts are linked. They could compare their list with the list of key terms given at the end of the topic in the Student Book. Discuss the checklist at the End of the topic and use questioning to see how much of the content they are comfortable with.

Students could make flashcards of the key content and then use the flashcards to quiz each other on the information.

Develop the learning episode

Ask students to work individually through the End of topic questions in the Student Book without looking at the text. As they work, walk around the classroom observing their answers and asking questions as necessary to find out which questions are causing difficulties.

Finish the learning episode

After a set period, ask the students to stop working and discuss any areas of difficulty you observed as you walked round the class. Students should complete any unanswered questions for homework, but you should stress that they should try to answer the questions without looking at the text, so that they can see how much they have remembered.

B4 Biological molecules

Contents

Introduction

This short section covers some of the important molecules in living organisms: carbohydrates, fats and proteins.

This learning episode provides the information needed for later sections on biological systems, particularly nutrition. If preferred, the content could be saved for these later sections where it can be introduced in context, for example protein structure in Section B5: *Enzymes,* or Section B7: *Human nutrition.*

Links to other topics

Topics	**Essential background knowledge**	**Useful links**
B2 Cells	B2a Cell structure and size of specimens	
B3 Movement into and out of cells		B3a Diffusion B3b Osmosis B3c Active transport
B5 Enzymes		B5a Enzyme structure
B6 Plant nutrition		B6a Photosynthesis B6b Leaf structure
B7 Human nutrition		B7a Diet B7b Digestion

Topic overview

B4a	**Carbohydrates, fats and proteins** In this learning episode, students learn about the basic structure of biological molecules and the elements from which they are made. This knowledge forms an essential basis for Section B7: *Human nutrition*.
B4b	**Food tests** This learning episode provides practical work on identifying starch, proteins, reducing sugars and lipids. This knowledge links closely to Learning episode B7a *Diet*.
B4c	**Consolidation and summary** This learning episode quickly recaps on the ideas encountered in the section. Students can answer the End of topic questions in the Student Book.

Careers links

These are some scientific careers that focus on this area of biology but careers in many other fields use the knowledge and skills gained studying science. **Food scientists** and **food technologists** use a knowledge of biological molecules in foods to design and plan the manufacture of food and drinks, making sure that they meet guidelines e.g., for sugar content.

Starting points

The Student Book section opener puts the ideas in the section into context and sets the scene. It also allows students to acknowledge and value their prior learning, and provides a benchmark against which future learning can be compared.

The questions provide a structure for introducing the section and can be used in a number of different ways:

- You could ask students to consider the questions as an introductory homework task.
- You could put students into groups to share their own ideas and understanding and then to report back to the whole class.
- Students could be given access to the Internet, preferably with a tight timescale, to find out the information required.

You could then use a spider chart or other form of wall chart to summarise everybody's ideas.

Recording these initial ideas allows you to retain them for reference as the individual topics are developed. In this way, your students' progress in learning can be readily acknowledged.

Learning episode B4a Carbohydrates, fats and proteins

Learning objectives

- List the chemical elements that make up: carbohydrates, fats, and proteins.
- State that large molecules are made from smaller molecules, limited to: starch, glycogen, and cellulose from glucose; proteins from amino acids; fats and oils from fatty acids and glycerol.

Common misconceptions

This section anticipates some of the content in Section B7: *Human nutrition,* and some students may be confused about the source of elements that form the molecules of our body. If they are uncertain, prompt for the correct answer by asking how nutrients get into our bodies.

Resources

Student Book Carbohydrates, fats and proteins Worksheet B4a.1 Elements in the human body

Resources for a class practical (see Technician's notes)

Approach

1. Introduction

Write the words *carbohydrate*, *protein* and *fat* on the board, and give students a few minutes to think of three really interesting facts that they remember about these substances. Take examples of facts from around the class and ask students to use the key points to construct a concept map about these biological molecules.

If any student mentions the word *lipid* to cover fats (lipids that are solid at room temperature, for example, butter or lard) and oils (lipids that are liquid at room temperature), explain what it means. Students may come across the term in any research they do. However, tell students that this term is beyond the core syllabus.

Students could add any other words they think are suitable to the concept map. Keep the concept map for the end of this learning episode.

2. Elements in the human body

Worksheet B4a.1 gives some information about the elements in the human body. Students should use the information to draw a pie chart and answer questions that reinforce their learning of elements present in the major groups of biological molecules.

Some students may need help with creating the pie chart to answer Question 1. Question 4 may confuse some students because it considers elements, not the molecules they form. Make sure any confusion is clarified before moving on.

3. The structure of carbohydrates, fats and proteins

Use the 'Carbohydrates, fats and proteins' section of the Student Book to introduce the different basic units of the three groups of food molecules. Students could use 'popper beads' or similar (for example, chemical models) to reinforce their understanding of the structure of carbohydrates and proteins. Use identical beads for making a 'carbohydrate', explaining that starch, glycogen and cellulose simply have different arrangements of identical glucose molecules. Use a selection of different beads for making a 'protein'. Students could suggest a different way of modelling to show how different fats and oils all contain glycerol, but each contain different fatty acids.

Students could make models using modelling putty if 'popper beads' or chemical models are not available.

4. Consolidation

Ask students to return to their concept maps from the introduction and add to, or amend, them using what they have learned in this learning episode. Take examples of changes from around the class to make sure all the key points are included.

Technician's notes

Be sure to check the latest safety notes on these resources before proceeding.

The following resources are needed for task 3, per group:

popper beads or similar

Learning episode B4b Food tests

Learning objectives

- Describe the use of:
  iodine solution test for starch; Benedict's solution test for reducing sugars; biuret test for proteins; ethanol emulsion test for fats and oils.

Common misconceptions

Students may have misconceptions about what types of nutrients are found in different food types. Use the food tests to show that different foods can have a variety of different nutrients.

Resources

Student Book Tests for food molecules

Worksheet B4b.1 Food tests

Resources for a class practical (see Technician's notes)

Approach

1. Introduction

Show students a selection of nutrition labels from a variety of foods. (These can be found on the back of food packaging. Examples can also be found on the web pages of food manufacturers.)

Ask students to look at the nutrition information and to note the amount of carbohydrate, protein and total fat in each food. Remind them to record only the value per 100 g, so that they can make fair comparisons of the data.

They should then order the foods for each nutrient, starting with the largest amount. Ask questions to check their results, such as:

- Which food contains the largest amount of carbohydrate?
- Which food contains the least fat?

2. Food tests

Worksheet B4b.1 guides students in the use of tests used to detect the presence of starch, reducing sugars (glucose), protein and fats and oils.

Check that students have drawn up a suitable table for their results before they carry out the tests.

In the evaluation question, it is suggested that Benedict's test is used to produce a quantitative test.

SAFETY INFORMATION: Safety goggles rather than safety spectacles should be worn because of the corrosive nature of potassium hydroxide solution. Students should be warned to take care when using it and wash off any spillages immediately.

There should not be any naked flames when ethanol is being used.

Remind students to be careful using the hot water and that foods must not be tasted in a science laboratory. Avoid foods that produce known allergies (such as peanuts). Take care when using potassium hydroxide solution. Wash off any spillages immediately.

3. Consolidation

Ask students to produce five quick quiz questions and answers based on what they have learned in this learning episode. They could then try out their questions on a neighbour. Check to see whether there are any recurring errors or weaknesses in knowledge or understanding, so that these can be addressed.

Technician's notes

Be sure to check the latest safety notes on these resources before proceeding.

The following resources are needed for the class practical B4b.1, per group:

samples of foods for testing
Benedict's solution
test tubes, measuring pipettes, heatproof tongs or test tube holder and small spoons or spatulas, 1 for each food sample
mortar and pestle
distilled water
white spotting tile
iodine solution
hot water from a kettle that has just boiled
beaker and insulation for beaker
5% potassium hydroxide solution
1% copper sulfate solution
absolute ethanol or industrial denatured ethanol (IDA)
eye protection (goggles, not safety spectacles)

Include a range of foods, some of which contain larger proportions of some food groups than others, such as biscuits, cake, plain crisps for carbohydrates, and yoghurt, cheese, cooking oil for fats. Dried pulses that have been soaked in water for 24 hours are suitable for proteins. If not already liquid, foods need to be easily broken down to a paste with a mortar and pestle.

You can use a water bath at 100 °C, but it is safer to use water that has just boiled and put it in an insulated beaker.

Labels on all bottles should comply with the manufacturer's instructions.

Learning episode B4c Consolidation and summary

Learning aims

- Review the learning points of the topic summarised in the End of topic checklist.
- Test understanding of the topic content by answering the End of topic questions.

Resources

Student Book End of topic checklist and End of topic questions

Approach

Introduce the learning episode

Ask students to work with a partner to make a list of key words from this topic. They could then work together to produce a spider diagram showing how the different concepts are linked. They could compare their list with the list of key terms given at the end of the topic in the Student Book. Discuss the checklist at the end of the topic and use questioning to see how much of the content they are comfortable with.

Students could make flashcards of the key content and then use the flashcards to quiz each other on the information.

Develop the learning episode

Ask students to work individually through the End of topic questions in the Student Book without looking at the text. As they work, walk around the classroom observing their answers and asking questions as necessary to find out which questions are causing difficulties.

Finish the learning episode

After a set period, ask the students to stop working and discuss any areas of difficulty you observed as you walked round the class. Students should complete any unanswered questions for homework, but you should stress that they should try to answer the questions without looking at the text, so that they can see how much they have remembered.

B5 Enzymes

Contents

Teacher Guide	Syllabus reference
B5 Enzymes	B5.1 Enzymes

Introduction

This section builds on the mention of enzymes in B4 *Biological molecules*. This section looks in more detail at the structure and function of enzymes, and their importance. Students also investigate the effect of temperature and pH on enzyme activity.

Links to other topics

Topics	Essential background knowledge	Useful links
B1 Characteristics of living organisms	B1a Characteristics of living organisms	
B4 Biological molecules	B4a Carbohydrates, fats and proteins	
B6 Plant nutrition		B6a Photosynthesis
B7 Human nutrition		B7a Diet B7b Digestive system B7c Digestion

Topic overview

B5a	**Enzyme structure** This learning episode looks at the role of enzymes in sustaining life. Students will learn about how enzymes function as biological catalysts. Supplement Students will also learn about enzyme-substrate complexes, complementary shape and enzyme specificity.
B5b	**Enzyme activity** This learning episode gives students several practical opportunities to explore the effects of temperature and pH on the rate of enzyme-catalysed reactions. Supplement The learning episode continues with an explanation of the effect of temperature and pH on the rate of enzyme-catalysed reactions.
B5c	**Consolidation and summary** This learning episode quickly recaps on the ideas encountered in the section. Students can answer the End of topic questions in the Student Book.

Careers links

These are some scientific careers that focus on this area of biology but careers in many other fields use the knowledge and skills gained studying science. Formulation **biochemists** work in companies to develop laundry products that use enzymes to clean clothes with less water and at lower temperatures.

Starting points

The Student Book section opener puts the ideas in the section into context and sets the scene. It also allows students to acknowledge and value their prior learning, and provides a benchmark against which future learning can be compared.

The questions provide a structure for introducing the section and can be used in a number of different ways:

- You could ask students to consider the questions as an introductory homework task.
- You could put students into groups to share their own ideas and understanding and then to report back to the whole class.
- Students could be given access to the Internet, preferably with a tight timescale, to find out the information required.

You could then use a spider chart or other form of wall chart to summarise everybody's ideas.

Recording these initial ideas allows you to retain them for reference as the individual topics are developed. In this way, your students' progress in learning can be readily acknowledged.

Learning episode B5a Enzyme structure

Learning objectives

- Describe enzymes as proteins that are involved in all metabolic reactions, where they function as biological catalysts.
- Supplement Describe and explain enzyme action with reference to: active site, enzyme-substrate complex, substrate, and product.
- Supplement Describe and explain the specificity of enzymes in terms of the complementary shape and fit of the active site with the substrate.

Common misconceptions

It is important to reinforce the fact that enzymes are proteins, and that all factors that affect proteins will also affect enzymes, and vice versa.

Resources

Student Book Enzymes as catalysts

Worksheet B5a.1 Know, want to know, learned

Approach

1. Introducing catalysts

Ask students whether they understand the word *catalyst*, and if so, to explain where they have heard it before and what it means. They may know it from catalytic converters in vehicle exhausts, or from their work in chemistry.

Explain that enzymes are catalysts found in living cells. Ask students to suggest what enzymes might do. It does not matter whether they answer correctly or not – this learning episode is to help you identify the limits of their knowledge and any misunderstandings they may have developed.

Tell students that humans have around 75 000 different enzymes in our bodies, and that each enzyme controls a different reaction. Give them a few minutes to work in pairs or small groups to write down as many reactions in the human body as they can think of. (You may need to remind them that a reaction changes reactants into products, so enzymes are not involved in processes such as diffusion and osmosis.)

Take examples from around the class. Make sure these include reactions that students should know from earlier work, such as in respiration, photosynthesis and the digestion of food molecules.

2. Enzyme structure

Worksheet B5a.1 is a KWL (Know, Want to know, Learned) grid that can be used to help support learning on the structure of enzymes. Some examples of what students need to know are included in the 'Know' column. Either students could decide what else they need to know using the learning objectives at the start of the section, or these could be added from class discussion. (Note that objectives based on practical work cannot be completed until after Learning episode B5b.)

Students should first complete the left-hand column and then identify in the middle column what else they need to find out in order to complete the knowledge more securely. If needed, check at this point that students have covered the main points. Note that students following the supplementary syllabus must include the term *active site* in their learning.

Students should then complete the right-hand column using the information in the 'Enzymes' section of the Student Book, or information from other sources such as textbooks or the Internet. Check that they are accessing information at a suitable level, as there is much content on this at A level and above, which could be confusing.

3. Consolidation

Ask students to prepare questions for which the following are the answers:

1. enzyme
2. catalyst
3. this makes enzymes essential for life
4. Supplement substrate
5. Supplement active site
6. Supplement enzyme-substrate complex
7. Supplement this makes enzymes specific to their substrate

They could then test their questions out on a neighbour to help identify any weaknesses in understanding.

For an extra challenge, ask students to write two or more questions per answer.

Learning episode B5b Enzyme activity

Learning objectives

- Investigate and describe the effect of changes in temperature and pH on enzyme activity.
- Supplement Explain the effect of changes in temperature on enzyme activity in terms of kinetic energy, shape and fit, frequency of effective collisions, and denaturation.
- Supplement Explain the effect of changes in pH on enzyme activity in terms of shape and fit, and denaturation.

Common misconceptions

Many students assume that all human enzymes work best at 'normal' human body temperature. Although this is the case for enzymes that work in the 'core' organs, such as the heart, lungs and digestive system, those that work in cells in the limbs and extremities may have lower optimum temperatures. Enzymes from other organisms may have very different optimum temperatures.

Some students may believe that enzymes are 'killed' by high temperatures. As enzymes are not living things, they cannot be killed – students should learn the correct term, *denaturation*, to explain why some enzymes no longer work at high temperatures. Some students may have the incorrect idea that enzymes can also be denatured by low temperatures – although their activity may be reduced, this is not a permanent change.

Resources

Student Book Enzymes and temperature; Investigating the effect of temperature on enzymes; The effect of temperature; Enzymes and pH; Investigating the effect of pH on enzymes; The effect of pH

Worksheet B5b.1 Gelatin, enzymes and temperature

Worksheet B5b.2 Trypsin, milk and temperature

Worksheet B5b.3 The effect of pH on amylase

Resources for a class practical (see Technician's notes)

Approach

1. Introduction

Write the following words on the board, and ask students to create sentences that use them. Ideally, they should use more than one of the words in the same sentence:

- enzyme
- metabolic reaction
- Supplement substrate
- Supplement complementary
- Supplement protein shape

Students could compare their sentences with a neighbour and select the ones combining most words into the same sentence correctly.

Supplement Briefly introduce the idea that the attraction or repulsion of amino acids within a protein/enzyme can be affected by external factors. Ask students what effect these changes may have on the shape of the active site and the way an enzyme works.

2. Enzyme activity

Ask students to take notes from the relevant sections of the Student Book on the effect of temperature and pH on enzyme action. Their notes should include the fact that enzymes speed up reactions and that they are affected by temperature and pH.

Supplement Students should include references to the *active site* and *denaturation* in their notes.

Supplement Students should include references to *kinetic energy*, *frequency of effective collisions* and *changes in shape* in their notes.

To test students' understanding, ask them to sketch a graph of reaction rate for an enzyme that digests food substances in the stomach, where the pH is around 2, and another for an enzyme that digests food substances in the small intestine, where the pH is near to 8.

3. Investigating enzyme activity

Students can investigate the effect of temperature on the rate of an enzyme-controlled reaction using Worksheet B5b.1 or B5b.2, depending on the resources available. It may be difficult to get hold of photographic film for Worksheet B5b.1. Worksheet B5b.2 provides an alternative if photographic film is not available (see Technician's notes).

The effect of pH on the rate of an enzyme-controlled reaction can be investigated using Worksheet B5b.3. Note that this practical includes the use of the starch test, which is covered in learning episode B4b specifically in relation to food tests. If students have not already used this test, describe how they should carry it out and explain what a change of colour, or lack of it, means in terms of the presence or absence of starch.

If necessary, discuss with students how they are going to record their results in a table and what the most suitable graph would be for them to draw. For Worksheet B5b.1, a suitable table would have temperature in one column and time taken for film to clear in another. A suitable graph would show temperature on the *x*-axis and time on the *y*-axis.

For Worksheet B5b.3, you may wish students to carry out a planning exercise for the investigation before giving them the worksheet.

SAFETY INFORMATION: Students should wear eye protection. Remind students to be careful when using the hottest water bath. Care should be taken with enzyme powders and concentrates (follow the precautions described by the supplier).

4. Consolidation: key points

Give students two minutes to write down the four most important points they think they have learned about enzymes. They should then spend another minute discussing their points with a partner and selecting the four best points from both sets.

Take selections from around the class and encourage discussion to produce a class list of key points to remember about enzymes.

Technician's notes

Be sure to check the latest safety notes on these resources before proceeding.

Enzymes purchased from suppliers may not produce the expected trend, or may not work under the conditions given. Please note the following general comments:

- Many purchased enzymes may not be affected by temperatures as high as 50 °C. The enzymes will have been developed for use in biological washing powders, and can still be fully functional up to 95 °C. Please consult suppliers' information before purchase.
- Many enzymes (particularly amylases) are very strongly affected by buffering chemicals. This is very frustrating for all concerned, particularly if the inhibition does not result in an end point. You should trial and perhaps repurchase enzymes for pH effects. Alternatively, the pH could be very carefully controlled and monitored during the experiments, removing the need for buffer chemicals entirely.
- Care should be taken with enzyme powders and concentrates (follow the precautions described by the supplier).
- Class practical B5b.1 may not work above 30–40 °C because some films may melt. Technicians should trial this practical before the class to test the film that will be used.

The following resources are needed for class practical B5b.1, per group:

5 test tubes
lengths of exposed and developed 35 mm black-and-white film
scissors
stopwatch or clock
0.1% protease solution, freshly prepared
eye protection
5 water baths set at e.g., 10 °C, 15 °C, 20 °C, 25 °C and 30 °C – only one set needed as all groups will use the same water baths

This is an investigation into the effect of temperature on an enzyme using exposed and developed black-and-white film. If you do not have any suitable film, Worksheet B5b.2 can be used instead.

Gelatin is digested by a wide range of protease enzymes. For the best results in the time available, it may be worth carrying out some test runs with different enzymes to see which performs best.

Some people are allergic to enzymes. If you suspect you are allergic, take great care when preparing the solution for students and wear disposable gloves and eye protection. Use a fume cupboard if preparing solutions from powdered enzymes. At the low concentration used in this experiment, students with an allergy to the enzyme are less likely to have a severe reaction, but gloves and eye protection are still recommended.

The following resources are needed for class practical B5b.2, per group:

5 test tubes
10% milk solution
stopwatch or clock
2% trypsin solution freshly prepared
eye protection
5 water baths set at 10 °C, 20 °C, 30 °C, 40 °C and 50 °C – only one set needed as all groups will use the same water baths
measuring pipettes

Please note that trypsin may not work well if it is made up with acidic distilled water (as is often the case). You may need to adjust the pH using alkali (such as sodium carbonate) to obtain an alkaline pH. The milk and trypsin should both be adjusted to pH 9 before starting, so that the pH optimum of the trypsin is met.

The investigation into the effect of temperature on an enzyme uses trypsin and milk. The milk can be fresh pasteurised milk, or made up into solution with water from powdered milk. Skimmed milk should be used, as the fat in milk will prevent the solution from clarifying.

The worksheet suggests a test run to identify the end point as effectively as possible in order to generate reliable results.

Students should find that trypsin has an optimum temperature at around 37 °C, similar to the core temperature of the human body.

Some people are allergic to enzymes. If you suspect you are allergic, take great care when preparing the solution for students and wear disposable gloves and eye protection. Use a fume cupboard if preparing solutions from powdered enzymes. At the low concentration used in this experiment, students with an allergy to the enzyme are less likely to have a severe reaction, but gloves and eye protection are still recommended.

The time required for this practical can be reduced by giving groups only one or two temperatures from the five, and then collating class results. Additional discussion time will be needed to cover the reliability of shared results.

The following resources are needed for class practical B5b.3, per group:

5 test tubes and rack
5 different pH solutions
1% amylase solution
10% starch solution
2 5 cm^3 syringes or pipettes
iodine solution in dropper bottle (The iodine solution used for testing should be no more than 0.01 M)
stopwatch or clock
marker pen
spotting tile
disposable gloves
eye protection

This practical investigates the effect of pH on amylase using the iodine test for starch to find the point at which all starch is digested. Different amylases have a different optimum pH. Amylase from the pancreas has an optimum between 6.7 and 7.0, but that from yeast has an optimum between 4.6 and 5.2. You may substitute a solution of pancreatin for amylase to give a higher result, more appropriate for the human body. The enzyme and starch solutions should be prepared fresh for the lesson.

The time required for this practical can be reduced by giving groups only one or two pH values from the six, and then collating class results. Additional discussion time will be needed to cover the reliability of shared results.

Different pH/buffer solutions can be made using disodium hydrogenphosphate and citric acid as follows:

pH	**Volume of 0.2 mol/dm^3 Na_2HPO_4**	**Volume of 0.1 mol/dm^3 citric acid**
3	20.55	79.45
4	38.55	61.45
5	51.50	48.50
6	63.15	36.85
7	82.35	17.65
8	97.25	2.75

Alternatively, some suppliers provide separate buffer solutions, or tablets, at the different pH values.

Some people are allergic to enzymes. If you suspect you are allergic, take great care when preparing the solution for students and wear disposable gloves and eye protection. Use a fume cupboard if preparing solutions from powdered enzymes. At the low concentration used in this experiment, students with an allergy to the enzyme are less likely to have a severe reaction, but gloves and eye protection are still recommended.

Learning episode B5c Consolidation and summary

Learning aims

- Review the learning points of the topic summarised in the End of topic checklist
- Test understanding of the topic content by answering the End of topic questions

Resources

Student Book End of topic Checklist and End of topic questions

Approach

Introduce the learning episode

Ask students to work with a partner to make a list of key words from this topic. They could then work together to produce a spider diagram showing how the different concepts are linked. They could compare their list with the list of key terms given at the end of the topic in the Student Book. Discuss the checklist at the end of the topic and use questioning to see how much of the content they are comfortable with.

Students could make flashcards of the key content and then use the flashcards to quiz each other on the information.

Develop the learning episode

Ask students to work individually through the End of topic questions in the Student Book without looking at the text. As they work, walk around the classroom observing their answers and asking questions as necessary to find out which questions are causing difficulties.

Finish the learning episode

After a set period, ask the students to stop working and discuss any areas of difficulty you observed as you walked round the class. Students should complete any unanswered questions for homework, but you should stress that they should try to answer the questions without looking at the text, so that they can see how much they have remembered.

B6 Plant nutrition

Contents

Teacher Guide	Syllabus reference
B6 Plant nutrition	B6.1 Photosynthesis B6.2 Leaf structure

Introduction

This section covers plant nutrition in terms of photosynthesis and leaf structure. Students should know about the role of photosynthesis for providing nutrition from B1: *Characteristics of living organisms*, they should also know the structure of plant cells from B2: *Cells*. The role of enzymes in photosynthesis can be highlighted. From the supplementary syllabus students should be reminded of the structure of root hair cells and palisade cells from B2: *Cells*.

Links to other topics

Topics	**Essential background knowledge**	**Useful links**
B1 Characteristics of living organisms	B1a Characteristics of living organisms	
B2 Cells	B2a Cell structure and size of specimens	
B3 Movement into and out of cells		B3a Diffusion B3b Osmosis B3c Active transport
B4 Biological molecules		B4a Carbohydrates, fats and proteins
B5 Enzymes		B5b Enzyme activity
B8 Transport in plants		B8a Xylem and phloem B8b Water uptake B8c Transpiration
B12 Respiration		B12a Respiration
B15 Organisms and their environment		B15a Energy flow, food chains and food webs B15b Carbon cycle

Topic overview

B6a	**Photosynthesis** This learning episode looks at the process and requirements for photosynthesis. It provides many opportunities for practical work. You may have to select from the suggestions offered, in order to fit the time and resources available. Some might also be presented as class demonstrations. Supplement Students need to be able to describe in more detail the conditions that effect photosynthesis, they also learn about the effect of light and dark conditions on gas exchange in aquatic plants
B6b	**Leaf structure** In this learning episode, students learn about the structure of the leaf.
B6c	**Consolidation and summary** This learning episode quickly recaps on the ideas encountered in the section. Students can answer the End of topic questions in the Student Book.

Careers links

These are some scientific careers that focus on this area of biology but careers in many other fields use the knowledge and skills gained studying science. **Agronomists** study the factors that affect crop growth, and work with farmers to improve plant yields. **Floriculturists** grow flowering and ornamental plants and produce cut flowers for decoration. Both of these jobs need to understand the conditions that promote healthy growth of plants.

Starting points

The Student Book section opener puts the ideas in the section into context and sets the scene. It also allows students to acknowledge and value their prior learning, and provides a benchmark against which future learning can be compared.

The questions provide a structure for introducing the section and can be used in a number of different ways:

- You could ask students to consider the questions as an introductory homework task.
- You could put students into groups to share their own ideas and understanding and then to report back to the whole class.
- Students could be given access to the Internet, preferably with a tight timescale, to find out the information required.

You could then use a spider chart or other form of wall chart to summarise everybody's ideas.

Recording these initial ideas allows you to retain them for reference as the individual topics are developed. In this way, your students' progress in learning can be readily acknowledged.

Learning episode B6a Photosynthesis

Learning objectives

- Describe photosynthesis as the process by which plants synthesise carbohydrates from raw materials using energy from light.
- State the word equation for photosynthesis as:
  carbon dioxide + water → glucose + oxygen
  in the presence of light and chlorophyll.
- State that chlorophyll is a green pigment that is found in chloroplasts.
- Investigate and understand the need for chlorophyll, light, and carbon dioxide for photosynthesis, using appropriate controls.
- Supplement State the balanced symbol equation for photosynthesis as:
  $6CO_2 + 6H_2O \rightarrow C_6H_{12}O_6 + 6O_2$.
- Supplement State that chlorophyll transfers energy from light into energy in chemicals, for the synthesis of carbohydrates.
- Supplement Understand and describe the effects of varying light intensity, carbon dioxide concentration, and temperature on the rate of photosynthesis.
- Supplement Understand and describe the effect of light and dark conditions on gas exchange in an aquatic plant using hydrogencarbonate indicator solution.

Common misconceptions

Some students may not recognise that light is essential for plant growth, but believe that sunlight helps plants to grow by keeping them warm. This is often reinforced by the idea that plants need 'plant food' in a similar way to animals needing food to provide energy and nutrients. In this first learning episode, it is crucial that students develop an understanding of the roles of light and photosynthesis in plant growth. A good way to test this misconception is to ask where plants get the molecules they are made from, many people (including science graduates) say the soil, but the carbon making up most of the plant molecules will have come from the carbon dioxide in the atmosphere.

Resources

Student Book Photosynthesis

Worksheet B6a.1 Testing leaves for starch

Worksheet B6a.2 Chlorophyll and photosynthesis

Worksheet B6a.3 Carbon dioxide and photosynthesis

Worksheet B6a.4 Light and the rate of photosynthesis

Resources for class practical (see Technician's notes)

Approach

1. Introduction to photosynthesis

Show students a potted plant and give them three minutes to jot down everything they can think of that relates to how the plant gets its food. Take examples from around the class, focusing on the answers that relate to photosynthesis.

Note any misconceptions that you will need to tackle during the work on this section.

Use the Student Book to introduce the word equation (and for Supplement content, the balanced symbol equation) for photosynthesis. Students should make suggestions for what is needed for photosynthesis to take place (and for Supplement content, the factors which might affect the rate of photosynthesis). Challenge students to use the equation for photosynthesis to justify their suggestions.

2. Investigating requirements for photosynthesis

There are several worksheets for this part of the section.

Worksheet B6a.1 explains how to prepare and test a leaf for starch. This method is needed for Worksheets B6a.2 and B6a.3, so students should be given the opportunity to do this. If you keep one plant in the dark and test a leaf from that at the end of the students' work, you will be able to show the 'need for light'.
Worksheet B6a.2 investigates the need for chlorophyll.
Worksheet B6a.3 investigates the need for carbon dioxide. Note that this might be better done as a demonstration, due to the apparatus used (see Technician's notes). You could also introduce the concept of a control in this investigation.

If time is limited, different groups of students could carry out different practicals at the same time, and then present their findings to the rest of the class.

SAFETY INFORMATION: Do not use open flames (such as that on a Bunsen burner) when heating ethanol. It is highly flammable. Students should wear eye protection when using ethanol or iodine solution. Remind students to take care to avoid scalding with very hot water. Students should rinse hands thoroughly at the end of the practical.

The soda lime should not be handled by the students.

You should supervise the pouring of very hot water from kettles to minimise the risk of scalding.

Supplement **3. Investigating the rate of photosynthesis**

Worksheet B6a.4 provides a method using a water plant to give quantitative data on the effect of light intensity on the rate of photosynthesis. The method assumes that the bubbles produced are all oxygen and of the same size. A more accurate method is to collect and measure the volume of oxygen produced over a fixed time, but this will take longer to produce measurable results.

The method could be adapted by adding different concentrations of sodium hydrogencarbonate to the water to produce quantitative data of the effect of carbon dioxide concentration on the rate of photosynthesis, or by changing the temperature of the water to investigate the effect of temperature on the rate of photosynthesis.

Check that students have drawn up suitable tables for their results before they carry out the tests.

SAFETY INFORMATION: Cover any cuts with a waterproof plaster. Students should wash their hands thoroughly after the practical work.

To provide a 'real-life' application, ask students to annotate a sketch of a glasshouse to explain how conditions could be adjusted to maximise the rate of growth of plants inside it.

Supplement **4. Net gas exchange**

The Developing Practical Skills box in the 'Gas exchange in plants' section of the Student Book describes a method, using hydrogencarbonate indicator, of investigating the effect of light and dark conditions on net gas exchange in an aquatic plant.

5. Consolidation: question tag

Ask students to jot down three key facts that they think the class should have learned in this learning episode on photosynthesis. They should then write a question for each of the facts.

Take examples from around the class, and select another student to answer it. If they answer it successfully, they should then read out one of their own questions.

Technician's notes

Be sure to check the latest safety notes on these resources before proceeding.

The method described on Worksheet for B6a.1 is also used in Worksheets B6a.2 and B6a.3, so the following applies to all three investigations.

The following resources are needed for the class practical B6a.1, per group:

Industrial denatured ethanol (IDA)
kettle of boiling water
water bath at 90 °C with boiling tube rack
marker pen
iodine solution in dropper bottle (a 0.1 M solution should be strong enough but you may need to use up to 0.5 M solution; you may need to carry out trials to check that the iodine solution is strong enough)
large beaker
leaf (best taken from the plant at the start of the practical so that it is fresh)
forceps
boiling tube
glass rod
Petri dish
white tile
eye protection

Please note that the plants need a large amount of daylight to produce starch. If this practical is being carried out in winter, the leaves will need to have several hours of strong, safe illumination to make sure that they have produced some starch.

IDA is highly flammable. All naked flames must be put out. IDA is also harmful (because of the methanol it contains), so eye protection should be worn.

The following resources are needed for the class practical B6a.2, per group, in addition to the apparatus for B6a.1:

destarched plant, either with variegated or green leaves
metal foil and paper clip (if green leaves) – students will need a sufficiently long strip to cover both sides of the leaf (if metal foil is not available, black card can be used instead)

The plants for use in this practical will need to be destarched, by placing them in the dark for about 48 hours. Test a leaf to check that the plants are fully destarched, otherwise this will affect the results.

The following resources are needed for the class practical B6a.3 per group, in addition to the apparatus for B6a.1:

2 bell jars
2 destarched plants
2 plastic trays
alternatively, glass sheets – to create a perfect seal with the bell jar using the jelly/grease
petroleum jelly or silicone grease
small beaker of soda lime
small beaker containing 10–20 marble chips
dilute hydrochloric acid
safety goggles (not safety spectacles)

The plants for use in this practical will need to be destarched, by placing them in the dark for about 48 hours. Test a leaf to check that the plants are fully destarched, otherwise this will affect the results.

Soda lime is corrosive – please consult safety guidance on how to handle it.

It would be best to have the soda lime beaker already under the bell jar, for about 24 hours. Students can then be supervised adding the plant. This will allow the atmospheric CO_2 to be removed before the plant is added. Otherwise, the plant will carry out some initial photosynthesis.

The following resources are needed for the class practical B6a.4, per group:

pondweed, for example *Elodea*
paper clip
boiling tube containing pond water
250 cm^3 beaker of water
metre rule
lamp – ideally low-voltage bulb or LED light
stopwatch or clock

Prepare the pondweed before the lesson by cutting off 10 cm lengths, one for each group. Check that there are no small invertebrates on the plants. Dispose of plant material safely to ensure non-native plants cannot contaminate the environment.

You will need to dim the lights in the classroom for this practical. Students should be reminded how to prepare for this (e.g., removing all bags, etc. from the floor) and how to behave appropriately in a darkened classroom.

If available, you could use a light meter, or sensor, to measure light intensity directly, instead of recording the distance between the light and the pondweed.

The voltage of the bulb should not exceed the safety limits of the bench lamp. Low-energy bulbs should have some sort of protection so that they do not protrude beyond the protective metal of the bench lamp; or they can be protected by a plastic or glass screen.

Low-energy bulbs contain mercury and so can be hazardous if the bulb is broken.

Be vigilant during this practical because of water being used close to electrical equipment.

LED lights emit a very narrow range of wavelengths of light, which might not be absorbed by the plant. Halogen light sources are fine for photosynthesis but they do get hot. A large beaker of tap water placed in front of the light source will absorb the heat.

Learning episode B6b Leaf structure

Learning objectives

- Identify in diagrams and images the following structures in the leaf of a dicotyledonous plant: chloroplasts, cuticle, guard cells and stomata, upper and lower epidermis, palisade mesophyll, spongy mesophyll, air spaces, vascular bundles, xylem, and phloem.

Resources

Student Book Leaf structure

Resources for class practical (see Technician's notes)

Approach

1. Introduction

Show students photos of different plants, from small grasses, to large trees, and ask them to say what they have in common. Then show them a photo of a solar panel and ask them what the two have in common. Try to encourage students to think of leaves as a way for plants to trap as much light as possible, which they use to make food.

2. Structure of a leaf

If available, ask students to look at prepared slides of transverse sections through a leaf (see Technician's notes). Alternatively, use a digital microscope to display such a slide, or display an example from the Internet. Ask students to identify the different tissues in the leaf. This could be done by annotating a sketch of a leaf section. Alternatively, ask students to draw a diagram of the micrograph of a plant leaf in Fig. B6.5 in the Student Book, and annotate it to show the different cells and tissues.

Note that the functions of xylem and phloem are covered in more detail in Section 8: *Transport in plants,* but it is helpful to identify these tissues in the leaf during this learning episode.

3. Consolidation

Ask students to write ten crossword clues for a crossword on photosynthesis. They could then test out their clues on a neighbour to identify any areas of learning that need further work.

Technician's notes

Be sure to check the latest safety notes on these resources before proceeding.

The following resources are needed for the demonstrations in the introduction and task 2:

potted plant
prepared slides of transverse sections through a leaf and microscope *or* digital microscope

Learning episode B6c Consolidation and summary

Learning aims

- Review the learning points of the topic summarised in the End of topic checklist.
- Test understanding of the topic content by answering the End of topic questions.

Resources

Student Book End of topic checklist and End of topic questions

Approach

Introduce the learning episode

Ask students to work with a partner to make a list of key words from this topic. They could then work together to produce a spider diagram showing how the different concepts are linked. They could compare their list with the list of key terms given at the end of the topic in the Student Book. Discuss the checklist at the end of the topic and use questioning to see how much of the content they are comfortable with.

Students could make flashcards of the key content and then use the flashcards to quiz each other on the information.

Develop the learning episode

Ask students to work individually through the End of topic questions in the Student Book without looking at the text. As they work, walk around the classroom observing their answers and asking questions as necessary to find out which questions are causing difficulties.

Finish the learning episode

After a set period, ask the students to stop working and discuss any areas of difficulty you observed as you walked round the class. Students should complete any unanswered questions for homework, but you should stress that they should try to answer the questions without looking at the text, so that they can see how much they have remembered.

B7 Human nutrition

Contents

Teacher Guide	Syllabus reference
B7 Human nutrition	B7.1 Diet B7.2 Digestive system B7.3 Digestion

Introduction

This section covers human nutrition in terms of diet, and digestion and absorption in the human digestive system. This topic adds more detail the nutrition part of B1: *Characteristics of living organisms.* Students will need to know both the food tests and different types of biological molecules from B4: *Biological molecules*. Students following the supplementary syllabus can also be reminded of the structure and function of enzymes from B5: *Enzymes*. You could also mention the role of the blood in carrying the absorbed nutrients around the body which will be followed up in topic B9: *Transport in animals*.

Links to other topics

Topics	**Essential background knowledge**	**Useful links**
B1 Characteristics of living organisms	B1a Characteristics of living organisms	
B2 Cells	B2a Cell structure and size of specimens	
B4 Biological molecules	B4a Carbohydrates, fats and proteins B4b Food tests	
B5 Enzymes		B5a Enzyme activity
B9 Transport in animals		B9b Blood vessels and blood
B12 Respiration		B12a Respiration

Topic overview

B7a	**Diet** This learning episode looks at what is meant by a balanced diet, and identifies the sources of some nutrients and some of the effects of nutrient deficiencies.
B7b	**Digestive system** In this learning episode, students study the structure of the digestive system and are introduced to the processes that take place there.

B7c	**Digestion** In this learning episode students learn about physical digestion and build on the work in Section B5: *Enzymes*, by looking at digestive enzymes. Some of the practical work from Section B5 could be used to support this learning. Supplement Students go into greater detail about how some digestive enzymes work, as well as the role of stomach acid.
B7d	**Consolidation and summary** This learning episode quickly recaps on the ideas encountered in the section. Students can answer the End of topic questions in the Student Book.

Careers links

These are some scientific careers that focus on this area of biology but careers in many other fields use the knowledge and skills gained studying science. **Dietitians** are health professionals who diagnose and treat people with nutrition problems. Some dietitians specialise in helping patients after illness or treatment for cancer. **Dental technicians** make dentures to replace broken or diseased teeth to allow people to chew and eat food properly.

Starting points

The Student Book section opener puts the ideas in the section into context and sets the scene. It also allows students to acknowledge and value their prior learning, and provides a benchmark against which future learning can be compared.

The questions provide a structure for introducing the section and can be used in a number of different ways:

- You could ask students to consider the questions as an introductory homework task.
- You could put students into groups to share their own ideas and understanding and then to report back to the whole class.
- Students could be given access to the Internet, preferably with a tight timescale, to find out the information required.

You could then use a spider chart or other form of wall chart to summarise everybody's ideas.

Recording these initial ideas allows you to retain them for reference as the individual topics are developed. In this way, your students' progress in learning can be readily acknowledged.

Learning episode B7a Diet

Learning objectives

- Describe what is meant by a balanced diet.
- State the principal dietary sources and describe the importance of:
  carbohydrates; fats and oils; proteins; vitamins, limited to C and D; mineral ions, limited to calcium and iron; fibre (roughage); water.

Common misconceptions

It is important that students understand that dietary advice is based on an idea of the 'average' person. Such advice is useful in guiding us towards healthy ideals, but everyone is different and has different needs. It is important to distinguish between the needs of the 'average' population and the needs of the individual.

Note that the term *malnutrition* is most commonly associated with deficiency diseases and starvation when people get insufficient food to eat. However, it correctly refers to any situation where the diet is not appropriate for health, and so relates to eating *too much* of some foods as well as too little.

Resources

Student Book Diet

Worksheet B7a.1 Government health advice

Approach

1. Introduction

Give students a few minutes to write a definition of the term *balanced diet.* Ask them to compare their definition with that of a neighbour and to work together to discuss what this means. Then ask pairs to work together to improve their definitions.

Finally, take examples from around the class and discuss with students any differences between what they think it means. It is not essential to reach a consensus on what the term means. However, this should raise some of students' misconceptions on the topic – make sure that these are tackled through this learning episode.

2. Food sources

Either provide images of plates of food, or ask students to research some. They should then annotate the images to show which foods are good sources of the main nutrients (carbohydrates, fats, proteins, vitamins C and D, calcium and iron, fibre and water). This could be extended with research on which foods are the best sources of these nutrients.

Students should also add annotations that identify the functions of these nutrients in the body. The text in the Student Book could be used to help them do this.

3. A balanced diet

This part of the learning episode provides opportunities for research work on many diet-related topics that link to what we mean by a balanced diet. For example, students could find out about:

- The association between obesity and diseases such as diabetes and heart disease – how concerned should we be with our weight?
- The huge market in vitamin pills – do we need them?
- Evidence for the role of fibre in preventing bowel cancer, diverticulitis and constipation – why should we eat fibre?
- The increase in cases of rickets in young children in some parts of the world over the past ten years, and its cause – is it preventable?

Be aware of students' sensitivities to ethnic and social differences and to concerns about individual body mass, and treat these with consideration.

Students could carry out their research individually, in pairs or in small groups, according to time and ability. They could use their findings to present a verbal or written report, or create a poster for display.

Alternatively, Worksheet B7a.1 provides a set of questions on UK government health information that covers advice on the amount of nutrients and energy we get from our food. The final question might be used as a class activity, to explore what a healthy diet is and how you persuade people to eat healthily.

Before starting on this work, it may help to discuss some of the terminology, such as saturated fats, calories and salt, and why these are focused on when thinking about healthy eating.

4. Consolidation: healthy eating notes

Ask students to imagine that they are part of a government health advice group. They should work in pairs or small groups to prepare a list of five points that cover how to make sure the population of the country eats healthily.

Take examples from around the class and group the responses according to whether they are advice to individuals, advice to the government department that makes sure sufficient quality food is available for everyone, or advice to farmers who grow the food, and shops and supermarkets who sell the food.

Learning episode B7b Digestive system

Learning objectives

- Identify in diagrams and images the main organs of the digestive system, limited to: alimentary canal: mouth, oesophagus, stomach, small intestine (duodenum and ileum), and large intestine (colon, rectum, anus); associated organs: salivary glands, pancreas, liver, and gall bladder.
- Describe the functions of the organs of the digestive system listed above, in relation to: ingestion – the taking of substances, e.g., food and drink, into the body; digestion – the breakdown of food; absorption – the movement of nutrients from the intestines into the blood; assimilation – uptake and use of nutrients by cells; egestion – the removal of undigested food from the body as faeces.

Common misconceptions

Remind students of the difference between *egestion* (the expelling of undigested food from the alimentary canal) and *excretion* (the expelling of waste substances from metabolism, such as urea in urine).

Some students may think that digestion is the release of energy from food molecules, confusing it with respiration, which takes place in all cells. It is important to clarify that digestion produces the small food molecules that enter the body before being assimilated.

Resources

Student Book Digestive system

Approach

1. Introduction

Write the following words on the board in alphabetic order: absorption, assimilation, digestion, egestion, ingestion. Ask students to identify the organ system in the body in which all these processes occur (they may not understand the meaning of some of the words but should be able to guess the answer from those they know), and then to put the terms in the order in which they occur, starting at the mouth.

Take suggestions from around the class – the answers may not be exact (for example, digestion covers chewing in the mouth as well as enzymes in the small intestine), but this will provide an opportunity to discuss the meaning of each term.

2. The digestive system

Learning the structure of the digestive system, and which process happens where, is mainly a case of learning the facts. The Student Book text and diagrams provide what students need to know.

Consider asking students to use the facts in a way that makes learning more fun, e.g., asking them to write the 'story' of the digestion of a meal as it passes through the alimentary canal, to explain the role of each organ in the digestion and assimilation of nutrients from large pieces of food.

3. Consolidation: crossword clues

Give students a few minutes to produce clues for a crossword on the human digestive system, to cover at least ten of the following 'answers': liver, stomach, physical digestion, chemical digestion, assimilation, absorption, egestion, pancreas, oesophagus, enzyme, duodenum, colon, rectum, anus. They should exchange their clues (without the answers) with a neighbour and attempt to answer the ones they have been given.

Ask students to make suggestions on how to improve any clues that they feel are not clear, and to discuss them with the original writer.

Learning episode B7c Digestion

Learning objectives

- Describe physical digestion as the breakdown of food into smaller pieces without chemical change to the food molecules.
- Describe chemical digestion as the breakdown of large insoluble molecules into small soluble molecules.
- Supplement State that physical digestion increases the surface area of food for the action of enzymes in chemical digestion.
- Supplement State the role of chemical digestion in producing small soluble molecules that can be absorbed.
- Supplement Describe the functions of enzymes as follows:
  amylase breaks down starch to simple reducing sugars; proteases break down protein to amino acids; lipases break down fats and oils to fatty acids and glycerol.
- Supplement State where, in the digestive system, amylase, protease, and lipase, are secreted and where they act.
- Supplement Describe the functions of hydrochloric acid in gastric juice, limited to killing harmful microorganisms in food and providing an acidic pH for optimum enzyme activity.

Common misconceptions

Some students may think that digestion is a purely chemical process involving enzymes. This learning episode should help them to realise that digestion by physical means is a useful first stage.

Supplement Some students may not realise that certain enzymes are released in the mouth, to begin the process of the chemical digestion of starch.

Supplement Reinforce the fact that enzyme names always end in *-ase*.

Resources

Student Book Different types of digestion; Chemical digestion; Stomach acid

Worksheet B7c.1 The digestion and absorption of carbohydrates

Resources for a class practical (see Technician's notes)

Approach

1. Introduction

Ask students the question 'What are teeth for?' and give them about a minute to think up one or more answers to the question. Most students should be able to answer that they are for breaking down large pieces of food. Extend the questioning to ask why teeth are in the mouth and not in another part of the digestive system. Take examples of useful answers, but do not elaborate further. Remind students of the work they did in Learning episode B7b about the role of the stomach. Explain that churning food is another example of physical digestion.

Supplement Reinforce that physical digestion is needed to increase surface area in order for enzymes to work more effectively.

2. Chemical digestion

Remind students of the molecules that make up their body and the work they did in Learning episode B4.a. Ask students why molecules that we eat in food need to be digested. Explain that most of the food we eat is made of large and/or insoluble molecules, and that chemical digestion helps break these down into small soluble molecules.

3. Supplement Enzymes in digestion

Write the following words on the board: protease, amylase, lipase. Then give students a minute or so to discuss in pairs or small groups what the words have in common and what they mean. They should associate *protease* and *lipase* with two of the major food groups that they learned about in Section B4: *Biological molecules*, and may remember amylase from the work on food tests. If needed, identify the -*ase* syllable as the part that means 'enzyme'.

Ask students to identify the smaller, simpler, molecules that each of the food groups is broken down into. Then give students a few minutes to jot down all that they remember about enzymes from Section B5: *Enzymes*. Take examples from around the class and note down any weaknesses that can be tackled in this learning episode.

Worksheet B7c.1 applies the methods from investigating the movement of molecules through visking tubing (dialysis tubing), from Section B3: *Movement into and out of cells* and investigating amylase from Section B5: *Enzymes*, to the digestion of starch and absorption of the products of digestion. As students should be familiar with these methods, this practical sheet is presented as a planning exercise.

If there is time, and students need to practise their investigative skills, allow them to carry out their plans after you have approved them.

Alternatively, students could carry out any of the practicals in Section B5 that they were unable to do earlier.

4. Supplement Sites of secretion

Give students a few minutes to look at Table B7.3 in the Student Book and to think of any way they can to remember the information. They might do this with a mnemonic, a brief sketch or a cartoon. They should then test each other by asking questions, trying to remember without the assistance of prompts.

5. Supplement Hydrochloric acid and gastric juice

Tell students that the acid produced in the stomach is as strong as the hydrochloric acid that is used in school laboratory experiments. Give them a minute or so to think about why the digestive system does this. Encourage them to make links to their work in Section B5: *Enzymes*, on the effect of pH on enzyme activity, and to identify the optimum pH of gastric enzymes (around pH 2). They should also consider the effect of acid on living cells and identify that acid will kill many microorganisms that we take in with our food.

(If students wonder why the acid does not attack the stomach lining cells, explain that some cells in the lining produce large amounts of mucus (slime), which protects the surface. However, if these cells are affected (e.g., by infection) so that they stop producing mucus, the acid will start to digest holes in the surface, causing painful stomach ulcers.)

6. Consolidation: concept map

Ask students to write the terms *physical digestion* and *chemical digestion* on two sides of a sheet of paper and then to add linked words to form a concept map on the section covering what they have learned in this learning episode. After a few minutes, ask students to compare their maps and to identify any improvements they could make.

Learning episode B7d Consolidation and summary

Learning aims

- Review the learning points of the topic summarised in the End of topic checklist.
- Test understanding of the topic content by answering the End of topic questions.

Resources

Student Book End of topic checklist and End of topic questions

Approach

Introduce the learning episode

Ask students to work with a partner to make a list of key words from this topic. They could then work together to produce a spider diagram showing how the different concepts are linked. They could compare their list with the list of key terms given at the end of the topic in the Student Book. Discuss the checklist at the end of the topic and use questioning to see how much of the content they are comfortable with.

Students could make flashcards of the key content and then use the flashcards to quiz each other on the information.

Develop the learning episode

Ask students to work individually through the End of topic questions in the Student Book without looking at the text. As they work, walk around the classroom observing their answers and asking questions as necessary to find out which questions are causing difficulties.

Finish the learning episode

After a set period, ask the students to stop working and discuss any areas of difficulty you observed as you walked round the class. Students should complete any unanswered questions for homework, but you should stress that they should try to answer the questions without looking at the text, so that they can see how much they have remembered.

B8 Transport in plants

Contents

Teacher Guide	Syllabus reference
B8 Transport in plants	B8.1 Xylem and phloem B8.2 Water uptake B8.3 Transpiration B8.4 Translocation

Introduction

In this section, students will learn about the transport systems in flowering plants, including the processes of water uptake and transpiration. Use what students learned in B2: *Cells* to remind them of plant cell structure, and supplementary students can be reminded of the function of root hair cells. The content from B3 will be further developed in the context of water uptake. The role of phloem can also be linked to transporting the products from photosynthesis learned about in B6: *Plant nutrition*.

Links to other topics

Topics	**Essential background knowledge**	**Useful links**
B2 Cells	B2a Cell structure and size of specimens	
B3 Movement into and out of cells	B3a Diffusion B3b Osmosis B3c Active transport	
B4 Biological molecules		B4a Carbohydrates, fats and proteins
B6 Plant nutrition	B6a Photosynthesis B6b Leaf structure	
B12 Respiration		B12a Respiration

Topic overview

B8a	**Xylem and phloem** In this learning episode, students will learn about the tissues involved in transporting materials around a plant.
B8b	**Water uptake** In this learning episode, students look at how water is gained through the root hair cells, and how it moves through the plant. Supplement Students look at the role of root hair cells and their large surface area.

B8c	**Transpiration** Students learn about the route of water loss through leaves. Supplement There is the opportunity to plan and carry out practical work on factors that affect the rate of transpiration from leaves.
B8d	**Consolidation and summary** This learning episode provides an opportunity for a quick recap on the ideas encountered in this section. Students can answer the End of topic questions in the Student Book.

Careers links

These are some scientific careers that focus on this area of biology but careers in many other fields use the knowledge and skills gained studying science. **Forestry managers** look after forests and woodland areas and need to ensure that newly planted trees will have sufficient space and water to grow. **Maple tree tappers** carefully remove some of the sap in the phloem of maple trees to produce maple syrup.

Starting points

The Student Book section opener puts the ideas in the section into context and sets the scene. It also allows students to acknowledge and value their prior learning, and provides a benchmark against which future learning can be compared.

The questions provide a structure for introducing the section and can be used in a number of different ways:

- You could ask students to consider the questions as an introductory homework task.
- You could put students into groups to share their own ideas and understanding and then to report back to the whole class.
- Students could be given access to the Internet, preferably with a tight timescale, to find out the information required.

You could then use a spider chart or other form of wall chart to summarise everybody's ideas.

Recording these initial ideas allows you to retain them for reference as the individual topics are developed. In this way, your students' progress in learning can be readily acknowledged.

Learning episode B8a Xylem and phloem

Learning objectives

- State the functions of xylem and phloem:
  xylem – transport of water and mineral ions, and support; phloem – transport of sucrose and amino acids.
- Identify in diagrams and images the position of xylem and phloem as seen in sections of roots, stems, and leaves of non-woody dicotyledonous plants.

Common misconceptions

Check that students learn the correct spelling *absorption*, as some spell it *absorbtion*.

This is a good opportunity to revise diffusion and absorption from earlier sections and check that students have not formed any misconceptions on that work.

Resources

Student Book Xylem and phloem

Resources for demonstrations and class practicals (see Technician's notes)

Approach

1. Introduction

Remind students of their work on Section B6: *Plant nutrition*, and give them a minute or so to jot down which substances the plant needs from the environment for growth. They should identify what these substances are used for, which provides an opportunity to briefly revise photosynthesis and the formation of proteins and other substances that need mineral ions from the environment. Take examples from around the class and note any weaknesses in understanding that need addressing during work on this section. Ask students which parts of the plants take in the substances they need. They should be able to answer: leaves for taking in carbon dioxide and roots for absorbing water and mineral ions.

2. Xylem and phloem

Students should use prepared slides of transverse sections of plant roots, stems and leaves to help identify the vascular bundles in the different plant organs. Make sure that the slides show non-woody, herbaceous, dicotyledonous plants. (If slides are not available, students could use the photomicrographs in Fig. B8.3 in the Student Book.)

Students should draw labelled diagrams from each slide, to show the main features. This is a good opportunity to remind students how to draw plan diagrams, which outline the main areas, and then add detail only to the areas of interest, in this case the xylem and phloem. Remind students that they should use a sharp pencil, only draw what they see, and should label the key features on their diagrams. Fig. B8.2 in the Student Book could help students identify and label the tissues correctly on their diagrams.

Students then use the relevant pages of the Student Book to describe the functions of each of xylem and phloem.

3. Consolidation

Ask students to start a concept map on plant transport tissues, which they will add to during other activities in this section. This could be done as a group activity to encourage discussion of what should be added. They should begin with the words *Transport in plants* in the middle of the page and add any terms and knowledge they have gained from this learning episode. Save the concept maps for adding to later.

Technician's notes

Be sure to check the latest safety notes on these resources before proceeding.

The following resources are needed for task 2 on xylem and phloem:

prepared slides of transverse sections of plant roots, stems and leaves The slides should show non-woody, herbaceous, dicotyledonous plants.

Learning episode B8b Water uptake

Learning objectives

- Identify in diagrams and images root hair cells and state their functions.
- Supplement State that the large surface area of root hairs increases the uptake of water and mineral ions.
- Outline the pathway taken by water through the root, stem and leaf as: root hair cells, root cortex cells, xylem, mesophyll cells.

Common misconceptions

Check that students learn the correct spelling *absorption*, as some spell it *absorbtion*.

This is a good opportunity to revise diffusion and absorption from earlier sections and check that students have not formed any misconceptions on that work.

Some students think that the ends of roots are open, like drinking straws. Ask students why water enters the roots – remind them of what they learned about osmosis in Section B3: *Movement into and out of cells.*

Resources

Student Book Water uptake

Worksheet B8b.1 Water movement through plants

Resources for demonstrations and class practicals (see Technician's notes)

Approach

Supplement **1. Introduction**

Remind students of their work on surface area and rate of diffusion from Worksheet B3a.3. If they did not carry out that practical work, present this here as a prepared demonstration (see Technician's notes).

If possible, show them a picture of a unicellular organism, such as *Paramecium* or *Amoeba*, from a slide or the Internet. Ask them to apply what they learned from the practical to the diffusion of nutrients into and out of the unicellular organism. They should compare this with the problem of getting materials to and from diffusion surfaces and the external environment in larger organisms. Give them a minute or so to discuss in pairs how large/multicellular organisms, such as plants, might solve this problem. Take ideas from round the class. (If needed, remind students of their work in Section B6: *Plant nutrition,* and ask them to think about which part of the plant absorbs mineral ions.) They should be able to suggest that the much-divided root system of the plant provides a large surface area for absorption.

2. Root hair cells

Students should look at a prepared slide of a plant root showing root hairs, or an image from the Internet, to identify the root hair cells. In addition, they should look at Fig. B8.5 in the Student Book, which shows the root hair cells on the outside of a root. They should identify the root hair cells as the site of absorption of water and mineral ions. This can be reinforced by explaining that if a plant is lifted out of the soil, these delicate root hairs are usually damaged. So, even if the plant is immediately replanted, it will take a few days before it can absorb water and mineral ions effectively again.

Supplement Link this work to the introductory task to help students understand the role of root hair cells in providing a much-increased surface area for absorption of water and mineral ions.

3. Water movement through plants

Worksheet B8b.1 provides a method for investigating the transport of water through the above-ground parts of a plant. Although knowledge of this method is beyond the requirements of the syllabus, it offers an interesting extension, and could simply be used as a demonstration. This takes a few minutes to set up, but then needs to be left for about 24 hours before looking at the results. Alternatively, this could be set up by a technician the day before the lesson.

Celery stalks work well for this, as they have large vascular bundles that can easily be removed from the stalk tissue. If celery is not available, use a plant stem that is wide and fairly transparent, such as busy Lizzie (*Impatiens*) or just wide, such as *Coleus*. If the plant stems have flowers, ideally choose white flowers as these will more clearly show the dye in the petals. If the plant stems have leaves, ideally choose those with pale or yellow leaves.

SAFETY INFORMATION: Remind students to take care with sharp knives.

Use the information on water uptake from the relevant pages of the Student Book to help describe the whole pathway taken by water through the plant from the root hair cells to the leaf mesophyll cells.

4. Consolidation

Students could develop the concept maps they started in the last learning episode, to include what they have learned in this learning episode. Alternatively, ask students following the supplementary syllabus to write a short paragraph for a web science encyclopaedia to explain how plants are adapted to take in water and mineral ions.

Technician's notes

Be sure to check the latest safety notes on these resources before proceeding.

The following resources are needed for the introduction:

picture / slide of a unicellular organism, such as *Paramecium* or *Amoeba* and the means to display it

The following resources are needed for task 2 on root hair cells, per student or group:

prepared slide of transverse section of a plant root, showing root hair cells, or a photograph of this

The following resources are needed for the class practical B8b.1, per group:

stem or stalk of plant
1 cm^3 food colouring
teat pipette
water
250 cm^3 beaker
sharp knife and cutting board
forceps
hand lens

See note above about suitable plants to use.

Keep the ends of the stalks/stems sitting in water until needed, to prevent air entering the vascular tissue.

SAFETY INFORMATION: Remind students to take care when using sharp knives.

Learning episode B8c Transpiration

Learning objectives

- Describe transpiration as the loss of water vapour from leaves.
- Supplement Investigate and describe the effects of variation of temperature and wind speed on transpiration rate.

Common misconceptions

Sometimes, when students describe transpiration as the evaporation of water from leaves, they do not appreciate that the evaporation takes place inside the leaf and the actual loss of water from the leaves is by diffusion of water vapour through the stomata.

Resources

Student Book Transpiration

Worksheet B8c.1 The rate of transpiration

Resources for class practicals (see Technician's notes)

Approach

1. Introduction

Give students the following words and ask them to write sentences that use as many of the words as possible: *root hair cell, absorption, water, root, stem, xylem, osmosis, mineral ions* and for students following the supplementary syllabus, *surface area*. Take examples from around the class and check that students have remembered all the key points from the previous learning episode.

2. Transpiration

On the day before you teach this section, place a potted plant in a plastic bag so that all the above-ground parts are enclosed, but the pot is not. Tie the bag loosely around the plant stem. Then place the plant in a bright place until the lesson. If the light is bright enough, and it is reasonably warm, condensation should be visible on the inside of the bag.

Show the plant in its bag to students and explain that the only water that plant has had was given to the roots. Ask them to explain as fully as they can why the condensation has formed inside the bag. They should be able to explain that the water in the pot has been absorbed by the roots and lost as water vapour from the leaves.

Introduce the term *transpiration* for the loss of water vapour from a leaf. Make sure that students identify *xylem* as the transport tissue through which water and dissolved mineral ions move through a plant.

If you have the equipment, consider allowing students to test their plans once you have checked them for safety and suitability. You will need to set up the potometer for them. Details are given in the Technician's notes below. The Developing Practical Skills activity in the 'Factors that affect the rate of transpiration' section of the Student Book relates directly to the experiment, and provides secondary data for students to analyse.

Supplement 3. Factors that affect transpiration

Worksheet B8c.1 is a planning sheet on the factors that affect the rate of transpiration. It suggests equipment for producing a simple potometer, but a commercial potometer is likely to give better results if set up correctly.

If you have the equipment, consider allowing students to test their plans once you have checked them for safety and suitability. You will need to set up the potometer for them. Details are given in the Technician's notes below.

This experiment could be extended to record the movement of the bubble when leaves are removed from the stem. Ask students to predict the effect of this when the apparatus is set up, and to explain any results from this extension.

If the practical set-up is not feasible, ask students to read the pages of the Student Book about the factors that affect the rate of transpiration. They should write a test question and answer asking how transpiration would change under different conditions (such as on a warm, still day or a cool, windy day).

Next, they should exchange their question and answer with another student who should first try to answer the question, then check their answer with the one given and identify ways of improving both the question and answer.

SAFETY INFORMATION: Wash hands thoroughly at the end of the practical. Be careful with the possibility of water being spilled close to the electrical fan.

4. Consolidation

Students could add to their concept maps from the last two activities, to include what they have learned in this learning episode. Alternatively, ask students to write a question about plant transport tissues and transpiration, that is worth 3 or 4 marks, and the mark scheme for it.

They should exchange their question and mark scheme with another student and look for weaknesses in the other student's answer. They should suggest any amendments to improve the question or answer, then return it to the other student to discuss and change.

Technician's notes

Be sure to check the latest safety notes on these resources before proceeding.

The following resources are needed for task 2, for demonstration:

potted plant
clear plastic bag
root section under a light microscope *or* an example displayed on a screen; transverse section of a plant leaf, either under a light microscope or displayed on a screen

On the day before the lesson, place a potted plant in a plastic bag so that all the above-ground parts are enclosed, but the pot is not. Tie the bag loosely around the plant stem. Then place the plant in a bright place until the lesson. If the light is bright enough, and it is reasonably warm, condensation should be visible on the inside of the bag.

The following resources are needed for the class practical B8c.1, per group:

potometer (see note above)
bright light (as light bank or next to bright window)
dim and/or dark conditions
light meter or sensor
electric fan with several settings
thermometer
access to warm area
access to cooler area

If a potometer is to be set up for the demonstration or for use by students in an investigation, use the following instructions.

Use woody shoots from a bush or tree that does not have glossy leaves. Assemble the potometer underwater to prevent air bubbles entering the apparatus. (Refer to the manufacturer's instructions.) The shoot should also be cut and inserted into the potometer under water.

Allow the leaves to dry before the lesson and allow the plant to adjust to the new conditions. Just before measurement starts, insert an air bubble into the tube as described in the manufacturer's instructions, and adjust its position to sit within the scale.

If students are to carry out their plans from the worksheet, the following apparatus may be needed. However, students may suggest other apparatus. The plans should be checked to make sure that the apparatus suggested is available.

If using lamps, please note the following:

- The voltage of the bulb should not exceed the safety limits of the bench lamp. Low-energy bulbs should have some sort of protection so that they do not protrude beyond the protective metal of the bench lamp. Or they can be protected by a plastic or glass screen.
- Low-energy bulbs contain mercury and so can be hazardous if the bulb is broken.
- You should be vigilant during this practical because of water being close to electrical equipment.
- LED lights emit a very narrow range of wavelengths of light, which might not be absorbed by the plant. Halogen light sources are fine for photosynthesis, but they do get hot. A large beaker of tap water placed in front of the light source will absorb the heat.
- Potometers should be prepared in a different part of the room from the light sources.

SAFETY INFORMATION: Wash hands thoroughly after handling the plants and ventilate the room after the activity.

Learning episode B8d Consolidation and summary

Learning aims

- Review the learning points of the topic summarised in the End of topic checklist
- Test understanding of the topic content by answering the End of topic questions

Resources

Student Book End of topic checklist and End of topic questions

Approach

Introduce the learning episode

Ask students to work with a partner to make a list of key words from this topic. They could then work together to produce a spider diagram showing how the different concepts are linked. They could compare their list with the list of key terms given at the end of the topic in the Student Book. Discuss the checklist at the end of the topic and use questioning to see how much of the content they are comfortable with.

Students could make flashcards of the key content and then use the flashcards to quiz each other on the information.

Develop the learning episode

Ask students to work individually through the End of topic questions in the Student Book without looking at the text. As they work, walk around the classroom observing their answers and asking questions as necessary to find out which questions are causing difficulties.

Finish the learning episode

After a set period, ask the students to stop working and discuss any areas of difficulty you observed as you walked round the class. Students should complete any unanswered questions for homework, but you should stress that they should try to answer the questions without looking at the text, so that they can see how much they have remembered.

B9 Transport in animals

Contents

Teacher Guide	Syllabus reference
B9 Transport in animals	B9.1 Circulatory systems B9.2 Heart B9.3 Blood vessels B9.4 Blood

Introduction

In this section, students will learn about the transport systems in animals. Students should remember the structure of cells from B2: *Cells*, and they should be reminded of this when looking at the cells found in blood. It would also be good to remind students of the importance of surface area to volume ratio.

Links to other topics

Topics	**Essential background knowledge**	**Useful links**
B1 Characteristics of living organisms	B1a Characteristics of living organisms	
B2 Cells	B2a Cell structure and size of specimens	
B3 Movement into and out of cells		B3a Diffusion B3b Osmosis
B5 Enzymes		B5b Enzyme activity
B7 Human nutrition	B7a Diet B7b Digestive system	
B10 Diseases and immunity		B10a Disease, defence and hygiene B10b The immune system
B11 Gas exchange in humans		B11a Gas exchange system
B12 Respiration		B12a Respiration

Topic overview

B9a	**Circulatory systems and the heart** This learning episode looks at the overall structure of circulatory systems and of the heart. It includes practical work on the effect of exercise on pulse rate, and a demonstration of the dissection of a heart. Supplement Students will also learn about the causes of coronary heart disease, some risk factors associated with it and how it may be prevented. Students will also learn why exercise affects pulse rate.

B9b	**Blood vessels and blood** This learning episode looks at the structure and function of arteries, veins and capillaries. It also looks at the structure and function of the different components in human blood. Supplement Students will learn how vessel structure is related to function, as well as learning further detail of the functions of some blood components, and blood clotting.
B9c	**Consolidation and summary** This learning episode provides an opportunity for a quick recap on the ideas encountered in the section. Students can answer the End of topic questions in the Student Book.

Careers links

These are some scientific careers that focus on this area of biology but careers in many other fields use the knowledge and skills gained studying science. Many health professionals need to have an understanding of the human circulatory system. **Phlebotomists** take blood from people for clinical testing, research or blood donations. **Cardiologists** are doctors who specialise in treating heart conditions.

Starting points

The Student Book section opener puts the ideas in the section into context and sets the scene. It also allows students to acknowledge and value their prior learning, and provides a benchmark against which future learning can be compared.

The questions provide a structure for introducing the section and can be used in a number of different ways:

- You could ask students to consider the questions as an introductory homework task.
- You could put students into groups to share their own ideas and understanding and then to report back to the whole class.
- Students could be given access to the Internet, preferably with a tight timescale, to find out the information required.

You could then use a spider chart or other form of wall chart to summarise everybody's ideas.

Recording these initial ideas allows you to retain them for reference as the individual topics are developed. In this way, your students' progress in learning can be readily acknowledged.

Learning episode B9a Circulatory systems and the heart

Learning objectives

- Describe the circulatory system as a system of blood vessels with a pump and valves to ensure one-way flow of blood.
- Identify in diagrams and images the structures of the mammalian heart, limited to: muscular wall, septum, left and right ventricles, left and right atria, one-way valves, and coronary arteries.
- State that blood is pumped away from the heart in arteries and returns to the heart in veins.
- State that the activity of the heart may be monitored by: ECG, pulse rate and listening to sounds of valves closing.
- Investigate and describe the effect of physical activity on the heart rate.
- Supplement Describe the functioning of the heart in terms of the contraction of muscles of the atria and ventricles, and the action of the valves.
- Supplement Explain the effect of physical activity on the heart rate.
- Supplement Describe coronary heart disease in terms of the blockage of coronary arteries and state the possible risk factors including: diet, lack of exercise, stress, smoking, genetic predisposition, age, and sex.
- Supplement Discuss the roles of diet and exercise in reducing the risk of coronary heart disease.

Common misconceptions

Some students think that the heart does not need its own blood supply – the heart is full of blood, so why does it need its own blood supply? Coronary blood vessels are needed to supply the tissues in the heart. This can be understood best when the role of different blood vessels is introduced, because substances in the blood diffuse most effectively into and out of capillaries, whose walls are formed from a single layer of cells.

Resources

Student Book Circulatory systems, Heart

Worksheet B9a.1 Structure of a mammalian heart

Worksheet B9a.2 Exercise and pulse rate

Resources for class practicals (see Technician's notes)

Approach

1. Introduction

Remind students of their earlier work on levels of organisation in Section B2: *Cells*. Using the circulatory system as an example, ask them to suggest an organ, a tissue and a cell in this system. List as many correct examples of each on the board and prompt for others that may not be immediately obvious to students. Examples are: organ – heart, artery, vein; tissue – blood, muscle in heart and blood vessel walls; cell – red blood cell, white blood cell, muscle cell. You could return to these lists at the end of this learning episode to see whether students can add more examples of each.

Note that the blood vessels content of the first learning objective in the syllabus, 'describing the term *circulatory system* as containing a pump (the heart) and valves that ensure one-way flow of blood', will be covered in the next episode.

2. Mammalian heart dissection demonstration

If you can get a mammalian heart from a butcher, and you are willing to do so, demonstrate the dissection of a heart (see Technician's notes). If possible, get a heart that still has the end of the blood vessels attached. You may want to wear disposable gloves during the dissection, and afterwards all surfaces and tools that have been in contact with the blood will need disinfecting, using a disinfectant such as 1% Virkon, for at least ten minutes. Wash your hands thoroughly after finishing the dissection and clearing up.

Start by pointing out where the blood vessels are attached (to the 'top' of the heart). Students will not be able to identify which vessel is which at this point. Use a sharp scalpel to cut down through each of the blood vessels into the heart, and then between the atria and ventricles, to open the heart out completely.

Help students to identify and distinguish the two atria and the two ventricles, by looking at differences in the thickness of the muscle walls. Also point out the valves and 'heart strings' in the ventricles that prevent the valves opening the wrong way when the heart contracts. Discuss how this ensures that blood only flows one way through the heart.

Students should make notes of the structures and their relationships during the dissection. Worksheet B9a.1 could be used to support this.

3. Exercise and pulse rate

Worksheet B9a.2 gives students an opportunity to investigate the effect of exercise on pulse rate. This work could be combined with the investigation of the effect of exercise on breathing in Worksheet B11b.1, Section B11: *Gas exchange in humans*. The data collected on breathing could then be saved for analysis in the appropriate lesson. If you intend to do this, please check the technician's notes for that investigation and explain to students how to carry out the additional measurements safely.

Ideally students should take repeat measurements at each level of exercise and calculate an average from their results. Also, if the same method is used by all students, the data could be collated and compared, to identify that different people have different pulse rates, but that they all respond to an increase in level of exercise with an increased pulse rate.

Check that students have drawn up a suitable table for their results before they carry out the tests.

The Developing Practical Skills activity in the 'Changing heart rate' section of the Student Book relates directly to the experiment, and provides secondary data for students to analyse, and an opportunity to evaluate a given method.

Supplement Students should be expected to explain these results in terms of delivering oxygen and glucose (food molecules) for respiration more rapidly to the actively respiring muscle cells, which need to use energy for contraction.

SAFETY INFORMATION: Be considerate of students' concerns about exercise. If some students are unable or unwilling to carry out the exercise, pair them with other students so that one is the test subject and the other is the recorder of results.

Over-exertion may be a hazard, especially for those with some medical conditions. Competitive situations can lead to careless behaviour and accidents.

Students who are exercising should be appropriately dressed, such as in gym kit (for example, shorts or tracksuit bottoms, T-shirt and training shoes), and the exercise should take place under supervision. If any student shows signs of difficulty, they should stop exercising immediately and sit quietly until they recover. For further advice on this, consult your PE department.

A suitable form of exercise is doing step-ups on stable equipment. This is much better than students running upstairs.

If taking the pulse rate in the neck, be careful that blood flow to the brain is not restricted.

Students should also research and report on other ways of monitoring heart activity using ECG or by listening to the sounds of valves closing. They could use the Internet or information from the 'Heart rate' section of the Student Book to collect their information.

Supplement **4. Coronary heart disease**

Discuss the importance of a separate, dedicated blood supply to the heart tissues through the coronary blood vessels. Although the heart is continually permeated with blood, the exchange of substances between the blood and tissues is most effective in through capillaries, which have very thin walls.

Ask students to consider the impact of a blockage of a coronary artery on the tissues that are beyond the blockage. From what students already know, they should be able to suggest that the tissues will be deprived of substances that they need, such as oxygen and glucose. They may also suggest that waste substances such carbon dioxide may build up in the tissues. The combined effects of this will quickly lead to heart damage or even death.

Ask students to read the 'Coronary heart disease' section of the Student Book to identify the direct cause of coronary heart disease (for example, deposition of cholesterol) and associated risk factors including diet, lack of exercise, stress, smoking, genetic predisposition, age and sex. They could extend this by completing research into the relationships between these risk factors and the causes of coronary heart disease, and then produce a poster or health leaflet giving guidance on how to reduce the risks.

4. Consolidation

Give students two pieces of paper. Ask them to write the word *true* on one piece and *false* on the other. Then give a series of true and false statements on what students have learned from this section, and ask students to hold up the correct piece of paper each time. Choose the statements to match the ability of the students. For example:

- The heart has four chambers. [true] (simple form)
- The mammalian heart has four chambers, two ventricles at the top and two atria at the bottom. [false] (more challenging)

The responses will help you identify any areas that students are still not sure about.

Technician's notes

Be sure to check the latest safety notes on these resources before proceeding.

The following resources are needed for the heart dissection demonstration:

heart
sharp knife or scalpel and dissecting board
disposable gloves
disinfectant such as 1% Virkon for disinfecting all tools and work surfaces
diagram of the heart

You will need one heart for this demonstration. Ideally the heart should still have the end of the blood vessels attached – they are best sourced from a butcher with a request to prepare them like this.

SAFETY INFORMATION: You may want to wear disposable gloves during the dissection, and afterwards all surfaces and tools that have been in contact with the blood will need disinfecting, using a disinfectant such as 1% Virkon, for at least 10 minutes. Wash your hands thoroughly after finishing the dissection and clearing up.

The following resources are needed for the class practical B9a.2, per group:

stopwatch

For this activity, students will need to exercise at different levels. Make sure that students are dressed appropriately and behave in a safe manner.

Learning episode B9b Blood vessels and blood

Learning objectives

- Describe the structure of arteries, veins, and capillaries, limited to: relative thickness of wall, diameter of the lumen, and the presence of valves in veins.
- Supplement Explain how the structure of arteries and veins is related to the pressure of the blood that they transport.
- List the components of blood as: red blood cells, white blood cells, platelets, and plasma.
- Identify red and white blood cells in photomicrographs and diagrams.
- State the functions of the following components of blood:
  red blood cells in transporting oxygen, including the role of haemoglobin; white blood cells in phagocytosis and antibody production; platelets in clotting (details are not required); plasma in the transport of blood cells, ions, nutrients, urea, hormones, and carbon dioxide.
- Supplement State the roles of blood clotting as preventing blood loss and the entry of pathogens.

Common misconceptions

Some students think that blood is a red liquid. Blood is more complex than this: it is composed of a straw-coloured liquid, called plasma, and many cells, including the red blood cells that give it its overall colour.

Resources

Student Book Blood vessels, Blood

Resources for class practicals (see Technician's notes)

Approach

1. Introduction

Give students one minute to write down three things they remember about blood and blood vessels. Take examples from around the class, and compile a list of remembered facts. Encourage discussion of points that have been misremembered, so as not to reinforce these, but do not worry if there is no resolution now. Instead, make sure these points are resolved correctly by the end of the lesson.

2. Blood vessels

Using the diagrams of the blood vessels from the Student Book, point out the distinction between arteries, which carry blood away from the heart, and veins, which carry blood back towards the heart (except the hepatic portal vein).

3. Structure of blood vessels

If available, students could look at, and draw, prepared slides of a vein, an artery and a capillary. Alternatively, display slides of these using a digital microscope, or use images from the Internet. Students should annotate their drawing to show the key features of each type of blood vessel.

Remind students that all these vessels are linked by asking them to describe the route blood might take from the heart through the kidneys and back to the heart.

If possible, show students a video clip of blood flowing through capillaries.

Supplement Students should explain how each of the key features of arteries and veins relates to their function.

4. Looking at blood

Use the section 'Blood' of the Student Book to introduce the components of blood. If possible, give students the opportunity to study prepared slides of human blood. Alternatively, display such a slide using a digital microscope, or use a suitable image from the Internet.

Ask students to draw examples of each of the different kinds of cell that they can see. This should include red blood cells, and white blood cells if visible. It is unlikely that they will identify platelets, but there may be more than one kind of white blood cell. Ask them to compare the size, structure and number of the different kinds of cell.

Students should also make notes to describe the functions of all the components of blood: red blood cells, white blood cells, platelets and plasma.

Supplement When discussing platelets, introduce the role of blood clotting as preventing blood loss and preventing the entry of pathogens.

Supplement 5. Red blood cells

Using the photograph of an electron micrograph of red blood cells from the Student Book, ask students to sketch one red blood cell and label it to explain how its structure is adapted to its function.

Students should use their knowledge of factors that affect the rate of diffusion from Section B3: *Movement into and out of cells*, to help explain the adaptations of red blood cells.

6. Consolidation: crossword clues

Ask students to write clues for a crossword on the structure and function of all of the components in blood, and the different kinds of blood vessel. They should write at least ten clues. Then ask them to test their clues on another student, and to amend the clues if they prove difficult to answer.

Technician's notes

Be sure to check the latest safety notes on these resources before proceeding.

The following resources are needed for the demonstration on blood vessels:

prepared slides of a vein, an artery and a capillary *or* slides of these using a digital microscope, *or* use images from the Internet
video clip of blood flowing through capillaries

The following resources are needed for a demonstration on blood:

prepared slides of human blood; display such a slide using a digital microscope *or* a suitable image from the Internet

Learning episode B9c Consolidation and summary

Learning aims

- Review the learning points of the topic summarised in the End of topic checklist
- Test understanding of the topic content by answering the End of topic questions

Resources

Student Book End of topic checklist and End of topic questions

Approach

Introduce the learning episode

Ask students to work with a partner to make a list of key words from this topic. They could then work together to produce a spider diagram showing how the different concepts are linked. They could compare their list with the list of key terms given at the end of the topic in the Student Book. Discuss the checklist at the end of the topic and use questioning to see how much of the content they are comfortable with.

Students could make flashcards of the key content and then use the flashcards to quiz each other on the information.

Develop the learning episode

Ask students to work individually through the End of topic questions in the Student Book without looking at the text. As they work, walk around the classroom observing their answers and asking questions as necessary to find out which questions are causing difficulties.

Finish the learning episode

After a set period, ask the students to stop working and discuss any areas of difficulty you observed as you walked round the class. Students should complete any unanswered questions for homework, but you should stress that they should try to answer the questions without looking at the text, so that they can see how much they have remembered.

B10 Diseases and immunity

Contents

Teacher Guide	Syllabus reference
B10 Diseases and immunity	B10.1 Diseases and immunity

Introduction

In this section, students will learn about transmissible diseases, hygiene and immunity.

A visit from a health professional could be a useful source of information for some of the tasks in Learning episode B10a on transmissible diseases and hygiene, and in Learning episode B10b on immunity and vaccination. Before the visit, work with students to produce suitable questions to give greatest benefit from this opportunity.

Links to other topics

Topics	Essential background knowledge	Useful links
B2 Cells	B2a Cell structure and size of specimens	
B5 Enzymes	B5a Enzyme structure	
B7 Human nutrition	B7b Digestive system	
B9 Transport in animals	B9b Blood vessels and blood	
B11 Gas exchange in humans		B11a Gas exchange system
B13 Drugs		B13a Drugs and medicinal drugs

Topic overview

B10a	**Disease, defence and hygiene** This learning episode covers transmissible diseases, their causes and methods of transmission. It also looks at the natural defences of the body. Supplement Students will learn about the structures of viruses and also how to prevent the spread of disease.
Supplement **B10b**	**The immune system** This learning episode looks in detail at how the immune system provides immunity to transmissible disease, through active and passive immunity, including the way in which vaccination can enhance immunity.
B10c	**Consolidation and summary** This learning episode provides an opportunity for a quick recap on the ideas encountered in the section. Students can answer the End of topic questions in the Student Book.

Careers links

These are some scientific careers that focus on this area of biology but careers in many other fields use the knowledge and skills gained studying science. **Epidemiologists** study patterns of health and disease in large populations to identify causes and suggest treatment methods. **Biotechnologists** and **pharmaceutical technicians** are involved in developing new vaccines.

Starting points

The Student Book section opener puts the ideas in the section into context and sets the scene. It also allows students to acknowledge and value their prior learning, and provides a benchmark against which future learning can be compared.

The questions provide a structure for introducing the section and can be used in a number of different ways:

- You could ask students to consider the questions as an introductory homework task.
- You could put students into groups to share their own ideas and understanding and then to report back to the whole class.
- Students could be given access to the Internet, preferably with a tight timescale, to find out the information required.

You could then use a spider chart or other form of wall chart to summarise everybody's ideas.

Recording these initial ideas allows you to retain them for reference as the individual topics are developed. In this way, your students' progress in learning can be readily acknowledged.

Learning episode B10a Disease, defence and hygiene

Learning objectives

- Describe a pathogen as a disease-causing organism.
- Describe a transmissible disease as a disease in which the pathogen can be passed from one host to another.
- State that a pathogen is transmitted:
  by direct contact, including through blood and other body fluids; indirectly, including from contaminated surfaces, food, animals, and air.
- Describe the body defences, limited to: skin, hairs in the nose, mucus, stomach acid and white blood cells.
- Supplement State the features of viruses, limited to a protein coat and genetic material.
- Supplement Explain the importance of the following in controlling the spread of disease:
  a clean water supply; hygienic food preparation; good personal hygiene; waste disposal; sewage treatment (details of the stages of sewage treatment are **not** required).

Common misconceptions

The term *germ* is commonly used to mean something microscopic that can cause disease. In a scientific sense, *germ* means something that gives rise to life and growth, such as the germ cells within a seed. So, it is important that students learn to use the correct term when talking about disease-causing microorganisms: *pathogen* (literally translated as 'something that gives rise to illness').

Many people confuse the relationship between microorganisms, bacteria and disease. Students should become aware that:

- microorganisms include many groups other than bacteria, such as viruses, and some fungi (e.g., yeasts);
- not all bacteria inside the body are pathogens – the majority have no effect on health and some are essential for health (particularly those in the digestive system.);
- many kinds of bacteria and other microorganisms live freely in the environment and play an important role in the recycling of nutrients.

Resources

Student Book Transmissible diseases; Methods of transmission; Controlling the spread of disease; Methods of defence

Worksheet B10a.1 Methods of transmission

Approach

1. Introduction

Ask students to define the term *disease*. Give them a minute or two to discuss their definitions in pairs or small groups and then take examples from around the class. Try to elicit a class definition of the word. Then write the word hyphenated like this: *dis-ease*. Explain that the word, quite literally, means the lack of ease or comfort. This meaning can be developed into 'something that causes the body to function abnormally, often causing symptoms (signs) of illness'.

Take examples of diseases from around the class and, without explaining why, write the names on the board in groups according to their types, such as transmissible or infectious diseases, disorders caused by malfunction of the body (e.g., cancer, diabetes, coronary heart disease), deficiency diseases (e.g., scurvy, rickets, genetic or inheritable diseases).

Challenge students to identify why the names are grouped as they are. Explain that this section will focus on diseases that can be passed from person to person, and introduce the term *transmissible* (or infectious) *disease* as the name for this group of diseases. (If students seem unsure about the term, make it clear that the other kinds of disease are not transmitted from person to person. Although genetic diseases are inherited from a parent, they are not considered transmissible diseases where infection can pass from an infected person to any uninfected person.) Also introduce the term *pathogen* (see common misconceptions above). For Supplement students include reference to viruses as examples of pathogens and include in the description that they are made of genetic material surrounded by a protein coat.

2. Transmissible diseases – methods of transmission

Worksheet B10b provides an outline diagram of Fig. B10.3 in the Student Book. Students could annotate the diagram to give examples of diseases that are transmitted by each of the methods shown. The 'Methods of transmission' section of the Student Book could help with this. Alternatively, students could use other sources, such as books or the Internet, to provide examples. (Suggest that students leave room for additional notes, which they will add in Task 4.)

Challenge students to explain the difference between direct and indirect contact methods of transmission.

3. Natural barriers to infection

Introduce this task by stating that we are surrounded by microorganisms, in the air and on every surface. (Take this opportunity, if needed, to reinforce the fact that only some of these are pathogens.) Ask why we are not ill all the time. Students should identify that the body is adapted to protect against infection much of the time.

Give students the headings: *Mechanical barriers*, *Chemical barriers*, *Immune system*. Then ask them to make notes of different examples under each of the headings. The 'Methods of defence' section of the Student Book can help with this, and supplemented by research in other books or the Internet if there is time. Note that detail of the immune system and vaccination is beyond the core syllabus. This detail will be covered in learning episode B10b for those students studying supplement content.

Supplement **4. Controlling the spread of disease**

Ask students to return to their annotated diagrams on transmissible diseases and to add notes that show how the risk of transmission by each method could be reduced. The 'Controlling the spread of disease' section of the Student Book can help with this.

Explain that methods that protect against the transmission of disease are known as good *hygiene*. Ask students to write a list of hygiene rules for students at the school, in order to reduce the risk of spreading disease.

5. Health poster

Give students the opportunity to bring together all that they have learned in this learning episode by focusing on one particular transmissible disease. Ask them to produce a health poster about the disease, which includes the following:

- identifies what causes the disease;
- describes how the disease is transmitted;
- Supplement describes how the body normally defends against the disease;
- Supplement explains how the risks of transmission can be reduced.

Allow time for students to swap posters and mark them. They should identify two points that are well covered, and two points that could be improved. Posters should be returned so that students can identify how they could improve the weaker points.

6. Consolidation

Ask students to compile a list of ten questions that could be used as crossword clues on what they have learned from this learning episode. They could then test out their clues on another student and use the responses to identify any weaknesses in their learning.

Supplement Learning episode B10b The immune system

Learning objectives

- Supplement Describe active immunity as defence against a pathogen by antibody production in the body.
- Supplement Explain that active immunity is gained after an infection by a pathogen or by vaccination.

Common misconceptions

It is important that students realise from this learning episode that immunity to one pathogen does not usually confer immunity to a different pathogen. For example, immunity to polio does not confer immunity to measles, or vice versa. The development of immunisation by Edward Jenner against smallpox, using a vaccine made from cowpox virus, was unusual and possible because the two viruses are very similar in structure. So, the immune response to one virus also protected against infection by the other.

There are many misconceptions about vaccination, many of which are beyond the requirements of the syllabus but may come up in discussion. One is that because a vaccine often contains the pathogen, then it must be dangerous. Students must realise that the form of the pathogen in a vaccine is harmless, or produces such a weak form of the illness that the person vaccinated may show few or no symptoms. In very rare circumstances, a person may react badly to vaccination, either to the weakened pathogen or to something else in the vaccine. The risk of a bad reaction to vaccination is far lower than the risk of harm from infection by the pathogen, which is why health professionals recommend vaccination.

Resources

Student Book The immune system and active immunity

Supplement Worksheet B10b.1 How antibodies defend the body

Approach

1. Introduction

Give students the saying, 'prevention is better than cure' and ask them to suggest as many reasons as they can think of how this could apply to transmissible diseases. They could do this individually, or in small groups to encourage discussion.

Take examples from around the class and try to encourage examples that relate not just to the individual, but also to their families, the health system and the country (which usually relates to income at a personal and national level, and the cost to healthcare).

2. The immune response – how antibodies defend the body

Worksheet B10b.1 provides a task and some images that will help students clarify their knowledge and understanding of how the immune system responds to infection by pathogens. Make sure the students identify the role of pathogens and antibodies, and the two possible outcomes for a pathogen tagged by antibodies.

3. Vaccination

Ask students to find out, either by questioning a health professional, looking at a health leaflet on vaccination, or by researching on the Internet, what the current recommendations for vaccination are where they live. They could extend their research to look at how recommendations vary because of the presence of different pathogens in different places, and how travellers abroad may need to consider vaccinations against additional diseases. They should realise that recommendations also vary at different times, as new vaccines become available or pathogens change.

Ask students to compare the effect of vaccination with the development of active immunity after infection, and to state the advantages of vaccination.

4. Consolidation

Ask students to write bullet notes as planning for a health leaflet for expectant mothers on protecting their baby against infection. The notes should cover the need for vaccination for conferring active immunity to common diseases, and the need for good hygiene to help prevent transmission of disease.

Learning episode B10c Consolidation and summary

Learning aims

- Review the learning points of the topic summarised in the End of topic checklist
- Test understanding of the topic content by answering the End of topic questions

Resources

Student Book End of topic checklist and End of topic questions

Approach

Introduce the learning episode

Ask students to work with a partner to make a list of key words from this topic. They could then work together to produce a spider diagram showing how the different concepts are linked. They could compare their list with the list of key terms given at the end of the topic in the Student Book. Discuss the checklist at the end of the topic and use questioning to see how much of the content they are comfortable with.

Students could make flashcards of the key content and then use the flashcards to quiz each other on the information.

Develop the learning episode

Ask students to work individually through the End of topic questions in the Student Book without looking at the text. As they work, walk around the classroom observing their answers and asking questions as necessary to find out which questions are causing difficulties.

Finish the learning episode

After a set period, ask the students to stop working and discuss any areas of difficulty you observed as you walked round the class. Students should complete any unanswered questions for homework, but you should stress that they should try to answer the questions without looking at the text, so that they can see how much they have remembered.

B11 Gas exchange in humans

Contents

Teacher Guide	Syllabus reference
B11 Gas exchange in humans	B11.1 Gas exchange in humans

Introduction

This section covers gas exchange surfaces in relation to the lungs in humans, and looks at the effect of exercise on breathing.

Links to other topics

Topics	Essential background knowledge	Useful links
B1 Characteristics of living organisms	B1a Characteristics of living organisms	
B2 Organisation of the organism	B2a Cell structure and size of specimens	
B3 Movement into and out of cells	B3a Diffusion	
B9 Transport in animals	B9a Circulatory systems and the heart B9b Blood vessels and blood	
B10 Diseases and immunity		B10a Disease, defence and hygiene
B12 Respiration		B12a Respiration

Topic overview

B11a	**Gas exchange system** This learning episode gives students an understanding of the structure of the human gas exchange system. Supplement Students will also learn how to describe the features of gas exchange systems, which provides good opportunities to revise and reinforce earlier learning on factors that increase the rate of diffusion (see Section B3: *Movement into and out of cells*).
B11b	**Investigating gas exchange** Students will investigate the effect of exercise on breathing.
B11c	**Consolidation and summary** This learning episode provides an opportunity for a quick recap on the ideas encountered in the section. Students can answer the End of topic questions in the Student Book.

Careers links

These are some scientific careers that focus on this area of biology but careers in many other fields use the knowledge and skills gained studying science. **Sports physiologists** work with athletes to improve their performance and need to understand how exercise affects gas exchange. **Anaesthetists** have to closely monitor patients' breathing, oxygen and carbon dioxide levels during operations.

Starting points

The Student Book section opener puts the ideas in the section into context and sets the scene. It also allows students to acknowledge and value their prior learning, and provides a benchmark against which future learning can be compared.

The questions provide a structure for introducing the section and can be used in a number of different ways:

- You could ask students to consider the questions as an introductory homework task.
- You could put students into groups to share their own ideas and understanding and then to report back to the whole class.
- Students could be given access to the Internet, preferably with a tight timescale, to find out the information required.

You could then use a spider chart or other form of wall chart to summarise everybody's ideas.

Recording these initial ideas allows you to retain them for reference as the individual topics are developed. In this way, your students' progress in learning can be readily acknowledged.

Learning episode B11a Gas exchange system

Learning objectives

- Identify in diagrams and images the following parts of the breathing system: lungs, diaphragm, ribs, intercostal muscles, larynx, trachea, bronchi, bronchioles, alveoli and associated capillaries.
- Supplement Describe the features of gas exchange surfaces in humans, limited to: large surface area, thin surface, good blood supply, and good ventilation with air.

Common misconceptions

It is important to reinforce the difference between the terms *breathing (ventilation)*, which is the movement of gases into and out of the gas exchange system, *gas exchange*, which is the exchange of gases across the gas exchange surface between the air and the body, and *respiration* (as in cellular respiration). Respiration will be covered in Section B12.

Students sometimes interpret diagrams of gas exchange to say that *all* the oxygen from air in the lungs moves into the blood, and *all* the carbon dioxide in the blood moves into the air in the lungs. If students are clear that the process of exchange is diffusion, then they should realise from their work in Section B3: *Movement into and out of cells*, that this is not possible. Work in this learning episode should help to clarify this.

Resources

Student Book The human breathing system; Gas exchange

Worksheet B11a.1 Structures of the human gas exchange system

Resources for demonstrations (see Technician's notes)

Approach

Supplement 1. Introduction

Remind students of their work on diffusion in Section B3: *Movement into and out of cells*, by asking them to remember what affects the rate of diffusion, and how this might influence what features an effective exchange surface has. Give them a short while to discuss these features in pairs or small groups, and then take examples from around the class. They should be able to remember a large surface area, thin surface across which diffusion occurs, and a good supply of the gases that are being exchanged.

Then get students to apply this knowledge to the human gas exchange system. They should be aware from earlier learning that gas exchange takes place between the air and the blood inside the lungs. Ask them what features they would expect to see in the lungs in order that the exchange of gases between the blood and air in the lungs by diffusion is as effective as possible. They should realise that the internal surface of the lung should be thin and have as large a surface area as possible.

2. Structures of the human gas exchange system

Use 'The human breathing system' section of the Student Book to introduce the structures of the gas exchange system in humans. This involves a lot of terms, some familiar, but also others that are new. An unlabelled version of Fig. B11.2 is given on Worksheet 11a.1. Students could use this to identify and label the key features of the human gas exchange system.

To help with learning, ask students to create their own glossary definitions for each of the bold words in the text. They could test out their definitions on other students to see whether they can guess the term correctly.

If available from a butcher, and if you are willing to do so, dissect lungs from a sheep to help students understand their three-dimensional structure.

SAFETY INFORMATION: You will need to wear disposable gloves while carrying out the demonstration, and wash down all surfaces, and soak all tools used in disinfectant (e.g., 1% Virkon) for at least ten minutes after the demonstration. Wash your hands thoroughly after removing the gloves.

If present, students should also identify the larynx, trachea, diaphragm, ribs and intercostal muscles.

3. Alveoli

If available, allow students to study prepared microscope slides of alveoli, and to draw and label what they see. Alternatively, display such a slide using a digital microscope, or a suitable image from the Internet, and discuss what is visible with the class.

Supplement Focus on the key adaptations that make alveoli well adapted for their role in gas exchange: the thin layer of tissue between the air in the lungs and the capillaries; the large number of capillaries supplying lots of blood; the large number of alveoli providing a large surface area for exchange.

Supplement **4. Gas exchange in the alveoli**

Ask students to sketch a simple outline diagram of the alveoli in an air sac and its related capillaries. They should then label the diagram to show what happens to the gases in air and blood at the alveolar surface.

Remind students of their work on diffusion in Section B3: *Movement into and out of cells*, and discuss why oxygen and carbon dioxide are exchanged between the blood and alveolar air in terms of diffusion. If students have difficulty remembering this, try reinforcing the idea with a simple model using beads or counters of two colours (one for oxygen and one for carbon dioxide) and a ruler or piece of string to model the alveolar surface. Arrange the beads so that there are some of each colour on each side, but that there are more of one colour on one side (i.e., oxygen on the alveolar side) and more of the other colour on the other side (i.e., carbon dioxide in the blood). Students should then move the beads to describe diffusion of the gases. They should remember that diffusion is the net movement of particles from a region of their higher concentration to their lower concentration, and if the system is static the two sides will end up with equal numbers of beads on each side. They could then consider the importance of ventilation and blood flow in maintaining the concentration differences that drive diffusion.

Ask students to write their own explanation on their diagram of how exchange of gases takes place in the lungs by diffusion.

5. Consolidation

Ask students to prepare a quick quiz of ten simple questions and answers on what they have learned from this learning episode. They should test out their questions on other students and identify those that they do not fully understand or could be improved. Allow a few minutes for students to check anything they are unsure of in the Student Book and discuss any areas that are showing general weakness.

Technician's notes

Be sure to check the latest safety notes on these resources before proceeding.

The following resources are needed for the demonstration on the structures of the human gas exchange system:

sheep lungs
sharp knife or scalpel
dissecting board
disposable gloves
disinfectant such as 1% Virkon for disinfecting all tools and work surfaces

If available from a butcher, and if you are willing to do so, dissect lungs from a sheep to help students understand their three-dimensional structure. You should wear disposable gloves while carrying out the demonstration. You will need to wash down all surfaces, and soak all tools used in disinfectant (e.g., 1% Virkon) for at least ten minutes after the demonstration. Wash your hands thoroughly after removing the gloves.

The following resources are needed for the demonstration on breathing:

If you have the equipment, use this apparatus to demonstrate the effect of the diaphragm on breathing.

rubber sheet for diaphragm (e.g., from a large rubber balloon)
Y tube
bung
2 small balloons
bell jar

Learning episode B11b Investigating gas exchange

Learning objectives

- Investigate and describe the effects of physical activity on the rate and depth of breathing.

Common misconceptions

Students may have misconceptions about breathing rate and fitness, it may be useful to show elite athletes after a race to show they also get out of breath.

Resources

Student Book Investigating the effects of exercise on breathing

Worksheet B11b.1 The effect of exercise on breathing

Resources for demonstration and class practical (see Technician's notes, following)

Approach

1. Introduction

Display a copy of Fig. B11.2 from the Student Book, or the diagram on Worksheet B11a.1, or ask them to look at the diagram in their book. If using the diagram from the worksheet, briefly revise the main structures, including the alveoli, capillaries, bronchioles, bronchi and trachea. Ask students to identify the main role of each of these structures. In particular, make sure they identify the alveolar surface as the gas exchange surface.

2. Modelling breathing demonstration

Although knowing how volume and pressure changes in the thorax lead to the ventilation of the lungs is beyond the requirements of the syllabus, it offers an interesting extension, and could simply be used as a demonstration.

3. The effect of exercise on breathing

Worksheet B11b.1 provides a method for investigating the effect of exercise on breathing. (This may be combined with gathering data for the investigation in Worksheet B9a.3 on the effect of exercise on pulse rate.) Students may have produced their own methods when answering questions in the Developing Practical Skills activity in the Student Book. If their plans are suitable, consider allowing them to carry those out instead of using the method on the worksheet. They should, however, complete the questions on the sheet using their own data.

Check that students have drawn up a suitable table for their results before they carry out the tests.

Students may need help with analysing their results, including identifying anomalous data. If students are unsure what is meant by 'quantitative', explain this before they answer the final question on the worksheet.

If a spirometer is not available, students can carry out the investigation measuring the rate of breathing only.

The method lets students decide how and when to take measurements, but guides them towards particular ways of carrying out the investigation. If class results are to be collated, you may wish to control the method so that all students work in the same way.

Students should find that the rate and depth of breathing increase as the level of exercise increases. However, there will probably be significant variation between students because of different levels of fitness and the way that students carry out their investigations. A quantitative method for measuring level of exercise would be using exercise equipment, such as a treadmill, which can be adjusted (for example, for speed).

SAFETY INFORMATION: Be considerate of students' concerns about exercise. If some students are unable or unwilling to carry out the exercise, pair them with other students so that one is the test subject and the other the recorder of results.

Over-exertion may be a hazard, especially for those with some medical conditions. Competitive situations can lead to careless behaviour and accidents.

Students who are exercising should be appropriately dressed, such as in gym kit, and the exercise should take place under supervision. If any student shows signs of difficulty, they should stop exercising immediately and sit quietly until they recover. For further advice on this, consult your PE department.

A suitable form of exercise is doing step-ups on stable equipment. This is much better than students running upstairs.

Mouthpieces should be sterilised before use and should not be shared.

4. Consolidation: making notes

Ask students to write five bullet point notes for a friend who has missed this work on how to investigate gas exchange. Give them two minutes to do this and then ask them to compare their notes with a neighbour and select the best five from the two lists. They could then work with another pair to produce the best five points from these. Then take examples from groups and ask students to select the best to create a class list of five key points.

Technician's notes

Be sure to check the latest safety notes on these resources before proceeding.

The following resources are needed for the class practical B11b.1, per group:

spirometer (see note above)
nose clip
stopwatch

If you do not have a commercial spirometer, the following simple apparatus can be constructed. This will not produce as reliable results, as it depends on the student breathing out normally once (rather than breathing in and out using the spirometer), and the tendency will be to breathe out more heavily than normal on each test. However, it may still produce useful results.

Calibrate a 2-litre plastic bottle by adding 500 cm^3 of water at a time, and marking the volume on the side of the bottle with a waterproof marker. When the bottle is full of water, invert it into a water trough without allowing any air into the bottle.

Insert a flexible plastic tube into the neck of the bottle and secure the bottle and tube in position. Clean the other end of the tubing with antiseptic solution. Alternatively, add a mouthpiece to the end of the tubing that can easily be removed and sterilised after each test. You will need a sterilised one of these for every student who is measured.

Sterilise mouthpieces using Milton® solution (sodium dichloroisocyanurate) made up according to the manufacturer's instructions. They should be soaked in the solution for at least 30 minutes before use.

Learning episode B11c Consolidation and summary

Learning aims

- Review the learning points of the topic summarised in the End of topic checklist
- Test understanding of the topic content by answering the End of topic questions

Resources

Student Book End of topic checklist and End of topic questions

Approach

Introduce the learning episode

Ask students to work with a partner to make a list of key words from this topic. They could then work together to produce a spider diagram showing how the different concepts are linked. They could compare their list with the list of key terms given at the end of the topic in the Student Book. Discuss the checklist at the end of the topic and use questioning to see how much of the content they are comfortable with.

Students could make flashcards of the key content and then use the flashcards to quiz each other on the information.

Develop the learning episode

Ask students to work individually through the End of topic questions in the Student Book without looking at the text. As they work, walk around the classroom observing their answers and asking questions as necessary to find out which questions are causing difficulties.

Finish the learning episode

After a set period, ask the students to stop working and discuss any areas of difficulty you observed as you walked round the class. Students should complete any unanswered questions for homework, but you should stress that they should try to answer the questions without looking at the text, so that they can see how much they have remembered.

B12 Respiration

Contents

Teacher Guide	Syllabus reference
B12	B12.1 Respiration

Introduction

This section covers the process and uses of aerobic cell respiration. Students following the supplementary syllabus will also need the balances chemical equation.

Links to other topics

Topics	Essential background knowledge	Useful links
B1 Characteristics of living organisms	B1a Characteristics of living organisms	
B2 Cells	B2a Cell structureand size of specimens	
B3 Movement into and out of cells		B3a Diffusion
B4 Biological molecules	B4a Biological molecules	
B5 Enzymes		B5b Enzyme activity
B9 Transport in animals	B9a Circulatory systems B9b Heart B9c Blood vessels B9d Blood	
B11 Gas exchange in humans	B11a Gas exchange system B11b Investigating gas exchange	
B15 Organisms and their environment		B15a Energy flow B15b Food chains and food webs B15c Carbon cycle

Topic overview

B12a	**Respiration** This learning episode gives students the opportunity to explore the importance of respiration, and find out in detail about aerobic respiration. Supplement Students will also encounter the balanced symbol equation for aerobic respiration.
B12b	**Consolidation and summary** This learning episode provides an opportunity for a quick recap on the ideas encountered in the section. Students can answer the End of topic questions in the Student Book.

Careers links

These are some scientific careers that focus on this area of biology but careers in many other fields use the knowledge and skills gained studying science. **Sports coaches** and **trainers** benefit from knowledge about anaerobic respiration and oxygen debt.

Starting points

The Student Book section opener puts the ideas in the section into context and sets the scene. It also allows students to acknowledge and value their prior learning, and provides a benchmark against which future learning can be compared.

The questions provide a structure for introducing the section and can be used in a number of different ways:

- You could ask students to consider the questions as an introductory homework task.
- You could put students into groups to share their own ideas and understanding and then to report back to the whole class.
- Students could be given access to the Internet, preferably with a tight timescale, to find out the information required.

You could then use a spider chart or other form of wall chart to summarise everybody's ideas.

Recording these initial ideas allows you to retain them for reference as the individual topics are developed. In this way, your students' progress in learning can be readily acknowledged.

Learning episode B12a Respiration

Learning objectives

- State the uses of energy in living organisms, including: muscle contraction, protein synthesis, cell division, active transport, growth and the maintenance of a constant body temperature.
- Describe aerobic respiration as the chemical reactions in cells that use oxygen to break down nutrient molecules to release energy.
- State the word equation for aerobic respiration as:
  glucose + oxygen → carbon dioxide + water.
- Supplement State the balanced chemical equation for aerobic respiration as:
  $C_6H_{12}O_6 + 6O_2 \rightarrow 6CO_2 + 6H_2O$.

Common misconceptions

In common language, the term *respiration* is used to mean breathing or ventilation. It is important to establish its correct scientific meaning: the release of energy from food molecules within cells.

There are many misconceptions around the understanding of the concept of energy. If students are also studying physics or chemistry, it can be helpful to develop a standard language and phraseology with colleagues. This can then be used when discussing aspects of energy and energy transfers to avoid the risk of developing misunderstandings about the conservation of energy in systems.

Note that the equation used to describe aerobic respiration is a summary of a complex series of reactions. At this level, students do not need to know any of the intermediate stages although it can help prepare students going on to higher-level courses to realise that there is much more going on than shown here.

Resources

Student Book Respiration, Aerobic respiration

Approach

1. Introduction

Following on from their work on human nutrition, ask students to explain why we need to eat. They should link food to the need for energy for life processes (if not, remind them of Section B1a: Characteristics of living organisms). Ask them to write down each of the life processes, and then an example to show how energy is used in each life process. Take examples from around the class for discussion. Try to get as wide a range of examples from students as you can.

2. Uses of energy in the human body

Give students a few minutes to list as many uses of energy in the human body as they can think of. Take examples from around the class and try to elicit all the examples in the syllabus (muscle contraction, protein synthesis, cell division, active transport, growth and maintenance of stable core body temperature). This may require prompting to think at the cellular level.

Ask students then to write a sentence for each example to show how the energy is used, for example:

- In muscles, energy is used when the muscle contracts.
- In protein synthesis, energy is used to form bonds between amino acids and so build a protein.

For examples that students have not yet come across in this course, provide sentences (or allow students to research on their own using later sections of the Student Book or on the Internet) such as:

- As energy is transferred from a warm object to its cooler surroundings, our body has to continually release energy from respiration to maintain a constant internal temperature.

Then ask how energy is released from food. From earlier work, students should be able to answer that it is respiration.

3. Aerobic respiration

Give students the terms *nutrient molecule*, *energy*, and *respiration*, and ask them to use the words to develop a definition for respiration. Students may well include *oxygen* in their definitions. In which case, point out that the general definition does not require this (although students will not be studying anaerobic respiration in the Combined Science course).

Use the text under the heading Aerobic respiration in the Student Book to introduce aerobic respiration and the word equation that summarises the reaction. Make sure students understand *aerobic* and how it is related to the equations.

Ask students to use their knowledge of body systems to explain how each of the reactants is delivered to each cell, and how waste products are removed. Also ask about why every human cell can carry out aerobic respiration.

Supplement Students should also learn the balanced symbol equation for aerobic respiration.

4. Consolidation

Give students a few minutes to write down three key points to answer a web query, 'What is aerobic respiration and what do we need it for?' They should compare their points with another student, selecting the three best, and repeat this in fours. Then take examples from around the class, and encourage discussion to select the best three for compiling a response to the query.

Learning episode B12b Consolidation and summary

Learning aims

- Review the learning points of the topic summarised in the End of topic checklist
- Test understanding of the topic content by answering the End of topic questions

Resources

Student Book End of topic checklist and End of topic questions

Approach

Introduce the learning episode

Ask students to work with a partner to make a list of key words from this topic. They could then work together to produce a spider diagram showing how the different concepts are linked. They could compare their list with the list of key terms given at the end of the topic in the Student Book. Discuss the checklist at the end of the topic and use questioning to see how much of the content they are comfortable with.

Students could make flashcards of the key content and then use the flashcards to quiz each other on the information.

Develop the learning episode

Ask students to work individually through the End of topic questions in the Student Book without looking at the text. As they work, walk around the classroom observing their answers and asking questions as necessary to find out which questions are causing difficulties.

Finish the learning episode

After a set period, ask the students to stop working and discuss any areas of difficulty you observed as you walked round the class. Students should complete any unanswered questions for homework, but you should stress that they should try to answer the questions without looking at the text, so that they can see how much they have remembered.

B13 Drugs

Contents

Teacher Guide	Syllabus reference
B13 Drugs	B13.1 Drugs

Introduction

This section covers drugs with a specific focus on antibiotics.

Links to other topics

Topics	Essential background knowledge	Useful links
B1 Characteristics of living organisms	B1a Characteristics of living organisms	
B2 Cells	B2a Cell structure and size of specimens	
B10 Diseases and immunity	B10a Disease, defence and hygiene	

Topic overview

B13a	**Drugs and medicinal drugs** This learning episode introduces the term *drug*, and then discusses antibiotics as examples of medicinal drugs. Supplement Students also look at how the development of antibiotic-resistant bacteria can be minimised. It may be helpful to arrange a visit from a health professional to discuss the correct use of antibiotics.
B13b	**Consolidation and summary** This learning episode provides an opportunity for a quick recap on the ideas encountered in the section, as well as time for the students to answer the End of topic questions in the Student Book.

Careers links

These are some scientific careers that focus on this area of biology but careers in many other fields use the knowledge and skills gained studying science. **Pharmacologists** study the effect of drugs on animals and humans. **Pharmacists** dispense drugs and provide people with treatment advice for some medical conditions.

Starting points

The Student Book section opener puts the ideas in the section into context and sets the scene. It also allows students to acknowledge and value their prior learning, and provides a benchmark against which future learning can be compared.

The questions provide a structure for introducing the section and can be used in a number of different ways:

- You could ask students to consider the questions as an introductory homework task.
- You could put students into groups to share their own ideas and understanding and then to report back to the whole class.
- Students could be given access to the Internet, preferably with a tight timescale, to find out the information required.

You could then use a spider chart or other form of wall chart to summarise everybody's ideas.

Recording these initial ideas allows you to retain them for reference as the individual topics are developed. In this way, your students' progress in learning can be readily acknowledged.

Learning episode B13a Drugs and medicinal drugs

Learning objectives

- Describe a drug as any substance taken into the body that modifies or affects chemical reactions in the body.
- Describe the use of antibiotics for the treatment of bacterial infections.
- State that some bacteria are resistant to antibiotics which reduces the effectiveness of antibiotics.
- State that antibiotics kill bacteria but do not affect viruses.
- Supplement Explain how using antibiotics only when essential can limit the development of resistant bacteria such as MRSA.

Common misconceptions

The common understanding of the term *drug* is often restricted to those drugs that are illegal or harmful. It is important that students understand the full meaning of the term, to cover legal drugs such as nicotine and caffeine as well as medicines.

Resources

Student Book Drugs

Approach

1. Introduction

Introduce the term *drug* to students and give them a few minutes to jot down any ideas related to the term. Take a few examples from the class, and use them to help draw together a definition of the word. Compare this definition with the one given in the Student Book, and ask students to comment on any similarities and differences. Students may hold a limited view of what a drug is, so use this opportunity to broaden it to the definition given in the Student Book.

Note: If you consider any comment from a student of concern in relation to safeguarding, do not respond directly but take your concern to a member of staff who has responsibility for matters of child protection.

2. Antibiotics

Use the example of antibiotics to introduce medicinal drugs – those used by doctors to treat the symptoms and causes of illness. Many students will have been given antibiotics at some time in their lives. Take a few minutes to collect examples of illnesses that have been treated. If necessary add other examples, such as throat infection, ear infection, infected skin wounds, kidney infection. Although students do not need to know the names of particular antibiotics, it may help some students if you mention some common types such as penicillin, ampicillin or erythromycin.

Ask students what these illnesses have in common. It is possible that they will not know, but lead them to the answer that these are all caused by bacteria. Remind them that antibiotics cannot be used to kill viruses and therefore are of no use against illnesses such as flu or HIV.

Supplement Remind students of their work in Section B2 on bacterial cells, and in Section B10 on the structure of viruses. Ask them to explain the differences between bacteria and viruses, and to use this to suggest possible reasons why antibiotics are not effective against viruses.

3. Problems with antibiotics

Ask students to make a list of questions about the use and misuse of antibiotics. These should include the problem of developing resistance, the need to complete a prescribed course of antibiotics, and their ineffectiveness against viruses.

Supplement Students should also look at the importance of only using antibiotics when absolutely essential – they may find it useful to read the information from the Antibiotic resistance section of the Student Book.

If possible, arrange a visit from a health professional to answer the questions. Alternatively, students could carry out research in books or on the Internet to find the answers.

Students could use their findings to produce a simple health leaflet about what antibiotics are and how they should be used.

4. Consolidation

Ask students to work in pairs to produce a brief answer for a web forum to the question, 'What is an antibiotic?' or 'Why should I follow what my doctor says about taking antibiotics?' Give them a few minutes to complete this and then ask them to exchange their answer with that of another pair and to identify one good point and one weakness in that answer. They should then return the answer, and consider how to improve the weakness identified in their own answer.

Learning episode B13b Consolidation and summary

Learning aims

- Review the learning points of the topic summarised in the End of topic checklist
- Test understanding of the topic content by answering the End of topic questions

Resources

Student Book End of topic checklist and End of topic questions

Approach

Introduce the learning episode

Ask students to work with a partner to make a list of key words from this topic. They could then work together to produce a spider diagram showing how the different concepts are linked. They could compare their list with the list of key terms given at the End of the topic in the Student Book. Discuss the checklist at the end of the topic and use questioning to see how much of the content they are comfortable with.

Students could make flashcards of the key content and then use the flashcards to quiz each other on the information.

Develop the learning episode

Ask students to work individually through the End of topic questions in the Student Book without looking at the text. As they work, walk around the classroom observing their answers and asking questions as necessary to find out which questions are causing difficulties.

Finish the learning episode

After a set period, ask the students to stop working and discuss any areas of difficulty you observed as you walked round the class. Students should complete any unanswered questions for homework, but you should stress that they should try to answer the questions without looking at the text, so that they can see how much they have remembered.

B14 Reproduction

Contents

Teacher Guide	Syllabus reference
B14 Reproduction	B14.1 Sexual reproduction in plants B14.2 Sexual reproduction in humans

Introduction

This section looks at reproduction in flowering plants and human reproduction. This can be linked to later topics in conservation when discussing captive breeding and seed banks

Links to other topics

Topics	**Essential background knowledge**	**Useful links**
B1 Characteristics of living organisms	B1a Characteristics of living organisms	
B2 Cells	B2a Cell structure and size of specimens	
B12 Respiration	B12a Respiration	
B16 Human influences on ecosystems		B16b Conservation

Topic overview

B14a	**Sexual reproduction in plants** In this learning episode, students will learn about the structure of an insect-pollinated flower and the method of pollination. This includes practical work looking at the structures in a flower. They will also study what happens during the fertilisation of a flower, and investigate the conditions that affect germination. Supplement Students will also describe the anthers and stigmas of a wind-pollinated flower.
B14b	**Sexual reproduction in humans** This learning episode introduces the structures and function of the organs of the human male and female reproductive systems. Students will also learn about the process of fertilisation and about the menstrual cycle.
B14c	**Consolidation and summary** This learning episode provides an opportunity for a quick recap on the ideas encountered in the section, as well as time for the students to answer the End of topic questions in the Student Book.

Careers links

These are some scientific careers that focus on this area of biology but careers in many other fields use the knowledge and skills gained studying science. **Horticultural technicians** grow, maintain and monitor plants for companies and research organisations. Medical staff specialising in **Assisted fertility** as well as **Midwives** and **Obstetrics** focus on this area of science.

Starting points

The Student Book section opener puts the ideas in the section into context and sets the scene. It also allows students to acknowledge and value their prior learning, and provides a benchmark against which future learning can be compared.

The questions provide a structure for introducing the section and can be used in a number of different ways:

- You could ask students to consider the questions as an introductory homework task.
- You could put students into groups to share their own ideas and understanding and then to report back to the whole class.
- Students could be given access to the Internet, preferably with a tight timescale, to find out the information required.

You could then use a spider chart or other form of wall chart to summarise everybody's ideas.

Recording these initial ideas allows you to retain them for reference as the individual topics are developed. In this way, your students' progress in learning can be readily acknowledged.

Learning episode B14a Sexual reproduction in plants

Learning objectives

- Identify in diagrams and images, and draw, the following parts of an insect-pollinated flower: sepals, petals, stamens, filaments, anthers, carpels, style, stigma, ovary and ovules.
- State the functions of the structures listed above.
- Describe pollination as the transfer of pollen grains from an anther to a stigma.
- State that fertilisation occurs when a pollen nucleus fuses with a nucleus in an ovule.
- Investigate and describe the environmental conditions that affect germination of seeds, limited to the requirement for: water, oxygen and a suitable temperature.
- Supplement Identify in diagrams and images, and describe, the anthers and stigmas of a wind-pollinated flower.

Common misconceptions

Many people confuse *pollination* and *fertilisation*. It is important for students to be able to distinguish properly between them. There can also be confusion between the dispersal of pollen and seed dispersal.

Resources

Student Book Sexual reproduction in plants

Worksheet B14a.1 Flower dissection

Worksheet B14a.2 Investigating conditions for germination

Resources for demonstration and class practicals (see Technician's notes, following)

Approach

1. Introduction

Give students one minute to write down three important facts about flowers. They should discuss their facts with another student, and choose together their three most important facts. Take examples from the class to cover the widest possible range of facts. Make sure at least one fact is related to pollination.

Use the opportunity to ask students to distinguish between pollination and fertilisation in plants. If they are not sure, leave this for now but make sure you return to it at the end of the learning episode.

2. The structure of flowers

If possible, provide flowers from wind-pollinated and insect-pollinated plants for students to dissect. Worksheet B14a.1 supports this task. Students will also need the diagram of flower structure (Fig.14.2) from the Student Book. Consider providing a photocopy of this diagram to avoid damage to the books.

If there is not enough time for students to dissect a flower of each type, then give one half of the class wind-pollinated flowers and the other half insect-pollinated flowers. Allow time for students to present their results to the other half at the end of the practical work.

If practical work is not possible, ask students to research examples on the Internet or from books, and to draw one example of each, clearly labelling each diagram to show how it is adapted for its particular mode of pollination.

Students could also research the shapes of pollen grains from different plants, to identify whether those from wind-pollinated plants differ from those of insect-pollinated plants.

SAFETY INFORMATION: Remind students to take care when using sharp knives. Wash hands thoroughly after the practical. Pollen from some plants can cause an allergic reaction in some students.

3. Investigating conditions for germination

Worksheet B14a.2 provides a method for investigating the conditions needed for germination, and supports the Developing Practical Skills activity from the Student Book. If students have suggested other possible conditions, add these to the investigation.

Note that class results will be analysed, taking averages of repeat tests from groups investigating the same set of conditions. So the more conditions investigated at the same time, the fewer groups and fewer repeats of a set of conditions will be available for analysis. Seeds do not always germinate successfully, so at least one repeat for each condition (light intensity, moisture level and temperature) is advisable.

It is important to control other variables as far as possible during the investigation. This will mean trying to provide a light place and a dark place at similar temperature, and a warm place and a cool place with similar light intensity. If any of these are not possible, remove that condition from the investigation.

Students should find that seeds need moisture and warmth to germinate successfully. Seeds in the dark should germinate as successfully as seeds in the light in similar moisture and temperature, particularly if they are large seeds. (Note: some very small seeds, such as poppy, need light for germination as they will only germinate when on or near the surface of soil.)

SAFETY INFORMATION: Remind students to wash hands thoroughly after handling seeds. Only keep germinating seeds for one week and dispose after this.

4. Consolidation: drawings

Ask students to sketch a flower with typical features of one type of pollination. They should exchange their sketch with a neighbour. Ask them to identify which type of pollination is involved and then annotate the features that are adaptations for this type of pollination. They should then discuss the sketch with the student who drew it to identify any other features to annotate.

Technician's notes

Be sure to check the latest safety notes on these resources before proceeding.

The following resources are needed for the class practical B14a.1, per group:

flowers from plants of different species
sharp knife or scalpel (only use scalpels when necessary) and cutting board
hand lens or magnifying glass
pencil and paper
optional: photocopy of Fig.14.2 from the Student Book

Offer a selection of flowers, some of which are wind-pollinated and some of which are insect-pollinated, but avoid complex 'flowers', such as daises and dandelions, which consist of a large number of true flowers within one flower head. Wind-pollinated flowers may be difficult to obtain, as they are very seasonal. It is easy to confuse the fruit of wind-pollinated plants with the flower.

Only use scalpels where necessary.

The following resources are needed for the class practical B14a.2, per group:

20 seeds
2 Petri dishes and 2 paper towels
light sensor or temperature sensor or thermometer
access to light
access to dark
access to bright, cold place
marker pen

Use seeds that quickly germinate such as *Arabidopsis* or cress.

Note that some seeds sold for growing in gardens or at home are treated with fungicide that may cause an allergic response.

Learning episode B14b Sexual reproduction in humans

Learning objectives

- Identify on diagrams and state the functions of the following parts of the male reproductive system: testes, scrotum, sperm ducts, prostate gland, urethra and penis.
- Identify on diagrams and state the functions of the following parts of the female reproductive system: ovaries, oviducts, uterus, cervix and vagina.
- Describe fertilisation as the fusion of the nuclei from a male gamete (sperm) and a female gamete (egg cell).
- Describe the menstrual cycle in terms of changes in the ovaries and in the lining of the uterus (knowledge of sex hormones is **not** required).

Common misconceptions

Some parts of the male reproductive system and the urinary system are shared – some diagrams can be confusing because of this. Also, some diagrams of the female reproductive system may show the bladder misplaced to one side – students must realise that this has been done to make the diagram more intelligible and does not reflect the bladder's true position.

Resources

Student Book Sexual reproduction in humans

Worksheet B14b.1 Male reproductive system

Worksheet B14b.2 Female reproductive system

Approach

1. Introduction

Ask students to give the scientific terms for the organs of the human male reproductive system, and then for the human female reproductive system.

Be aware of continuing embarrassment of any student and manage the class if needed by allowing students to work more individually than in pairs or groups.

2. Human male reproductive system

Ask students to read the text in the Student Book on the human male reproductive system. They should make notes on the different organs in the system, and identify as many tissues and cells as they can, explaining the role they play in reproduction.

Worksheet B14b.1 gives an outline of the male reproductive system that students can use to answer the Student Book question on labelling the human male reproductive system.

3. Human female reproductive system

Ask students to read the text in the Student Book on the human female reproductive system. They should make notes on the different organs in the system, and identify as many tissues and cells as they can, explaining the role they play in reproduction.

Worksheet B14b.2 gives an outline of the female reproductive system that students can use to answer the Student Book question on labelling the female reproductive system.

4. Fertilisation and egg development

Remind students that sexual reproduction involves the fusion of the nuclei of male and female gametes to form a zygote (fertilisation).

Tell students that if fertilisation doesn't occur then menstruation happens instead. Using the information from The menstrual cycle section of the Student Book ask students to make a time line showing the changes in the ovaries and changes in the lining of the uterus.

5. Consolidation: crossword clues

Ask students to write clues for a crossword on the human reproductive system and development of the zygote, embryo and fetus. They should write clues for at least ten of the scientific terms that have been used in the lesson. If there is time, they could test their clues out on another student (there are several good programs available on the Internet for generating crossword puzzles).

Learning episode B14c Consolidation and summary

Learning aims

- Review the learning points of the topic summarised in the End of topic checklist
- Test understanding of the topic content by answering the End of topic questions

Resources

Student Book End of topic checklist and End of topic questions

Approach

Introduce the learning episode

Ask students to work with a partner to make a list of key words from this topic. They could then work together to produce a spider diagram showing how the different concepts are linked. They could compare their list with the list of key terms given at the End of the topic in the Student Book. Discuss the checklist at the End of the topic and use questioning to see how much of the content they are comfortable with.

Students could make flashcards of the key content and then use the flashcards to quiz each other on the information.

Develop the learning episode

Ask students to work individually through the End of topic questions in the Student Book without looking at the text. As they work, walk around the classroom observing their answers and asking questions as necessary to find out which questions are causing difficulties.

Finish the learning episode

After a set period, ask the students to stop working and discuss any areas of difficulty you observed as you walked round the class. Students should complete any unanswered questions for homework, but you should stress that they should try to answer the questions without looking at the text, so that they can see how much they have remembered.

B15 Organisms and their environment

Contents

Teacher Guide	Syllabus reference
B15 Organisms and their environment	B15.1 Energy flow B15.2 Food chains and food webs B15.3 Carbon cycle

Introduction

This section introduces students to ecology, and gives them opportunities to study the flow of energy through food chains and food webs. This could be linked to B7a Diet to get students to place themselves in a food web. This is followed by work on the carbon cycle.

Links to other topics

Topics	**Essential background knowledge**	**Useful links**
B1 Characteristics of living organisms	B1a Characteristics of living organisms	
B6 Plant nutrition	B6a Photosynthesis	
B7 Human nutrition		B7a Diet
B12 Respiration	B12a Respiration	
B16 Human influences on ecosystems		B16a Habitat destruction B16b Conservation

Topic overview

B15a	**Energy flow, food chains and food webs** This learning episode introduces students to some of the terminology used in ecological studies, principally concerning food chains and webs. Supplement Students will consider the impacts humans have through overharvesting and the introduction of foreign species.
B15b	**The carbon cycle** Students will study the carbon cycle with particular attention to the roles of living (and once-living) organisms.
B15c	**Consolidation and summary** This learning episode provides an opportunity for a quick recap on the ideas encountered in this section, as well as time for the students to answer the End of topic questions in the Student Book.

Careers links

These are some scientific careers that focus on this area of biology but careers in many other fields use the knowledge and skills gained studying science. **Ornithologists** study birds and may be involved in conservation work. **Mycologists** and **plant pathologists** study fungi and the role they play in decomposition. **Countryside officers and rangers** manage open spaces, the organisms that live there and the people that use the spaces.

Starting points

The Student Book section opener puts the ideas in the section into context and sets the scene. It also allows students to acknowledge and value their prior learning, and provides a benchmark against which future learning can be compared.

The questions provide a structure for introducing the section and can be used in a number of different ways:

- You could ask students to consider the questions as an introductory homework task.
- You could put students into groups to share their own ideas and understanding and then to report back to the whole class.
- Students could be given access to the Internet, preferably with a tight timescale, to find out the information required.

You could then use a spider chart or other form of wall chart to summarise everybody's ideas.

Recording these initial ideas allows you to retain them for reference as the individual topics are developed. In this way, your students' progress in learning can be readily acknowledged.

Learning episode B15a Energy flow, food chains and food webs

Learning objectives

- State that the Sun is the principal source of energy input to biological systems.
- Describe the flow of energy through living organisms, including light energy from the Sun and chemical energy in organisms, and its eventual transfer to the environment.
- Describe a food chain as showing the transfer of energy from one organism to the next, beginning with a producer.
- Construct and interpret simple food chains.
- Describe a food web as a network of interconnected food chains and interpret food webs.
- Describe a producer as an organism that makes its own organic nutrients, usually using energy from sunlight, through photosynthesis.
- Describe a consumer as an organism that gets its energy by feeding on other organisms.
- State that consumers may be classed as primary, secondary and tertiary according to their position in a food chain.
- Describe a herbivore as an animal that gets its energy by eating plants.
- Describe a carnivore as an animal that gets its energy by eating other animals.
- Describe a decomposer as an organism that gets its energy from dead or waste organic material.
- Supplement Use food chains and food webs to describe the impact humans have through overharvesting of food species and through introducing foreign species to a habitat.

Common misconceptions

If students carry out research on food webs and food chains, they may meet statements such as 'All food webs begin with a plant', and 'All energy for life on Earth comes from the Sun'. Neither of these is completely true. There are many producers other than photosynthesising plants, most of which are bacteria and too small to be easily noticed, and/or they live in environments that are difficult to explore, such as in mid-ocean ridges or in deep caves. Some of these have unique biochemistry that takes energy from the breakdown of inorganic chemicals to make 'food' rather than using light energy from the Sun. However, it is fair to say that they contribute little to the food chains and food webs that we normally see.

Resources

Student Book Energy flow, Food chains and food webs

Approach

1. Introduction

Students should have some understanding of ecological studies from earlier work. Give them a minute or so to write down what they remember about the terms *food chain* and *food web*. Ask if they can remember any other terms, such as *habitat*, and, if so, also write down what they mean. Then take examples from around the class and encourage discussion to come up with class definitions for the terms. This should help you identify how much they remember from earlier work.

The following tasks assume that students have remembered little from previous work. The tasks may need adjusting if they have a lot of previous knowledge.

2. Energy inputs

Ask students what all organisms need in order to survive, grow and reproduce. Refer them back to their earlier work on the key characteristics of organisms from section B1 Characteristics of living organisms. They should be able to respond with 'food', and to describe sources of this for both animals and plants.

Ask them what the food is used for. Explain that in animals it is a source of energy as well as providing essential raw materials, while in plants 'food' is created in photosynthesis as their source of energy with additional raw materials (mineral ions) being absorbed from the ground with water. Make sure students make the link with respiration, the process for releasing energy from 'food molecules' for use in other life processes.

Reinforce the link between 'food', which they will have studied in earlier work, and 'energy stored in food molecules' to help them develop their understanding for this course.

3. Food chains and webs

Ask students to work in pairs and to write down a food chain that they remember. (If they are struggling, give them an example of organisms that could be used to make a food chain, such as grass, rabbit, fox.) They should use their food chain to write a set of 'rules' for drawing a food chain. Take examples from around the class to draw up a set of class rules. They should include the following:

- the chain is set out with the organisms in order of who eats what, placing the 'eater' on the right of what they eat
- an arrow should point from the organism being eaten to the organism that is the 'eater'.

Note the rules above are written in the simplest terms. Students may have a greater understanding and be able to give names for the feeding (trophic) levels in the chain (if not, do not introduce them yet). Extend their understanding to realise that chemical energy stored in the tissues of the organism being eaten is transferred to the animal eating it.

Ask students what information they would need to develop their food chain into a food web. If there is time and access to the Internet, they could look for examples of food webs, then choose one and identify as many food chains as possible within it.

4. Feeding levels

Ask students to read the Food chains and food webs section of the Student Book about the naming of producers and consumers in food chains and webs. They should then annotate one food chain and food web used in the task above, or the diagram of a food web given in the Student Book, to identify the producer and classify the consumers.

Make sure students can identify *producers*, *consumers* (*primary*, *secondary* and *tertiary*), *herbivores* and *carnivores*.

Alternatively, they could test each other by asking questions, such as: 'Name one primary consumer in the food web', or 'Which secondary consumer eats xxx and xxx?'

Note, this is also a good opportunity to introduce the role of *decomposers*, which are traditionally left out of food chains, although they may be included in some food webs. Students may like to suggest why this is. The role of decomposers will be covered in more detail in work on the carbon cycle later in this section.

Supplement 5. Interpreting human impacts

Ask students to read the text about Interpreting human impact on food chains and food webs in the Student Book. They could then use one of the examples given, or research their own example of overharvesting of a food species or introduction of a foreign species into a habitat. They should use what they learn to try to describe the changes in terms of food chain or food web diagrams.

If students carry out research on different examples, allow opportunity for them to present their results to the rest of the class, or to produce a poster that describes what has happened in text and diagrams.

6. Consolidation: definitions

Give students a few minutes to write notes for a web dictionary to explain food webs, feeding levels and energy flow through them. They should include definitions of the key words (in bold) in the Student Book. They should swap their notes with a neighbour to check them and see how they could be improved. They should then swap them back again and make any changes that are needed to improve their notes.

Learning episode B15b The carbon cycle

Learning objectives

- Describe the carbon cycle, limited to: photosynthesis, respiration, feeding, decomposition, formation of fossil fuels and combustion.

Common misconceptions

Combustion of fossil fuels rapidly contributes carbon to the atmosphere. This is easier to understand if students understand clearly that the fuels are formed from the remains of huge quantities of dead plant or animal material deposited over millions of years and changed by geological processes over this time. Since plant and animal tissue is rich in carbon, so are fossil fuels.

Resources

Student Book The carbon cycle

Worksheet B15b.1 Constructing the carbon cycle

Approach

1. Introduction

Revise work from section B6 Plant nutrition and section B12 Respiration, by asking students to write down equations for photosynthesis and respiration.

If they are not sure, remind them of the practical work they carried out to investigate photosynthesis and gas exchange. Focus on the measurement of the net release of carbon dioxide using hydrogencarbonate indicator. Ask them to write down (or give them) an equation for the burning of fuel (combustion).

Give them a minute or so to compare the equations and write down three key points from the comparison. Take examples from around the class to make sure that students grasp that combustion and respiration are similar in releasing carbon dioxide, and photosynthesis is opposite in taking in carbon dioxide.

2. Constructing the carbon cycle

Worksheet B15b.1 provides cards that students can cut out and use to construct a diagram of the carbon cycle. Explain that the term *stores of carbon* on the worksheet means 'something in which carbon is found'. Also make sure that students understand the terms *combustion* (burning, in this case, of organic material such as fossil fuels) and *decomposition* (decay of dead organisms and animal wastes).

Encourage students to make the link between the carbon dioxide released during decay and the respiration of the decomposers as they use the food materials to release energy for growth and reproduction.

Students should lay out the cards first in what they think is the right arrangement, and then compare their layout with the diagram of the carbon cycle from the Student Book before sticking them into their workbook.

You may wish to point out to some students that they will need several copies of the word *respiration*. For an extra challenge, students could be encouraged to identify the form of carbon in each of the stores shown in their diagram.

If there is time, students could research in books or on the Internet how much carbon there is in each carbon store given on the worksheet.

3. Consolidation: true or false

Give students two pieces of paper and ask them to write *true* on one and *false* on the other. Then read out some statements about the carbon cycle, some of which are true and some false, and ask students to hold up the correct answer each time. Suitable statements include:

- Fossil fuels are formed from dead plant or animal material. [true]
- Carbon is released from living organisms to the air by photosynthesis. [false]

Learning episode B15c Consolidation and summary

Learning aims

- Review the learning points of the topic summarised in the End of topic checklist
- Test understanding of the topic content by answering the End of topic questions

Resources

Student Book End of topic checklist and End of topic questions

Approach

Introduce the learning episode

Ask students to work with a partner to make a list of key words from this topic. They could then work together to produce a spider diagram showing how the different concepts are linked. They could compare their list with the list of key terms given at the End of the topic in the Student Book. Discuss the checklist at the End of the topic and use questioning to see how much of the content they are comfortable with.

Students could make flashcards of the key content and then use the flashcards to quiz each other on the information.

Develop the learning episode

Ask students to work individually through the End of topic questions in the Student Book without looking at the text. As they work, walk around the classroom observing their answers and asking questions as necessary to find out which questions are causing difficulties.

Finish the learning episode

After a set period, ask the students to stop working and discuss any areas of difficulty you observed as you walked round the class. Students should complete any unanswered questions for homework, but you should stress that they should try to answer the questions without looking at the text, so that they can see how much they have remembered.

B16 Human influences on ecosystems

Contents

Teacher Guide	Syllabus reference
B16 Human influences on ecosystems	B16.1 Habitat destruction B16.2 Conservation

Introduction

This section looks at the effects that human activities are having on ecosystems as a result of habitat destruction, pollution and conservation. Students will study a range of important ecosystems, and be expected to apply previous learning from several sections.

Where possible, introduce appropriate local examples of pollution and conservation, to help students fully appreciate the effect of human activity in their local area, and to help them consider what can be done to limit any damage.

Links to other topics

Topics	Essential background knowledge	Useful links
B1 Characteristics of living organisms	B1a Characteristics of living organisms	
B6 Plant nutrition	B6a Photosynthesis	
B15 Organisms and their environment	B15a Energy flow B15b Food chains and food webs B15c Carbon cycle	

Topic overview

B16a	**Habitat destruction** Students study how habitat destruction, including deforestation, may be caused, and how it can affect and damage ecosystems and the environment. Supplement Students look into more detail at the undesirable effects of deforestation.
B16b	**Conservation** In this learning episode, students learn about how organisms may become endangered or extinct. Supplement Students learn about how endangered species can be conserved, including by protecting their habitats.
B16c	**Consolidation and summary** This learning episode provides an opportunity for a quick recap on the ideas encountered in this section, as well as time for the students to answer the End of topic questions in the Student Book.

Careers links

These are some scientific careers that focus on this area of biology but careers in many other fields use the knowledge and skills gained studying science. **Environmental consultants** work to manage risks from floods, climate change and waste production. **Conservationists** work to protect both individual species and larger areas of habitat.

Starting points

The Student Book section opener puts the ideas in the section into context and sets the scene. It also allows students to acknowledge and value their prior learning, and provides a benchmark against which future learning can be compared.

The questions provide a structure for introducing the section and can be used in a number of different ways:

- You could ask students to consider the questions as an introductory homework task.
- You could put students into groups to share their own ideas and understanding and then to report back to the whole class.
- Students could be given access to the Internet, preferably with a tight timescale, to find out the information required.

You could then use a spider chart or other form of wall chart to summarise everybody's ideas.

Recording these initial ideas allows you to retain them for reference as the individual topics are developed. In this way, your students' progress in learning can be readily acknowledged.

Learning episode B16a Habitat destruction

Learning objectives

- Describe an ecosystem as a unit containing the community of organisms and their environment, interacting together.
- Describe biodiversity as the number of different species that live in an area.
- Describe the reasons for habitat destruction, including:
  (a) increased area for housing, crop plant production and livestock production
  (b) extraction of natural resources
  (c) freshwater and marine pollution (a detailed description of eutrophication is not required).
- State the undesirable effects of deforestation as an example of habitat destruction, to include: reducing biodiversity, extinction, loss of soil, flooding and increase of carbon dioxide in the atmosphere
- Supplement Explain the undesirable effects of deforestation as an example of habitat destruction, to include: reducing biodiversity, extinction, loss of soil, flooding and increase of carbon dioxide in the atmosphere.

Common misconceptions

Students may recognise the immediate impacts of habitat destruction, such as the killing of a particular plant or animal species, but may not appreciate the wider implications as a result of interdependency between organisms and their environment. Encourage them to look beyond the obvious, to see what else might happen.

Resources

Student Book Habitat destruction

Worksheet B16a.1 Effects of deforestation

Approach

1. Introduction

Ask students to imagine that a small local wood is about to be cleared to make space for some new buildings. Give them a few minutes, working individually or in pairs, to write down as many of the effects of this on the environment as they can think of.

Take examples from around the class and discuss any suggestions that might appear to conflict in order to identify a resolution. Encourage them to use terms learned in section B15 Organisms and their environment, including *habitat, food chain, food web, population* and *community*. In addition, introduce and explain the term *ecosystem* as a unit containing the community of organisms and their environment and *biodiversity* as a measure of how many species live in an area.

2. Habitat destruction

Use the Internet or books to provide students with a range of images of habitat destruction, such as:

- large-scale monocultures of crops or large-scale livestock production
- open-cast mining for rocks (e.g., limestone) or minerals
- freshwater or marine pollution (e.g., the results of eutrophication, oil spills or plastic waste).

(Avoid deforestation as an example, as this will be covered in greater detail later.) If possible, provide both local and global examples.

Ask students to choose one example, and to work in small groups to research details of why the destruction occurred, and what impact it had, or is having, on the environment and the organisms that lived there before destruction began. Encourage students to think beyond the immediate impact, to the long-term effects on food webs and how that might affect a larger area. Ask students to draw up a list of reasons for habitat destruction, and how it can affect organisms through changing food chains and food webs. Each group should present their findings to the rest of the class.

3. Reasons for large-scale deforestation

Give students access to the Internet or to books to research an area of large-scale deforestation, such as the Amazon Basin, parts of Indonesia and Malaysia, or a local example if appropriate. They should try to find answers to the following questions:

- Why is the deforestation taking place? What will replace the forest?
- Who will benefit from what happens after the deforestation?
- Are there any people who will not benefit from the deforestation? Explain your answer.
- Is anything being done to minimise the impact of the deforestation on the environment? If so, what, and how will it help?
- Is there anything that people outside the area could do to help minimise the impact of deforestation within the area? (Note: the answer to this should highlight the role of global trade in creating markets for products from these areas, and indicate that it is not just a local issue.)

Supplement **4. Effects of deforestation**

Ask students to read about the effects of deforestation from of the Student Book. Make appropriate links to previous learning, such as food chains and webs in section B15 Organisms and their environment, when discussing extinction, and photosynthesis in section B6 Plant nutrition, and the carbon cycle in section B15, when discussing the increase of carbon dioxide in the atmosphere.

Worksheet B16a.1 provides a set of cards with words relating to various aspects of deforestation and life processes. Students could use these in different ways, for example:

- to start a concept map on the topic, and then add words and annotations of their own
- to link aspects of deforestation to living processes, to help explain the impact of deforestation on the environment
- to organise ideas to use as the basis of a poster on the subject.

5. Consolidation

Ask students to write brief notes for a web article on habitat destruction and deforestation. They should organise their notes under main headings to cover the key points in this learning episode. Ask students to work in pairs to compare their headings and notes for content. They should decide on the best headings and notes. Then take examples around the class and encourage discussion to decide on which headings to use and which main points should be included in each section.

Learning episode B16b Conservation

Learning objectives

- Outline why organisms become endangered or extinct, including: climate change, habitat destruction, hunting, overharvesting, pollution and introduced species.
- Supplement Describe how endangered species can be conserved, limited to:
  (a) monitoring and protecting species and habitats
  (b) education
  (c) captive breeding programmes
  (d) seed banks.

Common misconceptions

Students may not appreciate that there are different levels of conservation, from international and national strategies for protecting organisms and ecosystems, down to the individual level, where there are things we can do in our daily lives to improve sustainability.

Resources

Student Book Conservation

Approach

1. Introduction

Show students an image of a species that is at risk of extinction as a result of human activity; for example, a tiger – tigers are now so limited in their habitats, and still hunted due to fear or for the trade in body parts for local medicine, that their population sizes are dangerously low. Identify the human-related activities that have put the tiger (or other species) at risk.

Give students a few minutes to suggest what we should do about it. Take examples from around the class and make a list of suggestions. Encourage discussion about how realistic each suggestion is and what more might be implied by each idea; for example, putting more animals into zoos for protection needs more room and more money to care for them.

Introduce the term *conservation* as referring to the methods used to protect species and the environment to prevent long-term damage.

2. Extinction

Extend the task in the introduction by discussing reasons why organisms become endangered or extinct. This could be introduced by reading through the bulleted list from the Student Book.

Ask students to research other examples of organisms that are endangered or have become extinct as a result of human activity. For each example, they should identify the human activity responsible. Gather the examples and reasons on the board to compile a class list for students to share.

Supplement 3. Conservation of a species and habitat

Ask students to select an example of either an endangered species or habitat, and to research what has been done in terms of conservation, why it has been done and how successful it has been in its aims. Encourage students to consider local examples.

They should quickly realise that the protection of either the species or a habitat leads to protection of the other. For example, there is little reason to protect a species if you cannot also bring about protection of a suitable habitat in which it can live permanently.

The Student Book provides an example of conservation of the Hawaiian goose in the Science in Context box. However, there are many examples that students could research, starting either from the perspective of a single species or from the perspective of a habitat (through protection of national parks, etc.).

Students could present their findings in a poster or a presentation so that other students can learn about the example. Encourage other students to ask questions to clarify the role and purpose of conservation.

4. Consolidation

Give students a few minutes to answer the question, 'What should we conserve and (Supplement) how should we conserve it?'

Take examples from around the class and encourage discussion to select the four best answers.

Alternatively, ask students to prepare a presentation using presentation software or a poster that answers the question 'What should we conserve and (Supplement) how should we conserve it?'

Learning episode B16c Consolidation and summary

Learning aims

- Review the learning points of the topic summarised in the End of topic checklist
- Test understanding of the topic content by answering the End of topic questions

Resources

Student Book End of topic checklist and End of topic questions

Approach

Introduce the learning episode

Ask students to work with a partner to make a list of key words from this topic. They could then work together to produce a spider diagram showing how the different concepts are linked. They could compare their list with the list of key terms given at the End of the topic in the Student Book. Discuss the checklist at the end of the topic and use questioning to see how much of the content they are comfortable with.

Students could make flashcards of the key content and then use the flashcards to quiz each other on the information.

Develop the learning episode

Ask students to work individually through the End of topic questions in the Student Book without looking at the text. As they work, walk around the classroom observing their answers and asking questions as necessary to find out which questions are causing difficulties.

Finish the learning episode

After a set period, ask the students to stop working and discuss any areas of difficulty you observed as you walked round the class. Students should complete any unanswered questions for homework, but you should stress that they should try to answer the questions without looking at the text, so that they can see how much they have remembered.

Biology Teaching sequence

The course can be taught in syllabus order and it will work well to do so. If you want to change the teaching sequence for the course, you should keep in mind the considerations below. An example alternative teaching sequence has been provided.

- There should be a logical progression in the development of concepts and knowledge in order to aid students' understanding.
- The course should have a 'storyline', so that wherever possible the sequence can be easily explained to students. You can achieve this in different ways for this syllabus, as some sections on body systems are covered in relation to animals (generally humans) and others focus on flowering plants. You might approach this from the view of the systems, or from the view of the organisms by covering all systems relating to humans and then covering plant systems at a different time.
- You should consider carefully how to introduce the more difficult concepts. For example, you should ensure that they are not introduced too early in the course, or put in a sequence where they all come one after another. Either of these approaches could result in demotivation. There should be time to consolidate more difficult concepts in the planned sequence.
- Maintain the emphasis on practical and investigative work throughout the course.
- You will need to allocate different amounts of teaching time to different sections, depending on levels of difficulty or extent of the content.
- It is important to retain some flexibility so that difficult ideas can be revisited, or new ideas applied in different contexts.
- Some topics may need consideration of seasons and geographical location of teaching, for example growing seedlings may be better in the spring/summer rather than the winter.
- You will need time for revision at the end of the course. It is also recommended that you encourage students to revisit material for revision throughout the course. This could be done using the flashcards created at the end of each topic.

Alternative teaching sequence

This example alternative teaching sequence is based on a two-year course taught in six terms. It is mapped over five terms on the assumption that much of the sixth term is spent on revision. For different numbers of terms or semesters, where the course is started early, or where the time allocated each week is either generous or limited, you will need to adjust this sequence.

First year, 1st term

B1 Characteristics of living organisms
B2 Cells
B3 Movement into and out of cells
B4 Biological molecules
B5 Enzymes

First year, 2nd term

B7 Human nutrition
B9 Transport in animals
B11 Gas exchange in humans
B12 Respiration

First year, 3rd term

B6 Plant nutrition
B8 Transport in plants
B13 Drugs
B15 Organisms and their environment

Second year, 1st term

B14 Reproduction

Second year, 2nd term

B10 Diseases and immunity
B16 Human influences on ecosystems

Biology Scheme of work

Learning episode	**Learning objectives and learning aims** Learning objectives are given which link to the syllabus. Learning aims reflect the intended outcomes of the consolidation and summary learning episodes.	**Syllabus reference**	**Student's Book heading/ subheading**
B1a Characteristics of living organisms	Learning objectives • Describe the characteristics of living organisms by defining: movement as an action by an organism or part of an organism causing a change of position or place. • Describe respiration as the chemical reactions in cells that break down nutrient molecules and release energy for metabolism. • Describe sensitivity as the ability to detect and respond to changes in the internal or external environment. • Describe growth as a permanent increase in size and dry mass. • Describe reproduction as the processes that make more of the same kind of organism. • Describe excretion as the removal of the waste products of metabolism and substances in excess of requirements. • Describe nutrition as the taking in of materials for energy, growth and development.	B1.1	Characteristics of living organisms
B1b Consolidation and summary	Learning aims • Review the learning points of the topic summarised in the End of topic checklist. • Test understanding of the topic content by answering the End of topic questions.	B1	End of topic checklist and End of topic questions

Learning episode	**Learning objectives and learning aims** Learning objectives are given which link to the syllabus. Learning aims reflect the intended outcomes of the consolidation and summary learning episodes.	**Syllabus reference**	**Student's Book heading/ subheading**
B2a Cell structure and size of specimens	Learning objectives • Describe and compare the structure of a plant cell with an animal cell, limited to: cell wall, cell membrane, nucleus, cytoplasm, chloroplasts, ribosomes, mitochondria, vacuoles. • Describe the structure of a bacterial cell, limited to: cell wall, cell membrane, cytoplasm, ribosomes, circular DNA, plasmids. • Identify the cell structures listed above in diagrams and images of plant, animal and bacterial cells. • Describe the functions of the structures listed above in plant and animal cells. • Supplement State that specialised cells have specific functions, limited to: root hair cells – absorption; palisade mesophyll cells – photosynthesis; red blood cells – transport of oxygen. • Describe the meaning of the terms: cell, tissue, organ, organ system and organism, as illustrated by examples given in the syllabus. • State and use the formula: magnification = image size ÷ actual size. • Calculate magnification and size of biological specimens using millimetres as units. • Supplement Convert measurements between millimetres (mm) and micrometres (μm).	B2.1, B2.2	Cell structure, Size of specimens
B2b Consolidation and summary	Learning aims • Review the learning points of the topic summarised in the End of topic checklist. • Test understanding of the topic content by answering the End of topic questions.	B2	End of topic checklist and End of topic questions
B3a Diffusion	Learning objectives • Describe diffusion as the net movement of particles from a region of their higher concentration to a region of their lower concentration (i.e., down a concentration gradient), as a result of their random movement. • State that some substances move into and out of cells by diffusion through the cell membrane. • Describe the importance of diffusion of gases and solutes in living organisms. • Supplement Investigate the factors that influence diffusion, limited to: surface area, temperature, concentration gradient, and distance.	B3.1	Diffusion

Learning episode	**Learning objectives and learning aims** Learning objectives are given which link to the syllabus. Learning aims reflect the intended outcomes of the consolidation and summary learning episodes.	**Syllabus reference**	**Student's Book heading/ subheading**
B3b Osmosis	Learning objectives • State that water diffuses through partially permeable membranes by osmosis. • State that water moves into and out of cells by osmosis through the cell membrane. • Investigate and describe the effects on plant tissues of immersing them in solutions of different concentrations. • Supplement Describe osmosis as the net movement of water molecules from a region of higher water potential (dilute solution) to a region of lower water potential (concentrated solution), through a partially permeable membrane. • Supplement Explain the effects on plant cells of immersing them in solutions of different concentrations by using the terms: turgid, turgor pressure, plasmolysis, flaccid.	B3.2	Osmosis
Supplement B3c Active transport	Learning objectives • Supplement Describe active transport as the movement of particles through a cell membrane from a region of lower concentration to a region of higher concentration (i.e., against a concentration gradient), using energy from respiration. • Supplement Explain the importance of active transport as a process for movement of molecules or ions across membranes, including ion uptake by root hairs.	B3.3	Active transport
B3d Consolidation and summary	Learning aims • Review the learning points of the topic summarised in the End of topic checklist. • Test understanding of the topic content by answering the End of topic questions.	B3	End of topic checklist and End of topic questions
B4a Carbohydrates, fats and proteins	Learning objectives • List the chemical elements that make up: carbohydrates, fats, and proteins. • State that large molecules are made from smaller molecules, limited to: starch, glycogen, and cellulose from glucose; proteins from amino acids; fats and oils from fatty acids and glycerol.	B4	Carbohydrates, fats and proteins
B4b Food tests	Learning objectives • Describe the use of: iodine solution test for starch; Benedict's solution test for reducing sugars; biuret test for proteins; ethanol emulsion test for fats and oils.	B4	Tests for food molecules
B4c Consolidation and summary	Learning aims • Review the learning points of the topic summarised in the End of topic checklist. • Test understanding of the topic content by answering the End of topic questions.	B4	End of topic checklist and End of topic questions

Learning episode	**Learning objectives and learning aims** Learning objectives are given which link to the syllabus. Learning aims reflect the intended outcomes of the consolidation and summary learning episodes.	**Syllabus reference**	**Student's Book heading/ subheading**
B5a Enzyme structure	Learning objectives • Describe enzymes as proteins that are involved in all metabolic reactions, where they function as biological catalysts. • Supplement Describe and explain enzyme action with reference to: active site, enzyme-substrate complex, substrate, and product. • Supplement Describe and explain the specificity of enzymes in terms of the complementary shape and fit of the active site with the substrate.	B5	Enzymes as catalysts
B5b Enzyme activity	Learning objectives • Investigate and describe the effect of changes in temperature and Ph on enzyme activity. • Supplement Explain the effect of changes in temperature on enzyme activity in terms of kinetic energy, shape and fit, frequency of effective collisions, and denaturation. • Supplement Explain the effect of changes in Ph on enzyme activity in terms of shape and fit, and denaturation.	B5	Enzymes and temperature; Investigating the effect of temperature on enzymes, The effect of temperature, Enzymes and Ph, Investigating the effect of pH on enzymes, The effect of Ph
B5c Consolidation and summary	Learning aims • Review the learning points of the topic summarised in the End of topic checklist. • Test understanding of the topic content by answering the End of topic questions.	B5	End of topic checklist and End of topic questions

Learning episode	**Learning objectives and learning aims** Learning objectives are given which link to the syllabus. Learning aims reflect the intended outcomes of the consolidation and summary learning episodes.	**Syllabus reference**	**Student's Book heading/ subheading**
B6a Photosynthesis	Learning objectives • Describe photosynthesis as the process by which plants synthesise carbohydrates from raw materials using energy from light. • State the word equation for photosynthesis as: carbon dioxide + water → glucose + oxygen in the presence of light and chlorophyll. • State that chlorophyll is a green pigment that is found in chloroplasts. • Supplement State that chlorophyll transfers energy from light into energy in chemicals, for the synthesis of carbohydrates. • Supplement State the balanced symbol equation for photosynthesis as: $6CO_2 + 6H_2O \rightarrow C_6H_{12}O_6 + 6O_2$. • Investigate and understand the need for chlorophyll, light and carbon dioxide for photosynthesis, using appropriate controls. • Supplement Understand and describe the effects of varying light intensity, carbon dioxide concentration and temperature on the rate of photosynthesis. • Supplement Understand and describe the effect of light and dark conditions on gas exchange in an aquatic plant using hydrogencarbonate indicator solution.	B6.1	Photosynthesis
B6b Leaf structure	Learning objectives • Identify in diagrams and images the following structures in the leaf of a dicotyledonous plant: chloroplasts, cuticle, guard cells and stomata, upper and lower epidermis, palisade mesophyll, spongy mesophyll, air spaces, vascular bundles, xylem and phloem.	B6.2	Leaf structure
B6c Consolidation and summary	Learning aims • Review the learning points of the topic summarised in the End of topic checklist. • Test understanding of the topic content by answering the End of topic questions.	B6	End of topic checklist and End of topic questions
B7a Diet	Learning objectives • Describe what is meant by a balanced diet. • State the principal dietary sources and describe the importance of: carbohydrates; fats and oils; proteins; vitamins, limited to C and D; mineral ions, limited to calcium and iron; fibre (roughage), water.	B7.1	Diet

Learning episode	**Learning objectives and learning aims** Learning objectives are given which link to the syllabus. Learning aims reflect the intended outcomes of the consolidation and summary learning episodes.	**Syllabus reference**	**Student's Book heading/ subheading**
B7b Digestive system	Learning objectives • Identify in diagrams and images the main organs of the digestive system, limited to: alimentary canal: mouth, oesophagus, stomach, small intestine (duodenum and ileum), and large intestine (colon, rectum, anus); associated organs: salivary glands, pancreas, liver, and gall bladder. • Describe the functions of the organs of the digestive system listed above, in relation to: ingestion – the taking of substances, e.g., food and drink, into the body; digestion – the breakdown of food; absorption – the movement of nutrients from the intestines into the blood; assimilation – uptake and use of nutrients by cells; egestion – the removal of undigested food from the body as faeces.	B7.2	Digestive system
B7c Digestion	Learning objectives • Describe physical digestion as the breakdown of food into smaller pieces without chemical change to the food molecules. • Describe chemical digestion as the breakdown of large insoluble molecules into small soluble molecules. • Supplement State that physical digestion increases the surface area of food for the action of enzymes in chemical digestion. • Supplement State the role of chemical digestion in producing small soluble molecules that can be absorbed. • Supplement Describe the functions of enzymes as follows: amylase breaks down starch to simple reducing sugars; proteases break down protein to amino acids; lipases break down fats and oils to fatty acids and glycerol. • Supplement State where, in the digestive system, amylase, protease and lipase are secreted and where they act. • Supplement Describe the functions of hydrochloric acid in gastric juice, limited to killing harmful microorganisms in food and providing an acidic Ph for optimum enzyme activity.	B7.3	Different types of digestion, Chemical digestion, Stomach acid
B7d Consolidation and summary	Learning aims • Review the learning points of the topic summarised in the End of topic checklist. • Test understanding of the topic content by answering the End of topic questions.	B7	End of topic checklist and End of topic questions

Learning episode	**Learning objectives and learning aims** Learning objectives are given which link to the syllabus. Learning aims reflect the intended outcomes of the consolidation and summary learning episodes.	**Syllabus reference**	**Student's Book heading/ subheading**
B8a Xylem and phloem	Learning objectives • State the functions of xylem and phloem: xylem – transport of water and mineral ions, and support; phloem – transport of sucrose and amino acids. • Identify in diagrams and images the position of xylem and phloem as seen in sections of roots, stems, and leaves of non-woody dicotyledonous plants.	B8.1	Xylem and phloem
B8b Water uptake	Learning objectives • Identify in diagrams and images root hair cells and state their functions. • Supplement State that the large surface area of root hairs increases the uptake of water and mineral ions. • Outline the pathway taken by water through the root, stem and leaf as: root hair cells, root cortex cells, xylem, mesophyll cells.	B8.2	Water uptake
B8c Transpiration	Learning objectives • Describe transpiration as the loss of water vapour from leaves. • Supplement Investigate and describe the effects of variation of temperature and wind speed on transpiration rate.	B8.3	Transpiration
B8d Consolidation and summary	Learning aims • Review the learning points of the topic summarised in the End of topic checklist. • Test understanding of the topic content by answering the End of topic questions.	B8	End of topic checklist and End of topic questions

Learning episode	**Learning objectives and learning aims** Learning objectives are given which link to the syllabus. Learning aims reflect the intended outcomes of the consolidation and summary learning episodes.	**Syllabus reference**	**Student's Book heading/ subheading**
B9a Circulatory systems and the heart	Learning objectives • Describe the circulatory system as a system of blood vessels with a pump and valves to ensure one-way flow of blood. • Identify in diagrams and images the structures of the mammalian heart, limited to: muscular wall, septum, left and right ventricles, left and right atria, one-way valves and coronary arteries. • State that blood is pumped away from the heart in arteries and returns to the heart in veins. • State that the activity of the heart may be monitored by: ECG, pulse rate and listening to sounds of valves closing. • Investigate and describe the effect of physical activity on the heart rate. • Supplement Describe the functioning of the heart in terms of the contraction of muscles of the atria and ventricles, and the action of the valves. • Supplement Explain the effect of physical activity on the heart rate. • Supplement Describe coronary heart disease in terms of the blockage of coronary arteries and state the possible risk factors including: diet, lack of exercise, stress, smoking, genetic predisposition, age, and sex. • Supplement Discuss the roles of diet and exercise in reducing the risk of coronary heart disease.	B9.1, B9.2	Circulatory systems, Heart
B9b Blood vessels and blood	Learning objectives • Describe the structure of arteries, veins and capillaries, limited to: relative thickness of wall, diameter of the lumen, and the presence of valves in veins. • Supplement Explain how the structure of arteries and veins is related to the pressure of the blood that they transport. • List the components of blood as: red blood cells, white blood cells, platelets and plasma. • Identify red and white blood cells in photomicrographs and diagrams. • State the functions of the following components of blood: red blood cells in transporting oxygen, including the role of haemoglobin; white blood cells in phagocytosis and antibody production; platelets in clotting (details are not required); plasma in the transport of blood cells, ions, nutrients, urea, hormones, and carbon dioxide. • Supplement State the roles of blood clotting as preventing blood loss and the entry of pathogens.	B9.3, B9.4	Blood vessels, Blood
B9c Consolidation and summary	Learning aims • Review the learning points of the topic summarised in the End of topic checklist. • Test understanding of the topic content by answering the End of topic questions.	B9	End of topic checklist and End of topic questions

Learning episode	**Learning objectives and learning aims** Learning objectives are given which link to the syllabus. Learning aims reflect the intended outcomes of the consolidation and summary learning episodes.	**Syllabus reference**	**Student's Book heading/ subheading**
B10a Disease, defence, hygiene	Learning objectives • Describe a pathogen as a disease-causing organism. • Describe a transmissible disease as a disease in which the pathogen can be passed from one host to another. • State that a pathogen is transmitted: by direct contact, including through blood and other body fluids; indirectly, including from contaminated surfaces, food, animals, and air. • Describe the body defences, limited to: skin, hairs in the nose, mucus, stomach acid and white blood cells. • Supplement State the features of viruses, limited to a protein coat and genetic material. • Supplement Explain the importance of the following in controlling the spread of disease: a clean water supply; hygienic food preparation; good personal hygiene; waste disposal; sewage treatment (details of the stages of sewage treatment are **not** required).	B10	Transmissible diseases, Methods of transmission, Controlling the spread of disease, Methods of defence
Supplement B10b The immune system	Learning objectives • Supplement Describe active immunity as defence against a pathogen by antibody production in the body. • Supplement Explain that active immunity is gained after an infection by a pathogen or by vaccination.	B10	The immune system and active immunity
B10c Consolidation and summary	Learning aims • Review the learning points of the topic summarised in the End of topic checklist. • Test understanding of the topic content by answering the End of topic questions.	B10	End of topic checklist and End of topic questions
B11a Gas exchange system	Learning objectives • Identify in diagrams and images the following parts of the breathing system: lungs, diaphragm, ribs, intercostal muscles, larynx, trachea, bronchi, bronchioles, alveoli, and associated capillaries. • Supplement Describe the features of gas exchange surfaces in humans, limited to: large surface area, thin surface, good blood supply, and good ventilation with air.	B11	The human breathing system, Gas exchange
B11b Investigating gas exchange	Learning objectives • Investigate and describe the effects of physical activity on the rate and depth of breathing.	B11	Investigating the effects of exercise on breathing

Learning episode	**Learning objectives and learning aims** Learning objectives are given which link to the syllabus. Learning aims reflect the intended outcomes of the consolidation and summary learning episodes.	**Syllabus reference**	**Student's Book heading/ subheading**
B11c Consolidation and summary	Learning aims • Review the learning points of the topic summarised in the End of topic checklist. • Test understanding of the topic content by answering the End of topic questions.	B11	End of topic checklist and End of topic questions
B12a Respiration	Learning objectives • State the uses of energy in living organisms, including: muscle contraction, protein synthesis, cell division, active transport, growth and the maintenance of a constant body temperature. • Describe aerobic respiration as the chemical reactions in cells that use oxygen to break down nutrient molecules to release energy. • State the word equation for aerobic respiration as: glucose + oxygen → carbon dioxide + water. • Supplement State the balanced chemical equation for aerobic respiration as: $C_6H_{12}O_6 + 6O_2 \rightarrow 6CO_2 + 6H_2O$.	B12	Respiration, Aerobic respiration
B12b Consolidation and summary	Learning aims • Review the learning points of the topic summarised in the End of topic checklist. • Test understanding of the topic content by answering the End of topic questions.	B12	End of topic checklist and End of topic questions
B13a Drugs and medicinal drugs	Learning objectives • Describe a drug as any substance taken into the body that modifies or affects chemical reactions in the body. • Describe the use of antibiotics for the treatment of bacterial infections. • State that some bacteria are resistant to antibiotics which reduces the effectiveness of antibiotics. • State that antibiotics kill bacteria but do not affect viruses. • Supplement Explain how using antibiotics only when essential can limit the development of resistant bacteria such as MRSA.	B13.1	Drugs
B13b Consolidation and summary	Learning aims • Review the learning points of the topic summarised in the End of topic checklist. • Test understanding of the topic content by answering the End of topic questions.	B13	End of topic checklist and End of topic questions

Learning episode	**Learning objectives and learning aims** Learning objectives are given which link to the syllabus. Learning aims reflect the intended outcomes of the consolidation and summary learning episodes.	**Syllabus reference**	**Student's Book heading/ subheading**
B14a Sexual reproduction in plants	Learning objectives • Identify in diagrams and images, and draw, the following parts of an insect-pollinated flower: sepals, petals, stamens, filaments, anthers, carpels, style, stigma, ovary and ovules. • State the functions of the structures listed above. • Describe pollination as the transfer of pollen grains from an anther to a stigma. • State that fertilisation occurs when a pollen nucleus fuses with a nucleus in an ovule. • Investigate and describe the environmental conditions that affect germination of seeds, limited to the requirement for: water, oxygen and a suitable temperature. • Supplement Identify in diagrams and images, and describe, the anthers and stigmas of a wind-pollinated flower.	B14.1	Sexual reproduction in plants
B14b Sexual reproduction in humans	Learning objectives • Identify on diagrams and state the functions of the following parts of the male reproductive system: testes, scrotum, sperm ducts, prostate gland, urethra and penis. • Identify on diagrams and state the functions of the following parts of the female reproductive system: ovaries, oviducts, uterus, cervix and vagina. • Describe fertilisation as the fusion of the nuclei from a male gamete (sperm) and a female gamete (egg cell). • Describe the menstrual cycle in terms of changes in the ovaries and in the lining of the uterus (knowledge of sex hormones is **not** required).	B14.2	Sexual reproduction in humans
B14c Consolidation and summary	Learning aims • Review the learning points of the topic summarised in the End of topic checklist. • Test understanding of the topic content by answering the End of topic questions.	B14	End of topic checklist and End of topic questions

Learning episode	**Learning objectives and learning aims** Learning objectives are given which link to the syllabus. Learning aims reflect the intended outcomes of the consolidation and summary learning episodes.	**Syllabus reference**	**Student's Book heading/ subheading**
B15a Energy flow, food chains and food webs	Learning objectives • State that the Sun is the principal source of energy input to biological systems. • Describe the flow of energy through living organisms, including light energy from the Sun and chemical energy in organisms, and its eventual transfer to the environment. • Describe a food chain as showing the transfer of energy from one organism to the next, beginning with a producer. • Construct and interpret simple food chains. • Describe a food web as a network of interconnected food chains and interpret food webs. • Describe a producer as an organism that makes its own organic nutrients, usually using energy from sunlight, through photosynthesis. • Describe a consumer as an organism that gets its energy by feeding on other organisms. • State that consumers may be classed as primary, secondary and tertiary according to their position in a food chain. • Describe a herbivore as an animal that gets its energy by eating plants. • Describe a carnivore as an animal that gets its energy by eating other animals. • Describe a decomposer as an organism that gets its energy from dead or waste organic material. • Supplement Use food chains and food webs to describe the impact humans have through overharvesting of food species and through introducing foreign species to a habitat.	B15.1, B15.2	Energy flow, Food chains and food webs
B15b The carbon cycle	Learning objectives • Describe the carbon cycle, limited to: photosynthesis, respiration, feeding, decomposition, formation of fossil fuels and combustion.	B15.3	The carbon cycle
B15c Consolidation and summary	Learning aims • Review the learning points of the topic summarised in the End of topic checklist. • Test understanding of the topic content by answering the End of topic questions.	B15	End of topic checklist and End of topic questions

Learning episode	**Learning objectives and learning aims** Learning objectives are given which link to the syllabus. Learning aims reflect the intended outcomes of the consolidation and summary learning episodes.	**Syllabus reference**	**Student's Book heading/ subheading**
B16a Habitat destruction	Learning objectives • Describe an ecosystem as a unit containing the community of organisms and their environment, interacting together. • Describe biodiversity as the number of different species that live in an area. • Describe the reasons for habitat destruction, including: (a) increased area for housing, crop plant production and livestock production (b) extraction of natural resources (c) freshwater and marine pollution (a detailed description of eutrophication is not required). • State that through altering food webs and food chains, humans can have a negative impact on habitats. • Supplement Explain the undesirable effects of deforestation as an example of habitat destruction, to include: reducing biodiversity, extinction, loss of soil, flooding and increase of carbon dioxide in the atmosphere.	B16.1	Habitat destruction
B16b Conservation	Learning objectives • Outline why organisms become endangered or extinct, including: climate change, habitat destruction, hunting, overharvesting, pollution and introduced species. • Supplement Describe how endangered species can be conserved, limited to: (a) monitoring and protecting species and habitats (b) education (c) captive breeding programmes (d) seed banks.	B16.2	Conservation
B16c Consolidation and summary	Learning aims • Review the learning points of the topic summarised in the End of topic checklist. • Test understanding of the topic content by answering the End of topic questions.	B16	End of topic checklist and End of topic questions

C1 States of matter

Contents

Teacher Guide	Syllabus reference
C1a States of matter	C1.1 Solids, liquids and gases

Overview of the section

This section may go back over ideas that students have already met, so it does not need to take up much teaching time. Nevertheless, it is worth making sure there are no misconceptions.

Starting points

The Student Book section opener puts the ideas in the section into context and sets the scene. It also allows students to acknowledge and value their prior learning, and provides a benchmark against which future learning can be compared.

The questions provide a structure for introducing the section and can be used in a number of different ways:

- You could ask students to consider the questions as an introductory homework task.
- You could put students into groups to share their own ideas and understanding and then to report back to the whole class.
- Students could be given access to the Internet, preferably with a tight timescale, to find out the information required.

You could then use a spider chart or other form of wall chart to summarise everybody's ideas.

Recording these initial ideas allows you to retain them for reference as the individual topics are developed. In this way, your students' progress in learning can be readily acknowledged.

C1a States of matter

Introduction

The kinetic particle theory links to work on a simple collision theory in Topic C6a *Rate of reaction*, and to an explanation of fractional distillation in Topic C11a *Fuels*.

Links to other topics

Section	**Essential background knowledge**	**Useful links**
C2 Atoms, elements and compounds		C2a Materials, atomic structure and the Periodic Table
C6 Chemical reactions		C6a Rate of reaction
C11 Organic chemistry		C11b Fuels

Topic overview

C1a.1	**Orientation** The purpose of this learning episode is to recap students' current knowledge and understanding of the three states of matter and the names of the processes involved when substances change states.
C1a.2	**Solids, liquids and gases** This learning episode introduces the particulate models of solids, liquids and gases. The focus is on the different arrangements and movements of the particles in the three states. The names of the processes that bring about a change in state are emphasised and it relates these changes to the energy involved. Supplement An explanation of the changes of state using the kinetic particle theory is then considered.
C1a.3	**Consolidation and summary** This learning episode allows for a quick recap of the key ideas of the topic and provides a link to subsequent work on atoms, elements and compounds.

Career links

These are some scientific careers that focus on this area of chemistry but careers in many other fields use the knowledge and skills gained studying science.

Many roles in chemistry rely on an understanding of the behaviour of different states of matter. For example, some paleoclimatologists study the interaction between ice and water in ice sheets or look at the composition of bubbles of air found in ice cores.

Learning episode C1a.1 Orientation

Learning objective

- State the distinguishing properties of solids, liquids and gases.

Resources

Student Book Solids, liquids and gases

Resources for demonstration (see Technician's notes, following)

Approach

Students should be familiar with classifying solids, liquids and gases. They may also know the names of the processes that cause changes of state. So this learning episode could be relatively short.

Ask students to list four solids, four liquids and four gases they are familiar with. Emphasise that substances with the same classification can have quite different properties. For example: a candle and a metal rod (hardness), water and liquid honey (viscosity), air and bromine gas (colour). Use the opening text in the Student Book to summarise the different properties of solids, liquids and gases. Explain that these distinguishing properties are generally true, but that students need to be careful. For example, sand is obviously a solid, but it takes the shape of any container it is put in.

Finally, ask students to name some of the processes involved in changing states, such as melting, boiling (evaporation) and freezing.

SAFETY INFORMATION
Wear eye protection - splash proof goggles when using bromine.
Wear chemical resistant gloves.
Keep bromine in fume cupboard.

Technician's notes

Be sure to check the latest safety notes on these resources before proceeding.

The following resources are needed for the demonstration on solids, liquids and gases:

a candle, a metal rod (similar diameter to a candle), a jar of liquid honey
a gas jar of bromine gas (in fume cupboard), eye protection (splash-proof goggles), chemical resistant gloves

Learning episode C1a.2 Solids, liquids and gases

Learning objectives

- Describe the structures of solids, liquids and gases in terms of particle separation, arrangement and motion.
- Describe changes of state in terms of melting, boiling, evaporating, freezing and condensing.
- Describe the effects of temperature and pressure on the volume of a gas.
- Supplement Explain changes of state in terms of kinetic particle theory.

Common misconceptions

Some students think that the particles in a liquid are not touching, that is, that the arrangement of particles in a liquid is closer to that of a gas than to that of a solid.

Resources

Student Book Solids, liquids and gases

Worksheet C1a.2 The particulate nature of matter

Approach

Refer to the particle arrangements in solids, liquids and gases. Emphasise the arrangements and movements of the particles in each model.

Ask students to speculate on the strength of the forces between the particles in the three models. Then ask them to use their ideas to explain why gases expand on heating more than solids and liquids. Refer to the changes in the arrangement of particles as a solid changes into a liquid and then into a gas.

Emphasise the names of the key processes: melting/freezing, evaporating/boiling/condensing. Ensure that students understand the different types of movement of the particles. They should also understand the average energy of the particles in each of the three states of matter – solid (vibration – relatively low energy), liquid (moving around – translation – more energy than in a solid hence the possibility of evaporation) and gas (rapid movement – translation – relatively high energy).

Supplement Use the flowchart (Fig C1.6) in the Student Book to emphasise the nature of the changes in particles as one state changes into another.

Finally, use Worksheet C1a.1 to check that students understand the key differences between the three states.

Learning episode C1a.3 Consolidation and summary

Learning aims

- Review the learning points of the topic summarised in the End of topic checklist.
- Test understanding of the topic content by answering the End of topic questions.

Resources

Student Book End of topic checklist; End of topic questions

Approach

Explain that for the rest of the course the term 'particle' will not be precise enough. Students will need to use terms like 'atoms' and 'molecules' (and 'ions') correctly to identify the type of particle. Emphasise that students will be learning more about these types of particles in the topics that follow.

Use the science in context box on *The States of Matter* (Student Book Solids, liquids and gases) to discuss plasma as a potential fourth state of matter. This content is beyond the syllabus but provides the opportunity to explore the content more deeply.

Ask students to work with a partner to make a list of key words from this topic. They could then work together to produce a spider diagram showing how the different concepts are linked. They could compare their list with the list of key terms given in the Student Book. Discuss the checklist and ask questions to see how much of the content students are comfortable with.

Students could make flashcards of the key content and then use the flashcards to quiz each other on the information.

Ask students to work individually through the End of topic questions in the Student Book without looking at the text. As they work, walk around the classroom observing their answers and questioning them as necessary to find out which questions are causing difficulties.

After a set period, ask students to stop working. Discuss any areas of difficulty you observed as you walked round the class.

Students should complete any unanswered questions for homework, but you should stress that they should answer the questions without looking at the text, so that they can see how much they have remembered.

C2 Atoms, elements and compounds

Contents

Teacher Guide	Syllabus reference
C2a Atoms, elements and compounds	C2.1 Elements, compounds and mixtures C2.2 Atomic structure and the Periodic Table
C2b Ions and ionic bonds	C2.3 Ions and ionic bonds
C2c Molecules and covalent bonds	C2.4 Simple molecules and covalent bonds

Overview of the section

As the list of contents suggests, this section is focused on chemical bonding. Fundamental to the study of chemical bonding are the substances in which bonding is critical, i.e. compounds and not mixtures. This leads on to the different types of bonding in elements. The nature of atomic structure underpins the study with the importance of electronic configuration a key feature.

Detailed study starts with ionic bonding and then, in the next subsection, covalent bonding. The differences in these types of bonding lead to the typical properties of ionic and covalent compounds.

The section as a whole is an essential starting point for much of the remaining content of the course. For example, ionic bonding is crucial in the study of electrochemistry, and covalent bonding in the study of organic compounds.

Starting points

The Student Book section opener puts the ideas in the section into context and sets the scene. It also allows students to acknowledge and value their prior learning, and provides a benchmark against which future learning can be compared.

The questions provide a structure for introducing the section and can be used in a number of different ways:

- You could ask students to consider the questions as an introductory homework task.
- You could put students into groups to share their own ideas and understanding and then to report back to the whole class.
- Students could be given access to the Internet, preferably with a tight timescale, to find out the information required.

You could then use a spider chart or other form of wall chart to summarise everybody's ideas.

Recording these initial ideas allows you to retain them for reference as the individual topics are developed. In this way, your students' progress in learning can be readily acknowledged.

C2a Atoms, elements and compounds

Introduction

This topic covers two syllabus sections namely, C2.1 Elements, compounds and mixtures, and C2.2 Atomic structure and the Periodic Table. It is the basis for future work on chemical bonding and more detailed work on the Periodic Table. It has strong links to Topic C2b *Ions and ionic bonds*, Topic C2c *Simple molecules and covalent bonds* and Section C8 *The Periodic Table*.

The work on atomic structure introduces the sub-atomic particles. This leads students to drawing and interpreting diagrams showing the numbers of protons and neutrons, and the arrangement of electrons in shells. There are no opportunities for practical work but plenty of opportunities for individual and group work.

Links to other topics

Section	Essential background knowledge	Useful links
C1 States of matter	C1a Solids, liquids and gases	
C2 Atoms, elements and compounds		C2b Ions and ionic bonds C2c Simple molecules and covalent bonds
C8 The Periodic Table		C8a Arrangement of elements

Topic overview

C2a.1	**Orientation** Students should be aware about the differences between elements, compounds and mixtures so this would be a good starting point. Students may have some knowledge of atomic structure and of particles called protons, neutrons and electrons. They may know something about how elements are arranged in the Periodic Table. The purpose of this learning episode is to find their level of understanding and to decide how much teaching time to spend on the topic.
C2a.2	**Sub-atomic particles** This learning episode starts with a brief consideration of Dalton's atomic theory. The learning episode then focuses on the protons, neutrons and electrons in an atom and how the standard representation of proton number and nucleon number gives information about the number of these particles in an atom. The arrangement of electrons is covered in the next learning episode.
C2a.3	**The arrangement of electrons in atoms** This learning episode introduces the arrangement of electrons in shells around the nucleus and the use of atom diagrams to represent the structure of an atom. The significance of the number of electrons in the outer electron shell is emphasised, along with the relationship between the number of electrons in the outer electron shell and the position of an element in the Periodic Table. This provides a very useful introduction to Section C8 *The Periodic Table*.

C2a.4	**Consolidation and summary** This learning episode provides a quick recap of the key ideas of the topic.

Career links

These are some scientific careers that focus on this area of chemistry but careers in many other fields use the knowledge and skills gained studying science.

Radiographers are trained medical professionals who use radioactive isotopes to diagnose and treat diseases.

Researchers, engineers and nuclear physicists at the CERN institute conduct experiments to study the fundamental particles making up all matter.

Learning episode C2a.1 Orientation

Learning objective

- Describe the differences between elements, compounds and mixtures.

Common misconceptions

Some students think that all atoms have the same number of protons and neutrons.

Resources

Student Book Elements, compounds and mixtures

Worksheet C2a.1 What do you know about atomic structure?

Approach

The orientation exercise can be based around the True/False quiz on Worksheet C2a.1. The 'not sure' column is meant to stop students from guessing. Collect students' responses and explore any areas of misunderstanding and lack of knowledge. You should then be able to determine how much time to devote to each of the activities that follow. You can then share the overall objectives of the topic.

Learning episode C2a.2 Sub-atomic particles

Learning objectives

- Describe the structure of the atom as a central nucleus containing neutrons and protons surrounded by electrons in shells.
- State the relative charges and relative masses of a proton, a neutron and an electron.
- Define proton number / atomic number as the number of protons in the nucleus of an atom.
- Define mass number / nucleon number as the total number of protons and neutrons in the nucleus of an atom.

Common misconceptions

Some students think that all the sub-atomic particles are found in the nucleus.

Resources

Student Book Atomic structure and the Periodic Table

Approach

This is a relatively brief learning episode starting with consideration of Dalton's atomic theory. All the key learning points are covered in the Student Book pages. Electron arrangements and atom diagrams are covered in learning episode C2a.3.

Once you have introduced the relative masses and charges of the proton, neutron and electron and the terms proton number/nucleon number (atomic/mass number), ask students to determine the numbers of protons, neutrons and electrons in various atoms, when given the mass number and name of the element. Students will need to look up the atomic number on the Periodic Table. With some students you may wish to introduce the notation as shown in Fig. C2.3, although this notation is not required. Ask students to use this notation to work out the structures of a number of atoms. Some possible examples include:

$^{56}_{26}Fe$ $^{127}_{53}I$ $^{108}_{47}Ag$ $^{197}_{79}Au$ $^{80}_{35}Br$ $^{35}_{17}Cl$ $^{37}_{17}Cl$

Refer to the Science in Context box on sub-atomic particles (at the end of the subsection on Proton number and mass number). The content of these Science in Context boxes is beyond the syllabus but provides the opportunity to explore the content more deeply.

Learning episode C2a.3 The arrangement of electrons in atoms

Learning objectives

- Determine the electronic configuration of elements, and their ions with proton number 1 to 20, e.g., 2,8,3.
- State that:
  (a) Group VIII noble gases have a full outer shell
  (b) The number of outer shell electrons is equal to the group number in Groups I to VII
  (c) The number of occupied electron shells is equal to the period number.

Resources

Student Book Electronic configuration: The first 20 elements of the Periodic Table Worksheet C2a.3 Atomic structures

Approach

This learning episode follows closely from the previous. Introduce the idea that electron shells can hold a maximum number of electrons, and that these numbers differ depending on which shell is being occupied by electrons.

Use the Student Book to illustrate the nature of atom diagrams and then the detailed electronic configuration of the first 20 elements. You can then make the link between the number of outer electrons and the position of the element in the Periodic Table. This idea will be revised at the start of Section C8 *The Periodic Table*.

Use Worksheet C2a.3 to reinforce the ideas introduced in this learning episode.

Learning episode C2a.4 Consolidation and summary

Learning aims

- Review the learning points of the topic summarised in the End of topic checklist.
- Test understanding of the topic content by answering the End of topic questions.

Resources

Student Book End of topic checklist; End of topic questions

Approach

Ask students to work with a partner to make a list of key words from this topic. They could then work together to produce a spider diagram showing how the different concepts are linked. They could compare their list with the list of key terms given in the Student Book. Discuss the checklist and ask questions to see how much of the content students are comfortable with.

Students could make flashcards of the key content and then use the flashcards to quiz each other on the information.

Ask students to work individually through the End of topic questions in the Student Book without looking at the text. As they work, walk around the classroom observing their answers and questioning them as necessary to find out which questions are causing difficulties.

After a set period, ask students to stop working. Discuss any areas of difficulty you observed as you walked round the class.

Students should complete any unanswered questions for homework, but you should stress that they should answer the questions without looking at the text, so that they can see how much they have remembered. Focus on the End of topic checklist and questions. Aim for a mix of individual and group work and provide plenty of opportunities for the students to articulate their ideas and understanding.

C2b Ions and ionic bonds

Introduction

This topic builds on previous ideas about atomic structure from Topic C2a *Materials, atomic structure and the Periodic Table*. It links to covalent bonding in Topic C2c *Simple molecules and covalent bonds*. There are also a number of other topics where knowledge of ions and ionic bonding is fundamental to understanding the content. These include Topic C4a *Electrolysis*; Topic C7a *Acids, bases and salts*; Topic C8b *Group I elements*; Topic C8c *Group VII elements*; Topic C12b *Identification of ions and gases*. For the supplementary syllabus there are also links to Topic C6b *Redox reactions*.

This topic has three main aspects: the nature of ionic bonding and the associated properties of ionic compounds, and for those following the supplementary syllabus, an extension to the idea of dot and cross diagrams and the lattice structure.

Links to other topics

Section	**Essential background knowledge**	**Useful links**
C2 Atoms, elements and compounds	C2a Materials, atomic structure and the Periodic Table	C2c Simple molecules and covalent bonds
C4 Electrochemistry		C4a Electrolysis
C6 Chemical reactions		C6b Redox reactions
C7 Acids, bases and salts		C7a Acids, bases and salts
C8 The Periodic Table		C8b Group I elements C8c Group VII elements
C12 Experimental techniques and chemical analysis		C12b Identification of ions and gases

Topic overview

C2b.1	**Orientation** This learning episode recaps previous work on atomic structure, with emphasis on electron arrangements. Considering the properties of ionic compounds provides an insight into the nature of the bonding in these compounds.
C2b.2	**The formation of ions** This learning episode involves modelling the process of ionic bonding and then provides plenty of practice for producing dot-and-cross diagrams. Reference to the relationship between ion charges and the Periodic Table will provide a link to future work in that area. Supplement Consideration is given to compounds formed by elements other than those in Group I and Group VII.

C2b.3	**Properties of ionic compounds** This learning episode focuses on the description of the characteristic properties of ionic compounds and contrasts them with other compounds (essentially covalent compounds). Supplement The idea of a giant ionic lattice structure as a representation of how ions are arranged in an ionic compound is covered.
C2b.4	**Consolidation and summary** This learning episode provides a quick recap of the ideas encountered and an opportunity to answer the End of topic questions in the Student Book.

Career links

These are some scientific careers that focus on this area of chemistry but careers in many other fields use the knowledge and skills gained studying science.

Geochemists study the composition of ionic compounds in the form of minerals in the Earth's mantle and crust to find resources such as metals and help with the understanding of environmental issues such as soil depletion.

Oceanographers study the presence of ions dissolved in seawater to measure changes in salinity which are influenced by climate change.

Learning episode C2b.1 Orientation

Learning objectives

- To revise ideas on atomic structure.
- To map out the main purposes of the topic.
- Know the structure of an atom by interpreting proton and nucleon numbers.
- Know the arrangement of electrons in shells for the first 20 elements.

Common misconceptions

Some students think that all compounds are made up of molecules.

Resources

Student Book Sub-atomic particles; Proton number and mass number

Approach

Start by revising the ideas on atomic structure, emphasising the electronic configurations of atoms and the arrangement of electrons in shells around the nucleus.

Ask students to research the melting points of some common substances, including a range of ionic compounds and covalent compounds. Ask them to see if there is a pattern in the compounds that have melting points in the higher ranges. The students should identify the inclusion of a metal as a common feature.

Explain that these properties must be a result of how metals and non-metals combine together – a process that produces particles known as ions.

Learning episode C2b.2 The formation of ions

Learning objectives

- Describe the formation of:
  a) positive ions, known as cations, by the loss of electrons
  b) negative ions, known as anions, by the gain of electrons
- State that an ionic bond is a strong electrostatic attraction between oppositely charged ions.
- Describe the formation of ionic bonds between elements from Group I and Group VII, including the use of dot-and-cross diagrams.
- Supplement Describe the formation of ionic bonds between ions of metallic and non-metallic elements, including the use of dot-and-cross diagrams.

Common misconceptions

Some students think that all compounds are made up of molecules.

Resources

Student Book The formation of ions; Electron transfer in ionic bonding

Worksheet C2b.2 Ionic bonding: dot-and-cross diagrams

Approach

Revise the idea of atom diagrams. Use the examples in the Student Book to introduce dot- and-cross diagrams.

Emphasise the importance of electron transfer from metal to non-metal. In the examples that students are likely to meet, both the metal and non-metal acquire a full outer shell of electrons (that is, 2 or 8). Take time to explain that the resulting particles are no longer atoms (no longer neutral) and are known as ions.

Emphasise that the ion charges in the chemical formula of an ionic compound cancel each other out. Use the questions in the Student Book to allow for practice producing simple dot-and-cross diagrams.

Ask students if they can see a pattern between the charge on the resulting ions and their position in the Periodic Table.

Supplement Students consider the formation of ionic bonds where more than one electron is transferred (that is, compounds other than those formed from Group I and Group VII elements). Worksheet C2b.2 gives some examples of the transfer of more than one electron.

Learning episode C2b.3 Properties of ionic compounds

Learning objectives

- Describe the properties of ionic compounds:
  - (a) high melting points and boiling points
  - (b) good electrical conductivity when aqueous or molten and poor when solid
  - (c) generally soluble in water.
- Supplement Describe the giant lattice structure of ionic compounds as a regular arrangement of alternating positive and negative ions, exemplified by sodium chloride.

Resources

Student Book Properties of ionic compounds

Resources for a demonstration (see Technician's notes, following)

Approach

Use the text in the Student Book to summarise the properties of sodium chloride. Mention that these properties apply to all ionic compounds. Refer to the Science in Context box on ionic crystals in the Student Book. The content of these Science in Context boxes is beyond the syllabus but provides the opportunity to explore the content more deeply.

Supplement If possible, use models to illustrate the 3D structure of ionic compounds such as sodium chloride. Explain that the attractions between positive and negative ions are called electrostatic attractions and that multiple attractions in all directions produce a very strong structure that is not easily broken down; for example, the melting point of sodium chloride is 801 °C.

Technician's notes

Be sure to check the latest safety notes on these resources before proceeding.

The following resources are needed for the demonstration of models of ionic structure:

molecular model, made up to show two or three ionic compounds such as sodium chloride

Learning episode C2b.4 Consolidation and summary

Learning aims

- Review the learning points of the topic summarised in the End of topic checklist.
- Test understanding of the topic content by answering the End of topic questions.

Resources

Student Book End of topic checklist; End of topic questions

Approach

Ask students to work with a partner to make a list of key words from this topic. They could then work together to produce a spider diagram showing how the different concepts are linked. They could compare their list with the list of key terms given in the Student Book. Discuss the checklist and ask questions to see how much of the content students are comfortable with.

Students could make flashcards of the key content and then use the flashcards to quiz each other on the information.

Ask students to work individually through the End of topic questions in the Student Book without looking at the text. As they work, walk around the classroom observing their answers and questioning them as necessary to find out which questions are causing difficulties.

After a set period, ask students to stop working. Discuss any areas of difficulty you observed as you walked round the class.

Students should complete any unanswered questions for homework, but you should stress that they should answer the questions without looking at the text, so that they can see how much they have remembered.

If you have chosen to teach Topic C2c *Simple molecules and covalent bonds* as the next topic, then the link can be emphasised.

C2c Simple molecules and covalent bonds

Introduction

This topic builds on previous work on atomic structure. Students must understand the nature of covalent bonding and the properties of covalent compounds in order to study a number of topics later in the course.

Topic C2a *Materials, atomic structure and the Periodic Table* gave important information about electronic configurations. Covalent bonding features in all the topics in Section C11 *Organic chemistry*. For the supplementary syllabus, bond breaking and forming is an aspect of Topic C5a *Exothermic and endothermic reactions*.

This topic has two main aspects: the nature of covalent bonding and the typical properties of covalent compounds and how they compare with the properties of ionic compounds.

Links to other topics

Section	Essential background knowledge	Useful links
C2 Atoms, elements and compounds	C2a Materials, atomic structure and the Periodic Table	C2b Ions and ionic bonds
C5 Chemical energetics		C5a Exothermic and endothermic reactions
C11 Organic chemistry		C11b Fuels C11c Alkanes C11d Alkenes C11e Polymers

Topic overview

C2c.1	**Orientation** This learning episode builds on earlier work on atomic structure and ionic bonding. Covalent bonding is contrasted with that of ionic bonding and the differences in typical properties of covalent and ionic compounds are highlighted.
C2c.2	**The formation of covalent molecules** This learning episode involves modelling the process of covalent bonding and then provides plenty of practice producing dot-and-cross diagrams. Reference to the relationship between the number of covalent bonds an atom can form and the Periodic Table will provide a link to future work in that area. Supplement The formation of multiple covalent bonds is considered through examples such as ethene, oxygen, carbon dioxide and nitrogen.
C2c.3	**Properties of simple covalent molecules** This learning episode focuses on the characteristic properties of simple covalent molecules and contrasts them with other compounds (essentially ionic compounds).

C2c.4	**Consolidation and summary** This learning episode provides an opportunity for a quick recap of the ideas encountered in the topic as well as time for students to answer the End of topic questions in the Student Book.

Career links

These are some scientific careers that focus on this area of chemistry but careers in many other fields use the knowledge and skills gained studying science.

A team of professors and academics at the University of Manchester, from backgrounds in material science, engineering, chemistry and physics, work together to research the properties and uses of graphite and graphene.

Cryogenicists use liquid nitrogen to study the effects of very cold temperatures on materials.

Learning episode C2c.1 Orientation

Learning aims

- To revise ideas on atomic structure and ionic bonding.
- To map out the main purposes of the topic.
- Know the structure of an atom from its proton and nucleon numbers.
- Know the arrangement of electrons in shells for the first 20 elements.

Approach

If you choose to teach this topic soon after Topic C2b *Ions and ionic bonds*, you will be able to revise atomic structure very quickly. Refer to the exercise on melting points from Topic C2b.1 Orientation, and the differences between compounds made from only non-metals and those that contain a metal.

Explain that these differences can be explained in terms of the relative strengths of the attractive forces between the ions (in ionic compounds) and between the molecules in covalent compounds. Tell students that it is important to distinguish between the attractive forces **within** molecules (covalent bonds) and the attractive forces **between** molecules.

Explain that the topic will explore another way that atoms/elements can combine to form compounds. It will then look in more detail at the properties of compounds with covalent bonds (made up of molecules) compared with those with ionic bonds (made up of ions).

Learning episode C2c.2 The formation of covalent molecules

Learning objectives

- State that a covalent bond is formed when a pair of electrons is shared between two atoms leading to noble gas electronic configurations.
- Describe the formation of covalent bonds in simple molecules, including H_2, Cl_2, H_2O, CH_4, NH_3 and HCl. Use dot-and-cross diagrams to show the electronic configurations in these and similar molecules.
- Supplement Describe the formation of covalent bonds in simple molecules, including CH_3OH, C_2H_4, O_2, CO_2 and N_2. Use dot-and-cross diagrams to show the electronic configurations in these and similar molecules.

Resources

Student Book Electron sharing in covalent bonding

Approach

Use examples from the Student Book to illustrate how non-metals (which do not like to give up electrons) share electrons to achieve the electronic configuration of the nearest noble gas (full outer electron shells). Emphasise the importance of showing shared electrons in shared pairs in single (and multiple) covalent bonds.

It is also a good idea to show non-bonding electrons as electron pairs. This will help students to avoid mistakes in showing the correct number.

Introduce the structure of covalent molecules represented by a displayed formula. At this level it is acceptable to draw methane with a square planar displayed formula, as opposed to a pyramidal one.

Use the questions in the Student Book to give students practice at producing dot-and-cross diagrams.

Ask students if they notice a relationship between the number of covalent bonds an atom can form and its position in the Periodic Table.

Supplement Use the Student Book to look at the formation of molecules with more than two different types of atoms (methanol) and then with double covalent bonds (ethene, oxygen, carbon dioxide) and triple covalent bonds (nitrogen).

Learning episode C2c.3 Properties of simple covalent molecules

Learning objectives

- Describe the properties of simple molecular compounds:
  (a) low melting points and boiling points
  (b) poor electrical conductivity.

Common misconceptions

Some students think that a covalent bond is weaker than an ionic bond and that covalent bonds are weak bonds.

Resources

Student Book Properties of covalent compounds

Approach

When comparing the type of forces present in covalent compounds you could use the terms 'within molecules' and 'between molecules'. Many students are confused by the difference, so it is worth spending a little time explaining this.

Use the table in the Student Book to identify the characteristic properties of simple covalent compounds. Emphasise the lack of electrical conductivity due to the absence of ions.

Contrast the properties of these simple molecular structures with those that are characteristic of ionic compounds.

Learning episode C2c.4 Consolidation and summary

Learning aims

- Review the learning points of the topic summarised in the End of topic checklist.
- Test understanding of the topic content by answering the End of topic questions.

Resources

Student Book End of topic checklist; End of topic questions

Approach

Ask students to work with a partner to make a list of key words from this topic. They could then work together to produce a spider diagram showing how the different concepts are linked. They could compare their list with the list of key terms given in the Student Book. Discuss the checklist and ask questions to see how much of the content students are comfortable with.

Students could make flashcards of the key content and then use the flashcards to quiz each other on the information.

Ask students to work individually through the End of topic questions in the Student Book without looking at the text. As they work, walk around the classroom observing their answers and questioning them as necessary to find out which questions are causing difficulties.

After a set period, ask students to stop working. Discuss any areas of difficulty you observed as you walked round the class.

Students should complete any unanswered questions for homework, but you should stress that they should answer the questions without looking at the text, so that they can see how much they have remembered.

Ask students to produce a table summarising the differences between the bonding and properties of ionic and covalent compounds.

C3 Stoichiometry

Contents

Teacher Guide	Syllabus reference
C3a Formulas	C3.1 Formulas

Overview of the section

The work on chemical formulas is a necessary part of generating chemical equations. Previously students may have only been introduced to word equations. If so, mastering the process will be essential if progress is to be made. It introduces chemical equations as the chemist's 'shorthand' for describing reactions.

Starting points

The Student Book section opener puts the ideas in the section into context and sets the scene. It also allows students to acknowledge and value their prior learning, and provides a benchmark against which future learning can be compared.

The questions provide a structure for introducing the section and can be used in a number of different ways:

- You could ask students to consider the questions as an introductory homework task.
- You could put students into groups to share their own ideas and understanding and then to report back to the whole class.
- Students could be given access to the Internet, preferably with a tight timescale, to find out the information required.

You could then use a spider chart or other form of wall chart to summarise everybody's ideas.

Recording these initial ideas allows you to retain them for reference as the individual topics are developed. In this way, your students' progress in learning can be readily acknowledged.

C3a Formulas

Introduction

The work on formulas and equations builds on the ideas encountered in Topic C2a *Materials, atomic structure and the Periodic Table*, Topic C2b *Ions and ionic bonds* and Topic C2c *Simple molecules and covalent bonds*.

Supplement Ionic equations are only required for students following the supplementary syllabus.

This work links to Topic C6b *Redox reactions*, Topic C7a *Acids, bases and salts* and Topic C12b *Identification of ions and gases*.

The first three learning episodes focus on writing chemical formulas and balancing chemical equations.

Supplement Constructing symbol equations and ionic equations are only required for the supplementary syllabus.

There are opportunities for practical work, particularly quantitative practical work. These provide practice of the full range of practical skills.

Links to other sections

Section	Essential background knowledge	Useful links
C2 Atoms, elements and compounds	C2a Materials, atomic structure and the Periodic Table C2b Ions and ionic bonds C2c Simple molecules and covalent bonds	
C6 Chemical reactions		C6b Redox reactions
C7 Acids, bases and salts		C7a Acids, bases and salts
C12 Experimental techniques and chemical analysis		C12b Identification of ions and gases

Note: All topics after this one will make use of chemical formulas and word and chemical equations. The topics listed are the strongest links.

Topic overview

C3a.1	**Orientation** The purpose of this learning episode is to revise and reinforce students' abilities to write chemical formulas. This includes deducing formulae from diagrammatic representations. Supplement The focus here is on the use of ion charges to write chemical formulas.

C3a.2	**Writing chemical equations** Students should be familiar with word equations and this learning episode is intended to help students to become more familiar with state symbols, symbol equations and how they can be balanced. Supplement Students construct symbol equations with state symbols. Ionic equations are introduced as a way of generalising reactions by removing 'spectator' ions.
C3a.3	**Consolidation and summary** This learning episode provides a quick recap of the key ideas encountered in the topic.

Career links

These are some scientific careers that focus on this area of chemistry but careers in many other fields use the knowledge and skills gained studying science.

Synthetic chemists and process chemists work with process engineers to develop the equipment and manufacturing plants that take reactions developed in a research lab and scale them up so that they can be used commercially.

Learning episode C3a.1 Orientation

Learning objectives

- State the formulas of the elements and compounds named in the subject content.
- Define the molecular formula of a compound as the number and type of different atoms in one molecule.
- Deduce the formula of a simple compound from the relative number of atoms present in a model or diagrammatic representation.
- Supplement Deduce the formula of an ionic compound from the relative numbers of the ions present in a model or a diagrammatic representation or from the charges on the ions.

Resources

Student Book Formulas

Approach

Part 1

Students may already be familiar with rules on combining powers (and ion charges – see below) and/or the use of the Periodic Table to write chemical formulas. However, if they are not, then this learning episode will provide the practice needed.

The Student Book sets out how combining powers can be derived from the position of the element in the Periodic Table and then presents a systematic approach to writing chemical formulas. Radicals are introduced and the use of brackets in formulas is explained. Questions 3 and 5 on pages 263 and 264 in the Student Book test understanding of working out a formula from a diagram.

Supplement The use of ion charges to write chemical formulas links to the work covered in Topic C2b *Ions and ionic bonds*. A short section in the Student Book highlights this approach to writing formulas. Emphasise that this approach only applies to ionic compounds, that is, compounds formed between a metal and a non-metal.

Part 2

Explain that the rest of the topic concentrates on writing chemical equations.

Learning episode C3a.2 Writing chemical equations

Learning objectives

- Construct word equations to show how reactants form products, including state symbols.
- Balance and interpret simple symbol equations, including state symbols.
- Supplement Construct symbol equations with state symbols, including ionic equations.

Resources

Student Book Writing chemical equations

Worksheet C3a.2 Balancing equations

Approach

You will need to decide just how much time you want to devote to this activity. Clearly there will be opportunities throughout the rest of the course to practise writing chemical equations, but establishing the basic principles and methodology is equally important.

The Student Book gives examples of how word equations can be translated into chemical equations. The process of balancing equations is then developed through the use of worked examples. Worksheet C3a.2 provides examples for the students to try. Students will need to know the non-metallic elements that exist as diatomic molecules and not as single atoms.

The use of state symbols is introduced. Time will need to be devoted to consideration of the section on balancing more complex equations that contain radicals.

The second set of questions in the Student Book includes some more difficult examples for extra challenge.

Supplement Use the Student Book to focus on balancing ionic equations. The supplementary questions on Worksheet C3a.2 (9 and 10) require the balancing of ionic equations.

Learning episode C3a.3 Consolidation and summary

Learning aims

- Review the learning points of the topic summarised in the End of topic checklist.
- Test understanding of the topic content by answering the End of topic questions.

Resources

Student Book End of topic checklist; End of topic questions

Approach

Ask students to work with a partner to make a list of key words from this topic. They could then work together to produce a spider diagram showing how the different concepts are linked. They could compare their list with the list of key terms given in the Student Book. Discuss the checklist and ask questions to see how much of the content students are comfortable with.

Students could make flashcards of the key content and then use the flashcards to quiz each other on the information.

Ask students to work individually through the End of topic questions in the Student Book without looking at the text. As they work, walk around the classroom observing their answers and questioning them as necessary to find out which questions are causing difficulties.

After a set period, ask students to stop working. Discuss any areas of difficulty you observed as you walked round the class.

Students should complete any unanswered questions for homework, but you should stress that they should answer the questions without looking at the text, so that they can see how much they have remembered.

C4 Electrochemistry

Contents

Teacher Guide	Syllabus reference
C4a Electrochemistry	C4.1 Electrolysis

Overview of the section

Electrochemistry is a key element of physical chemistry and its applications are continually growing. The use of electricity to generate chemical reactions is of significant industrial importance, whether to produce metals such as aluminium or non-metals such as chlorine.

The section starts with the characteristics of electrolytes and electrolysis and introduces the electrolysis of a range of electrolytes. Supplementary material focuses on the prediction of the products produced by the electrolysis of binary compounds in the molten state.

Starting points

The Student Book section opener puts the ideas in the section into context and sets the scene. It also allows students to acknowledge and value their prior learning, and provides a benchmark against which future learning can be compared.

The questions provide a structure for introducing the section and can be used in a number of different ways:

- You could ask students to consider the questions as an introductory homework task.
- You could put students into groups to share their own ideas and understanding and then to report back to the whole class.
- Students could be given access to the Internet, preferably with a tight timescale, to find out the information required.

You could then use a spider chart or other form of wall chart to summarise everybody's ideas.

Recording these initial ideas allows you to retain them for reference as the individual topics are developed. In this way, your students' progress in learning can be readily acknowledged.

C4a Electrochemistry

Introduction

This topic builds on the understanding of chemical bonding from Topics C2b *Ions and ionic bonds* and Topic C2c *Simple molecules and covalent bonds*. There are clear links to Topic C9a *Metals*, because the reactivity series can be used to predict the preferential discharge of ions.

The topic starts with an introduction to electrolytes and electrolysis. You will then consider the electrolysis of molten compounds and aqueous solutions.

The concepts of electrolysis can be reinforced by practical demonstrations. This also gives students a good opportunity to develop their practical investigative skills, particularly *Observing, measuring and recording* and *Handling experimental observations and data*.

Links to other topics

Section	Essential background knowledge	Useful links
C2 Atoms, elements and compounds	C2b Ions and ionic bonds C2c Simple molecules and covalent bonds	
C9 Metals		C9a Metals

Topic overview

C4a.1	**Orientation** This learning episode revises ideas on chemical bonding, especially ionic bonding, and then introduces some of the terms associated with electrolysis. Demonstrations highlight the aspects to be covered in the topic.
C4a.2	**Electrolysis of molten electrolytes** This learning episode introduces the electrolysis of molten electrolytes using anhydrous zinc chloride as an example. The need for the ions in an electrolyte to be able to move is emphasised.
C4a.3	**Electrolysis of aqueous electrolytes** This is a relatively long learning episode and offers the opportunity for two practical activities for students.
C4a.4	**Consolidation and summary** This learning episode provides an opportunity for a quick recap of the ideas encountered in the topic.

Career links

These are some scientific careers that focus on this area of chemistry but careers in many other fields use the knowledge and skills gained studying science.

Some electrochemists are involved in the development of medical diagnostic devices which use electrochemistry techniques. Fuel cell material scientists develop new materials that can be used in fuel cell manufacture.

Learning episode C4a.1 Orientation

Learning objectives

- Define electrolysis as the decomposition of an ionic compound, when molten or in aqueous solution, by the passage of an electric current.
- Identify in simple electrolytic cells
  (a) the anode as the positive electrode
  (b) the cathode as the negative electrode
  (c) the electrolyte as the molten or aqueous substance that undergoes electrolysis.

Resources

Student Book The process of electrolysis

Resources for demonstration (see Technician's notes, following)

Approach

Demonstrate the passage of electricity through a solid conductor such as a strip of copper **(see Technician's notes)**. Contrast it with the fact that a strip of plastic acts as an insulator.

Ask why a metal conducts electricity but plastic does not. Students may be aware from studying physics that electrons are able to flow through metals (the structure of metals, with delocalised electrons, is beyond the scope of this syllabus).

Then demonstrate the passage of electricity through a solution (such as aqueous sodium chloride) and compare it with the lack of conductivity of solid sodium chloride. Ask students what they can remember about the bonding in sodium chloride. Elicit the idea that the mobility of the ions in sodium chloride might have something to do with whether an electric current can pass through it.

Ask what the difference between the ions in solid sodium chloride will be and in sodium chloride solution. Use this demonstration to introduce some terms and definitions for electrode, anode, cathode, electrolyte and electrolysis.

Tell students that the topic will explore further the involvement of ions in electrolysis and why the ions must be able to move. They should have observed gases being produced in the electrolysis of sodium chloride solution.

SAFETY INFORMATION

Wear eye protection (splash-proof goggles).

Turn off the electricity when the teacher smells chlorine.

Chlorine is oxidising, toxic and hazardous to the environment.

Small amounts of Cl_2, which can cause respiratory distress in some people, may be produced. ***The laboratory must be well ventilated.***

Finally, for those students following the supplementary syllabus explain how the charge is transferred in electrolysis – in the electrodes (electrons added or removed), wires (electrons) and electrolyte (ions).

Technician's notes

Be sure to check the latest safety notes on these resources before proceeding.

The following resources are needed for the demonstration of conduction:

copper foil, about 5 cm long, cleaned with abrasive paper such as emery cloth; avoid sharp edges
strip of plastic, about 5 cm long
solid sodium chloride, about 10 g
0.5 mol / dm^3 sodium chloride solution, 50 cm^3
100 cm^3 beaker
6–8 V power supply, bulb, leads, crocodile clips, carbon electrodes

Learning episode C4a.2 Electrolysis of molten electrolytes

Learning objectives

- Identify the products formed at the electrodes and describe the observations made during the electrolysis of molten lead(II) bromide using inert electrodes made of platinum or carbon / graphite.
- Supplement State that metals or hydrogen are formed at the cathode and that non-metals (other than hydrogen) are formed at the anode.
- Supplement Predict the identity of the products at each electrode for the electrolysis of a binary compound in the molten state.

Resources

Student Book Electrolysis of molten lead(II) bromide; Predicting the products of electrolysis

Resources for demonstration (see Technician's notes, following)

Approach

Although lead(II) bromide is in the syllabus, the demonstration should be of anhydrous zinc chloride **(see Technician's notes)**. This is due to the specific hazards of demonstrating lead(II) bromide in the laboratory.

Revise the conditions for electrolysis discussed in the previous learning episode. The section Conditions for electrolysis in the Student Book should help. Then introduce the electrolysis of molten anhydrous zinc chloride. You must set this up in a working fume cupboard. If possible, have a lamp in the circuit.

Demonstrate that an electric current cannot pass through the anhydrous zinc chloride as a solid, but once it is molten, an electric current does flow. It may be possible to see effervescence at the anode as the chlorine is formed and, after cooling, it might be possible to see the zinc at the bottom of the crucible.

If you need to reinforce the fact that not all molten substances conduct electricity, you could repeat the experiment using candle wax (a covalently bonded compound). Ask students to predict the products of the electrolysis of other molten binary ionic compounds, for example sodium chloride, lead(II) bromide, magnesium chloride and aluminium oxide.

Supplement Ask students to predict the changes that occur when some other molten ionic compounds are electrolysed, for example sodium chloride, lead(II) bromide, magnesium chloride and aluminium oxide.

SAFETY INFORMATION

Wear eye protection (splash-proof goggles) and use safety screens.

Use a fume cupboard for the electrolysis of molten salts, e.g., zinc chloride. Ensure that the fume cupboard door is closed so that air is gently drawn in. These salts are appreciably volatile at melting temperatures.

Impure lead bromide may give off bromine on melting and during electrolysis; bromine is very toxic and corrosive and must not be inhaled. Pure lead iodide starts to decompose at just above melting point. Use anhydrous zinc chloride as a safer alternative.

Chlorine is oxidising, toxic and hazardous to the environment.

Small amounts of Cl_2, which can cause respiratory distress in some people, may be produced. ***The laboratory must be well ventilated.***

Technician's notes

Be sure to check the latest safety notes on these resources before proceeding.

The following resources are needed for the demonstration of electrolysis of anhydrous zinc chloride:

The apparatus should be set up in a fume cupboard.

carbon electrodes in a holder (pencil 'lead' can be used to allow smaller quantities to be used)
leads, crocodile clips, lamp, 6–8 V power pack
tripod, pipeclay triangle, stand and clamp, crucible, heatproof mat
solid anhydrous zinc chloride
possibly also candle wax
eye protection (splash-proof goggles)

Learning episode C4a.3 Electrolysis of aqueous electrolytes

Learning objectives

- Identify the products formed at the electrodes and describe the observations made during the electrolysis of:
  (a) concentrated aqueous sodium chloride
  (b) dilute sulfuric acid.
  using inert electrodes made of platinum or carbon / graphite.
- Supplement State that metals or hydrogen are formed at the cathode and that non-metals (other than hydrogen) are formed at the anode.

Common misconceptions

Some students think that water is an electrolyte (perhaps arising from knowledge of the dangers associated with electrical appliances in the bathroom).

Resources

Student Book The electrolysis of dilute sulfuric acid; Electrolysis of concentrated sodium chloride solution; Predicting the products of electrolysis.

Worksheet C4a.3(1) Identifying and investigating electrolytes

Worksheet C4a.3(2) The electrolysis of concentrated sodium chloride solution

Resources for class practical (see Technician's notes, following)

Approach

1. Investigating electrolytes

Students test a range of liquids and solutions as indicated in Worksheet C4a.3(1), using their knowledge gained from the first two activities. They make predictions before they undertake each test.

They will become aware that effervescence occurs at one electrode at least with all four electrolytes. They should also be able to identify copper as a pink deposit on the cathode when aqueous copper(II) sulfate is electrolysed.

SAFETY INFORMATION
Wear eye protection (splash-proof goggles).
Ethanol is highly flammable.
Chlorine is oxidising, toxic and hazardous to the environment.
Small amounts of Cl_2, which can cause respiratory distress in some people, may be produced. ***The laboratory must be well ventilated.***

2. Electrolysis of sodium chloride solution

Students can now use Worksheet C4a.3(2) to look in more detail at the electrolysis of concentrated sodium chloride solution. This learning episode introduces the tests for hydrogen and chlorine. If these gases have not been covered previously, you will need to provide some guidance.

SAFETY INFORMATION
Wear eye protection (splash-proof goggles).
Chlorine is oxidising, toxic and hazardous to the environment.
Small amounts of Cl_2, which can cause respiratory distress in some people, may be produced. ***The laboratory must be well ventilated.***

3. Electrolysis of dilute sulfuric acid

The Developing Practical Skills box in the Student Book can be used instead of further practical activity.

Refer back to the observations that the students made for this reaction in Worksheet C4a3(1).

Technician's notes

Be sure to check the latest safety notes on these resources before proceeding.
The following resources are needed for the class practical C4a.3(1), per group:

6–8 V power pack, leads, crocodile clips, lamp, carbon electrodes
100 cm^3 beaker
50 cm^3 ethanol (industrial denatured ethanol) + container for collection of used ethanol
50 cm^3 dilute hydrochloric acid (0.5 mol / dm^3)
50 cm^3 dilute sulfuric acid (0.4 mol / dm^3)
50 cm^3 copper(II) sulfate solution (0.4 mol / dm^3)
50 cm^3 sodium chloride solution (0.4 mol / dm^3)
eye protection (splash-proof goggles)

The following resources are needed for the class practical 4a.3(2), per group:

electrolysis cell with carbon electrodes
6–8 V power supply, leads, crocodile clips
ignition tubes
sodium chloride solution (0.4 M) to fill electrolysis cell
eye protection (splash-proof goggles)

Learning episode C4a.4 Consolidation and summary

Learning aims

- Review the learning points of the topic summarised in the End of topic checklist.
- Test understanding of the topic content by answering the End of topic questions.

Resources

Student Book End of topic checklist and End of topic questions

Approach

Ask students to work with a partner to make a list of key words from this topic. They could then work together to produce a spider diagram showing how the different concepts are linked. They could compare their list with the list of key terms given in the Student Book. Discuss the checklist and ask questions to see how much of the content students are comfortable with.

It is worth stressing the importance of electrolysis in the manufacture of some important chemicals – aluminium and chlorine.

Students could make flashcards of the key content and then use the flashcards to quiz each other on the information.

Ask students to work individually through the End of topic questions in the Student Book without looking at the text. As they work, walk around the classroom observing their answers and questioning them as necessary to find out which questions are causing difficulties.

After a set period, ask students to stop working. Discuss any areas of difficulty you observed as you walked round the class.

Students should complete any unanswered questions for homework, but you should stress that they should answer the questions without looking at the text, so that they can see how much they have remembered.

C5 Chemical energetics

Contents

Teacher Guide	Syllabus reference
C5a Chemical energetics	C5.1 Exothermic and endothermic reactions

Overview of the section

This section is an important element of physical chemistry. Students will be very familiar with exothermic reactions, such as burning fuels, although they may not be familiar with the term 'exothermic'.

Endothermic reactions, as they are far less common, may not be at all familiar.

Supplement For supplementary study, reaction pathway diagrams are introduced as a way of summarising in diagrammatic form the key energy components in a chemical reaction. The energy associated with making and breaking of bonds is introduced.

Starting points

The Student Book section opener puts the ideas in the section into context and sets the scene. It also allows students to acknowledge and value their prior learning, and provides a benchmark against which future learning can be compared.

The questions provide a structure for introducing the section and can be used in a number of different ways:

- You could ask students to consider the questions as an introductory homework task.
- You could put students into groups to share their own ideas and understanding and then to report back to the whole class.
- Students could be given access to the Internet, preferably with a tight timescale, to find out the information required.

You could then use a spider chart or other form of wall chart to summarise everybody's ideas.

Recording these initial ideas allows you to retain them for reference as the individual topics are developed. In this way, your students' progress in learning can be readily acknowledged.

C5a Chemical energetics

Introduction

This topic does not need much prior knowledge, although students must understand covalent bonding before they consider bond breaking and bond forming in reactions.

There are many links to other topics; for example, at supplementary level, energy change in neutralisation reactions links to Topic C7a *Acids, bases and salts* and the energy of combustion links to Topic C11a *Fuels*, Topic C11c *Alkanes* and Topic C11d *Alkenes*. There are also links at supplementary level to Topic C6a *Rate of Reaction* (the activation energy of reactions).

This topic starts with classifying reactions as exothermic and endothermic and then focuses on two types of calorimetric techniques – first using an insulated cup (for reactions such as displacement and neutralisation), and then a metal container (for combustion reactions). At supplementary level, reaction pathway diagrams (energy level diagrams) are introduced and the origin of these energy transfers is explained qualitatively in terms of bond breaking and bond forming.

Students have good opportunities to develop their practical/investigative skills, including performing accurate quantitative work.

Links to other topics

Section	Essential background knowledge	Useful links
C2 Atoms, elements and compounds	C2c Simple molecules and covalent bonds	
C6 Chemical reactions		C6a Rate of reaction
C7 Acids, bases and salts		C7a Acids, bases and salts
C11 Organic chemistry		C11a Fuels C11b Alkanes C11c Alkenes

Topic overview

C5a.1	**Orientation** Limited background knowledge is required for this topic, other than an understanding of covalent bonding. This learning episode introduces the definitions of exothermic and endothermic reactions and then provides students with a number of practical examples of each type of reaction. Reactions include acid/alkali neutralisation reactions and displacement reactions. The physical change of dissolving potassium chloride in water is also introduced as work on 'cold packs' is developed in the next learning episode.

C5a.2	**Measuring energy transfers (1)** This learning episode introduces the calorimetric method for measuring energy changes. Students learn that thermal energy transfer to and from a reaction has to be controlled if accurate measurements have to be made. The choice of the reaction vessel is therefore crucial. There is an investigation into 'cold packs'. Students design their own investigation to find out what quantities of potassium chloride and water must be mixed to provide a temperature change of –10 °C. This learning episode provides a good opportunity for students to learn about this simple calorimetric technique.
C5a.3	**Measuring energy transfers (2)** This learning episode is in two parts. The first part involves using the calorimetric method to compare the energy produced on burning different fuels. Supplement The second part introduces the terminology of reaction pathway diagrams to distinguish between exothermic and endothermic reactions. Students use a calorimetric technique to find out the energy transferred when a fuel burns. The Developing Practical Skills box in the Student Book can be used to reinforce the ideas encountered or could be used later in the consolidation and summary learning episode.
C5a.4	**Reaction pathway diagrams, bond breaking and bond forming** Supplement This learning episode uses the concept of a reaction pathway diagram or energy level diagram to explain the differences between exothermic and endothermic reactions. The use of the term activation energy is reinforced and a reaction pathway diagram is explained in terms of how it illustrates the difference in the energy required to break bonds and the energy released when bonds are formed.
C5a.5	**Consolidation and summary** This learning episode recaps on the ideas encountered in the topic. The Developing Practical Skills box in the Student Book could be used as a revision activity if it has not already been used as part of Learning episode C5a.3. Students can also do the End of topic questions.

Career links

These are some scientific careers that focus on this area of chemistry but careers in many other fields use the knowledge and skills gained studying science.

Pyrotechnic engineers design and test fireworks for firework displays. **Combustion chemists** can be involved in the development of new fuels. **Health and safety officers and firefighters** look at how to limit risks from fires and how to deal with fires safely and efficiently.

Learning episode C5a.1 Orientation

Learning objectives

- State that an exothermic reaction transfers thermal energy to the surroundings leading to an increase in the temperature of the surroundings.
- State than an endothermic reaction takes in thermal energy from the surroundings leading to a decrease in the temperature of the surroundings.

Common misconceptions

Students may think that dissolving is a chemical reaction – dissolving potassium chloride in water is a physical change, not a chemical change.

Resources

Student Book Thermal energy transfer in chemical reactions

Worksheet C5a.1 Exothermic and endothermic reactions

Resources for demonstration and class practical (see Technician's notes, following)

Approach

Demonstrate the burning of magnesium ribbon in a Bunsen flame (see Technician's notes). Warn students not to look at the flame directly. If possible, provide sunglasses/blue glass. Explain that the reaction generates a lot of thermal/heat energy – energy is transferred from the reaction to the surroundings.

Such reactions are known as exothermic reactions.

SAFETY INFORMATION

Wear eye protection (splash-proof goggles).

For longer lengths of magnesium ribbon burnt by the teacher or burning in oxygen, do not look directly at the flame. A welding filter or even viewing it through a 1mm gap between the fingers of one hand will offer protection. Magnesium burns with such brightness that viewing it directly may cause retinal damage.

Raising and lowering the crucible lid should not allow a direct view of the burning magnesium.

Introduce the practical activity (Worksheet C5a.1). This includes one reaction that results in a fall in the temperature of the surroundings (citric acid and sodium hydrogen carbonate) and a physical change (potassium chloride and water) which also absorbs thermal energy from the surroundings. The other reactions (metal + acid; displacement of a metal ion; neutralisation) are exothermic.

Students may be familiar with metal + acid and neutralisation reactions. This is an opportunity to recap or to probe their knowledge. Explain that later in the topic they will study techniques for comparing the amount of thermal energy released in exothermic and endothermic reactions, and at supplementary level, the reasons behind these energy changes.

After completing the practical activity ask the question 'Do all chemical reactions generate thermal energy – are they all exothermic?' In fact, this is not the case. Introduce the terminology of an endothermic reaction – a reaction that absorbs thermal energy from the surroundings.

SAFETY INFORMATION
Wear eye protection (splash-proof goggles).
Sodium hydroxide is corrosive.
Copper(II) sulfate is corrosive, hazardous to the aqueous environment and a moderate hazard.
Sulfuric acid and citric acid are irritants.
Wear disposable gloves.

Technician's notes

Be sure to check the latest safety notes on these resources before proceeding.

The following resources are needed for the demonstration of an exothermic reaction:

magnesium ribbon – about 5 cm (supply students with pre-cut lengths)
tongs, Bunsen burner, heatproof mat
eye protection (splash-proof goggles)

The following resources are needed for the class practical, C5a.1, per group:

boiling tubes, boiling tube rack, thermometer, spatula, droppers, stirring rods
potassium chloride
magnesium ribbon (1 cm per group)
1 mol / dm^3 sulfuric acid, 1 mol / dm^3 citric acid, 1 mol / dm^3 sodium hydroxide – about 8 cm^3 of each
1 mol / dm^3 hydrochloric acid, 1 mol / dm^3 copper(II) sulfate – about 8 cm^3 of each
sodium hydrogen carbonate, iron filings
eye protection (splash-proof goggles)

SAFETY INFORMATION
Only the quantities required should be dispensed.

Learning episode C5a.2 Measuring energy transfers (1)

Learning aims

- To investigate ‘cold packs’.
- To measure temperature changes in a reaction using an insulated cup as a simple calorimeter.
- Describe how an insulated cup can be used to carry out simple calorimetric experiments to determine thermal energy changes (such as in dissolving solids or neutralisation reactions).

Resources

Student Book Thermal energy transfer in chemical reactions

Worksheet C5a.2 Designing a ‘cold pack’

Resources for class practical (see Technician’s notes, below)

Approach

This learning episode introduces the calorimetric technique for measuring energy changes using an insulated beaker.

Designing a ‘cold pack’ is intended to provide some open-ended practical/investigational work, with the students being required to plan and design their own approach to identifying the quantities of potassium chloride and water in a ‘cold pack’ that will produce a drop in temperature of 10 °C.

Explain that cold packs can be used to limit the impact of sports injuries. Some manufacturers use potassium chloride and water in these products. (See also the Science in Context box in the Student Book on How common are endothermic reactions?, which provides some more information on cold packs. The content of these Science in Context boxes is beyond the syllabus but provides the opportunity to explore the content more deeply.)

Worksheet C5a.2 provides some structure for the practical/investigation work.

Encourage students to consider what a systematic approach might require – not just to use trial and error. One approach is to repeat the experiment three or four times using different quantities of potassium chloride with a fixed volume of water, then plot results on a graph and extrapolate to identify the quantities required for the target temperature drop.

Students are required to evaluate the procedure they used and to suggest how it could be improved.

SAFETY INFORMATION
Wear eye protection (splash-proof goggles).

Technician’s notes

Be sure to check the latest safety notes on these resources before proceeding. The following resources are needed for the class practical C5a.2, per group:

insulated cup
thermometer accurate to 0.1 °C
measuring cylinder
Spatula
glass rod
access to electronic balance
potassium chloride, water
eye protection (splash-proof goggles)

Learning episode C5a.3 Measuring energy transfers (2)

Learning objectives

- Supplement Interpret reaction pathway diagrams showing exothermic and endothermic reactions.
- Supplement Define activation energy, E_a, as the minimum energy that colliding particles must have in order to react.
- Supplement Draw and label reaction pathway diagrams for exothermic and endothermic reactions using information provided, to include: reactants; products; overall energy change of reaction; activation energy, E_a.

Resources

Student Book Thermal energy transfer in chemical reactions

Worksheet C5a.3 Measuring energy transfers

Resources for class practical (see Technician's notes, below)

Approach

Part 1

This learning episode uses a different calorimetric technique – one that will provide a comparison of the energy released when a series of fuels burn. This provides a link to the topics in the *Organic Chemistry* section, C11.

A metal container is used as the calorimeter because the metal is a good conductor of heat and will transfer the energy from the flame to the water. The technique involves considerable energy transfer to the environment but can provide comparative results for a number of different fuels.

The technique is explained on Worksheet C5a.3.

Warn students to be extremely careful when handling the spirit burners. Often the tops of the burners are loose fitting and the fuel can readily escape. Make sure that the wick fits tightly in the holder and that the holder fits tightly in the container. If a spirit burner is knocked over when burning, it could cause a substantial fire. Tell students that if this happens, you will extinguish the flames with a fire blanket.

In the *Handling experimental observations and data* section of the worksheet, students will compare the temperature rise per g of fuel burned, and then in *Planning and evaluating investigations* they will consider why their results are likely to be very different from those quoted in a data book.

SAFETY INFORMATION

Wear eye protection (splash-proof goggles).

Ethanol, propanol and pentanol are highly flammable.

Ensure that the laboratory is well ventilated.

Part 2

Supplement Introduce the convention of the overall energy change in the reaction (using ΔH is not included in this syllabus) by referring to the diagrams and explanations in the Student Book.

Introduce the use of a reaction pathway diagram or energy level diagram to illustrate the differences between exothermic and endothermic reactions.

Finally, the Developing Practical Skills box in the Student Book could be used to reinforce learning on calorimetric techniques. Alternatively, it could be used later in the Consolidation and Summary learning episode.

Technician's notes

Be sure to check the latest safety notes on these resources before proceeding.

The following resources are needed for class practical C5a.3, per group:

tin or copper can with a capacity of 100–200 cm^3
Thermometer
heat resistant boards to use as bench protection and draught screens
50–100 cm^3 measuring cylinder
access to electronic balance accurate in grams to 2 decimal places
boss, clamp and stand
small ($\leq$ 100 cm^3) spirit burners and lids containing ethanol, propanol and pentanol
eye protection (splash-proof goggles)

Check that the lids on the spirit burners are secured and that the wicks are not loose. Fill the spirit burners or fill the extra space with cotton wool.

Spirit burners should be dedicated to use with a specific alcohol.

Learning episode C5a.4 Reaction pathway diagrams, bond breaking and bond forming

Learning objectives

- Supplement Interpret reaction pathway diagrams showing exothermic and endothermic reactions.
- Supplement Draw and label reaction pathway diagrams for exothermic and endothermic reactions using information provided, to include: reactants products overall energy change of reaction; activation energy, E_a.
- Supplement State that bond breaking is an endothermic process and bond making is an exothermic process.

Resources

Student Book Reaction pathway diagrams; Activation energy; Where does the energy come from

Approach

Supplement Reaction pathway diagrams were introduced in the previous learning episode. Explain the origin of the overall energy changes in exothermic and endothermic reactions in terms of the energy needed to break bonds (an endothermic process) and the energy released on forming bonds (an exothermic process).

Learning episode C5a.5 Consolidation and summary

Learning aims

- Review the learning points of the topic summarised in the End of topic checklist.
- Test understanding of the topic content by answering the End of topic questions.

Resources

Student Book End of topic checklist and End of topic questions

Approach

It is worth spending some time to consolidate the learning in this topic, particularly, at supplementary level, the concept of reaction pathway diagrams.

Supplement You may need to revise bond breaking and bond forming with students studying the supplementary syllabus.

If you have not used it earlier, the Developing Practical Skills box in the Student Book could be used here. The Science in Context box on endothermic reactions in the Student Book may also be used as a basis for emphasising certain aspects of the topic. The content of these Science in Context boxes is beyond the syllabus but provides the opportunity to explore the content more deeply

Ask students to work with a partner to make a list of key words from this topic. They could then work together to produce a spider diagram showing how the different concepts are linked. They could compare their list with the list of key terms given in the Student Book. Discuss the checklist and ask questions to see how much of the content students are comfortable with.

Students could make flashcards of the key content and then use the flashcards to quiz each other on the information.

Ask students to work individually through the End of topic questions in the Student Book without looking at the text. As they work, walk around the classroom observing their answers and questioning them as necessary to find out which questions are causing difficulties.

After a set period, ask students to stop working. Discuss any areas of difficulty you observed as you walked round the class.

Students should complete any unanswered questions for homework, but you should stress that they should answer the questions without looking at the text, so that they can see how much they have remembered. Consider the End of topic checklist as a basis for revising ideas. Give students opportunities to answer the End of topic questions.

C6 Chemical reactions

Contents

Teacher Guide	Syllabus reference
C6a Rate of reaction	C6.1 Physical and chemical changes C6.2 Rate of reaction
C6b Redox reactions	C6.3 Redox

Overview of the section

This section does not feature all chemical reactions as key chemical reactions feature in many other sections, such as the reactions of acids and the reactions covered in the organic chemistry section. Students should be familiar with the differences between physical and chemical changes and so only a brief revision is included. Investigating the factors that govern the rate of a reaction has great industrial importance and have widescale applicability across very many reactions. The topic on redox reactions emphasises the importance of the inseparable nature of the processes of reduction and oxidation.

Starting points

The Student Book section opener puts the ideas in the section into context and sets the scene. It also allows students to acknowledge and value their prior learning, and provides a benchmark against which future learning can be compared.

The questions provide a structure for introducing the section and can be used in a number of different ways:

- You could ask students to consider the questions as an introductory homework task.
- You could put students into groups to share their own ideas and understanding and then to report back to the whole class.
- Students could be given access to the Internet, preferably with a tight timescale, to find out the information required.

You could then use a spider chart or other form of wall chart to summarise everybody's ideas.

Recording these initial ideas allows you to retain them for reference as the individual topics are developed. In this way, your students' progress in learning can be readily acknowledged.

C6a Rate of reaction

Introduction

The core material focuses on describing how concentration, surface area of solids, catalysts and temperature affect the rate of reactions. A study of catalysts provides a link to catalytic cracking in Topic C11d *Alkenes*. You will find opportunities to practise writing and balancing chemical equations.

Supplement The supplement material extends the idea of the particulate nature of matter to develop a simple collision theory that explains how the rate of a reaction is affected by the factors mentioned above. The concept of a reaction pathway diagram/energy profile can be used to explain the effect of a catalyst on a reaction. Activation energy provides a link with Topic C5a *Chemical energetics*.

The topic lends itself to a range of different investigative activities. Once you have introduced the experimental techniques for measuring the rate of a reaction, the factors – particle size/surface area, concentration, temperature and catalysts – can be investigated.

Links to other topics

Section	Essential background knowledge	Useful links
C1 States of matter	C1a States of matter	
C2 Atoms, elements and compounds	C2a Atoms, elements and compounds	C2c Simple molecules and covalent bonds
C5 Chemical energetics		C5a Chemical energetics
C11 Organic chemistry		C11c Alkenes

Topic overview

C6a.1	**Orientation** This learning episode brings together a number of different ideas: • A recap on the students' current knowledge and understanding of the particulate nature of matter. • A quick demonstration of reactions that occur at different speeds. • An explanation of the difference between rate, speed and time. • Supplement An introduction to a simple collision theory.
C6a.2	**Monitoring the rate of a reaction** This learning episode involves practical work based on the reaction of marble chips with dilute hydrochloric acid and the reaction of magnesium with dilute hydrochloric acid. The rate of the two reactions is monitored either by measuring the loss in mass of the reaction mixture (as the carbon dioxide escapes) or by measuring the volume of (hydrogen) gas produced. Supplement The collision theory can be used to explain why the rate of reaction changes.

C6a.3	**Concentration of solutions** This learning episode involves practical work using the reaction between marble chips and dilute hydrochloric acid. The concentration of the hydrochloric acid is changed and the time taken to collect 40 cm^3 of carbon dioxide is measured. Supplement The collision theory can be used to explain why the rate of reaction is affected by the concentration of solutions.
C6a.4	**Temperature** This involves practical work using one of the reactions encountered in the topic. Supplement The collision theory explains why the rate of reaction is affected by temperature.
C6a.5	**Surface area of solids** This learning episode involves practical work using apparatus similar to that used in Learning episode C6a.2. The reaction of 'large' pieces of marble chip with dilute hydrochloric acid can be compared with the same reaction involving 'small' pieces of marble chip. Supplement The collision theory explains why the rate of reaction is affected by particle size/surface area of solids.
C6a.6	**Use of a catalyst** This involves practical work focused on identifying a good catalyst for the decomposition of hydrogen peroxide solution. In addition, the effect of changing the amount of the catalyst on the rate of the reaction is demonstrated. Supplement The effect of a catalyst on the reaction is illustrated using a reaction pathway diagram.
C6a.7	**Consolidation and summary** This learning episode allows for all the learning in the topic to be consolidated and for key ideas to be applied through tackling the End of topic questions in the Student Book.

Career links

These are some scientific careers that focus on this area of chemistry but careers in many other fields use the knowledge and skills gained studying science.

Chemical manufacturers need to understand the ways in which they can control the rate of reactions for safety and for maximising profits.

Learning episode C6a.1 Orientation

Learning objective

- Identify physical and chemical changes and understand the differences between them.

Common misconceptions

Some students think that rate/speed and time are the same.

Some students think that all collisions between reacting particles result in a reaction.

Resources

Student Book Physical and chemical changes; Rate of reaction; Collision theory

Worksheet C6a.1 The particulate nature of matter

Resources for demonstration (see Technician's notes, following)

Approach

1. Recap on students' current knowledge and understanding of the particulate nature of matter

Introduce the topic using the Introduction, knowledge check and learning objectives in the Student Book. Start with the card sort activity on Worksheet C6a.1 where small groups of students arrange the cards into 'true', 'false' and 'don't know'.

Join each group of students with another to discuss any differences in their card sorts. Take feedback and summarise the key differences between the arrangement and movement of the particles in solids, liquids and gases.

Check there is good understanding of the differences between physical and chemical changes.

If necessary, use of the opening text in the Student Book to illustrate the differences between physical changes and chemical changes (reactions).

2. Quick demonstration of reactions that occur at different speeds.

Demonstrate a few reactions to show the different speeds at which the reactions take place (see Technician's notes):

a)	Almost instantaneous	Copper(II) nitrate solution and sodium carbonate solution (or alternative precipitation reaction)
b)	Rapid once started	Burning magnesium ribbon in air (oxygen)
c)	Quite slow	Calcium carbonate (marble chips) and dilute hydrochloric acid
d)	Very slow	A shiny steel nail and moist air (no observable change during the lesson) Ask the question: 'Why do some reactions happen very quickly and others are very slow?'.

SAFETY INFORMATION

Wear eye protection.

For longer lengths of magnesium ribbon burnt by the teacher or burning in oxygen, do not look directly at the flame. A welding filter or even viewing it through a 1mm gap between the fingers of one hand will offer protection. Magnesium burns with such brightness that viewing it directly may cause retinal damage.

Raising and lowering the crucible lid should not allow a direct view of the burning magnesium.

3. Explanation of the difference between speed, rate and time

Ask students what 'rate' is. Refer to a car: its speed in kilometres/hour is in fact the rate of travel of the car. Refer to Table C6.1 in the Student Book on speed, rate and time.

Supplement 4. Introduction to a simple collision theory

Although the idea of a collision theory is only required in the supplementary syllabus it is probably worth referring to it with all students, varying the detail to match the needs of the particular group that you are teaching. Explain the collision theory model that scientists use to explain the different rates at which reactions occur. Emphasise that there are two requirements:

- The reacting particles must collide.
- There must be sufficient energy involved in the collision to break chemical bonds. These collisions are called effective collisions.

Technician's notes

Be sure to check the latest safety notes on these resources before proceeding. The following resources are needed for the demonstration of reaction speeds:

0.01 mol / dm^3 copper(II) nitrate solution (50 cm^3) – Moderate hazard
0.01 mol / dm^3 sodium carbonate solution (50 cm^3)
magnesium ribbon (2–5 cm)
5 g calcium carbonate (as marble chips)
0.2 mol / dm^3 hydrochloric acid (50 cm^3)
4 beakers (100 cm^3), crucible, tongs, Bunsen burner and heatproof mat
eye protection

Learning episode C6a.2 Monitoring the rate of reaction

Learning objectives

- Describe practical methods for investigating the rate of a reaction including change in mass of a reactant or product and the formation of a gas.
- Interpret data, including graphs, from rate of reaction experiments.
- Supplement Describe collision theory in terms of:

  number of particles per unit volume; frequency of collisions between particles; kinetic energy of particles; activation energy, E_a.

Common misconceptions

Some students think that reactions proceed at a constant rate.

Resources

Student Book Collision theory; Monitoring the rate of a reaction; Interpreting graphs from rate of reaction experiments

Worksheet C6a.2 Monitoring the rate of a reaction

Resources for demonstration and class practical (see Technician's notes, following)

Approach

Your choice of approach will depend on the apparatus available. You will need an electronic balance reading in grams accurate to two decimal places to accurately measure the loss in mass during the reaction between marble chips and dilute hydrochloric acid.

When measuring the volume of gas produced in a reaction, a gas syringe is ideal but an inverted measuring cylinder in a trough of water is an acceptable alternative. There are opportunities to practise all the practical skills.

One approach is to:

1. Demonstrate how the rate of the reaction between marble chips and hydrochloric acid can be monitored by measuring the loss in mass as the carbon dioxide escapes. The apparatus and method are given in the Student Book in Fig. C6.5. Record the class results and ask students to show them graphically. Discuss the different regions of the graph and link the steepness of the gradient to the rate of the reaction.
2. Use a class practical to explore the reaction between dilute hydrochloric acid and magnesium ribbon by measuring the volume of hydrogen produced. Full details are provided on Worksheet C6a.2.

SAFETY INFORMATION
Wear eye protection.
Hydrogen gas is extremely flammable.
Ensure that there are no naked flames in the room.

3. You may use the graph from the class practical to reinforce the interpretation of the changing rate of reaction as a reaction proceeds.

Supplement Use simple collision theory to explain why the rate decreases – as the reactant particles are used up in the reaction, there will be fewer effective collisions in a given time.

Technician's notes

Be sure to check the latest safety notes on these resources before proceeding.

The following resources are needed for the demonstration of the reaction between marble chips and dilute hydrochloric acid:

5 g of marble chips (an excess)
30 cm^3 of 0.25 mol / dm^3 hydrochloric acid
measuring cylinder
electronic balance (reading in grams and accurate to 2 decimal places)
conical flask and cotton wool plug
Stopwatch
eye protection

The following resources are needed for the class practical C6a.2, per group:

8 cm magnesium ribbon
25 cm^3 1 mol / dm^3 hydrochloric acid
conical flask, bung and delivery tube
gas syringe or 100 cm^3 measuring cylinder and trough of water
stand and clamp
eye protection
stopwatch

Learning episode C6a.3 Concentration of solutions

Learning objectives

- Describe the effect on the rate of a reaction of changing the concentration of solutions.
- Supplement Explain the effect on the rate of reaction of: changing the concentration of solutions using collision theory.
- Describe practical methods for investigating the rate of a reaction including change in mass of a reactant or product and the formation of a gas.
- Interpret data, including graphs, from rate of reaction experiments.

Common misconceptions

Check that the difference between time and rate is understood.

Resources

Student Book What can change the rate of a reaction?; Concentration

Resources for class practical (see Technician's notes, following)

Approach

The Developing Practical Skills activity in the Student Book relates directly to the experiment used in this learning episode.

Ask the student groups to base their experiment on the Developing Practical Skills experiment using exactly the same method and quantities. They could use the five different concentrations as in the Student Book activity or, if time is a consideration, four would be acceptable. Check whether measuring the production of 40 cm^3 of carbon dioxide is a sensible amount given the time allowed for the experiment. Change this if necessary. The students will need to create their own table to record their results.

Supplement Explain the results in terms of the collision theory: the more particles there are in a certain volume of solution the more effective collisions there will be.

SAFETY INFORMATION
Wear eye protection.

Technician's notes

Be sure to check the latest safety notes on these resources before proceeding.

The following resources are needed for the class practical C6a.3, per group:

200 cm^3 0.2 mol / dm^3 hydrochloric acid
25 g calcium carbonate (marble chips)
50 cm^3 measuring cylinder or burette (hydrochloric acid)
50 cm^3 measuring cylinder or burette (water)
Stopwatch
eye protection
conical flask, delivery tube and gas syringe (as in Student Book diagram)

Learning episode C6a.4 Temperature

Learning objectives

- Describe the effect on the rate of reaction of changing the temperature.
- Interpret data, including graphs, from rate of reaction experiments.
- Describe practical methods for investigating the rate of a reaction including change in mass of a reactant or product and the formation of a gas.
- Supplement Explain the effect on the rate of reaction of changing the temperature using collision theory.

Common misconceptions

The dramatic impact of temperature may come as a surprise to students – a 10 °C rise in temperature can double the rate of a reaction.

Resources

Student Book What can change the rate of a reaction?; Temperature

Worksheet C6a.4 Reaction rate and temperature

Resources for class practical (see Technician's notes, following)

Approach

This learning episode is a very good opportunity for students to devise and plan their own investigation. They will be familiar with the approach used in the previous learning episode, but will have to decide how to adapt it to allow an investigation of temperature. It is important to allow students to produce a plan without too much prompting.

Choosing the concentration of hydrochloric acid to use is a key aspect. If they choose the fastest reaction from the previous learning episode, they will find the reaction is too rapid to measure accurately. Choosing the slowest will be much more successful.

Supplement Once students have completed their investigation, their results can be explained in terms of the collision theory. A crucial point is that increasing temperature will not only ensure more collisions; it will significantly increase the energy of the collisions.

Make sure that students' plans have been checked before they start to carry them out.

SAFETY INFORMATION
Wear eye protection.

Technician's notes

Be sure to check the latest safety notes on these resources before proceeding.

150 cm^3 0.2 mol / dm^3 hydrochloric acid
25 g of calcium carbonate (marble chips)
50 cm^3 measuring cylinder or burette (hydrochloric acid)
50 cm^3 measuring cylinder or burette (water)
Stopwatch
conical flask, delivery tube, gas syringe (as in Student Book)
eye protection (splash-proof goggles)
Thermometer

Learning episode C6a.5 Surface area of solids

Learning objectives

- Describe the effect on the rate of reaction of changing the surface area of solids.
- Describe practical methods for investigating the rate of a reaction including the formation of a gas.
- Interpret data, including graphs, from rate of reaction experiments.
- Supplement Explain the effect on the rate of reaction of changing the surface area of solids using collision theory.

Common misconceptions

Students may need convincing that the same mass of smaller particles will have a greater surface area than bigger particles. A sliced loaf of bread can be used to model why smaller pieces of a large lump will have a greater surface area.

Resources

Student Book What can change the rate of a reaction?; Surface area

Worksheet C6a.5 Reaction rate and surface area

Resources for the demonstration and class practical (see Technician's notes, following)

Approach

Whichever practical approach you use, students will be familiar with the apparatus from a previous learning episode. The reaction between marble chips and hydrochloric acid will be slower than that of magnesium ribbon and dilute hydrochloric acid, so you have to decide for how long students should take readings. If you choose to demonstrate the loss of mass method, then a video camera connected to a projector showing the balance display and the stopwatch allows students to take their own readings. This helps to keep them interested and motivated especially while the mass changes are slow.

It will be sufficient to compare the initial rates or early stages of the two reactions rather than waiting for the reaction to go to completion. Ideally, students should not be taking readings for more than 5 or 6 minutes for each experiment. (About 90 cm^3 of carbon dioxide should be produced at room temperature. If you want a quicker reaction you can increase the concentration of the acid, but make sure that the syringe is disconnected once the volume reaches 100 cm^3.)

The results can be shown graphically and the gradients of the two curves compared for the early stages of the reactions.

Supplement You may choose to use the collision theory to explain why the reaction has a greater rate with smaller particles. More marble surface and calcium carbonate 'particles' (ions) will be exposed to the hydrochloric acid particles (H^+ ions), so there will be more effective collisions in a given time.

SAFETY INFORMATION
Wear eye protection.
Ensure that the laboratory is well ventilated.

Technician's notes

Be sure to check the latest safety notes on these resources before proceeding.

The reaction between marble chips and dilute hydrochloric acid can be monitored by measuring the loss in mass or the volume of carbon dioxide produced as the reaction proceeds. The availability of accurate electronic balances (accurate to 0.01 g) will determine whether the loss of mass method is used, either as a demonstration (see Technician's notes) or a class practical. The method involving the collection of gas might be easier to resource. The marble chips can be categorised roughly into 'large' and 'small'.

The following resources are needed for the demonstration of the loss of mass method:

5 g of 'large' marble chips (an excess)
5 g of 'small' marble chips (an excess)
2 × 30 cm^3 of 0.25 M hydrochloric acid (allow 100 cm^3)
measuring cylinder
access to an electronic balance in grams accurate to 2 decimal places
conical flask and cotton wool plug
stopwatch
eye protection

The following resources are needed for the class practical C6a.5, per group:

5 g 'large' marble chips (an excess)
5 g 'small' marble chips (an excess)
2 × 30 cm^3 0.25 M hydrochloric acid (allow 100 cm^3)
conical flask, bung and delivery tube
gas syringe or 100 cm^3 measuring cylinder and trough of water
stand and clamp
stopwatch
eye protection

Learning episode C6a.6 Use of a catalyst

Learning objectives

- Describe the effect on the rate of reaction of adding or removing a catalyst.
- State that a catalyst increases the rate of a reaction and is unchanged at the end of a reaction.
- Supplement Explain the effect on the rate of reaction of adding or removing a catalyst using collision theory.
- Interpret data, including graphs, from rate of reaction experiments.

Common misconceptions

Some students may have encountered inhibitors or negative catalysts. However, as far as this syllabus is concerned the focus is only on catalysts defined as increasing the rate of a reaction.

Resources

Student Book What can change the rate of a reaction?; Catalysts

Worksheet C6a.6 Catalysts

Resources for class practical and demonstration (see Technician's notes, below)

Approach

Students should now be familiar with the experimental methods used to investigate the rate of various reactions. It may be that they do not need any further practice or experience. A suggested approach is as follows:

1. Students do a quick test tube investigation of possible catalysts for the decomposition of hydrogen peroxide (see Worksheet C6a.6). They add spatula measures of the solids to approximately 5 cm^3 of hydrogen peroxide in a series of test-tubes. They observe the tubes to see if effervescence occurs (oxygen formation) and, if it does, at what rate.

SAFETY INFORMATION

Wear eye protection.

Hydrogen peroxide solution is an irritant – using 10 volume maximum solution is low hazard.

Copper(II) oxide and manganese(IV) oxide are both harmful. Take care not to inhale.

Copper(II) oxide is also hazardous to the aquatic environment and so needs to be carefully disposed of.

Calcium oxide is a moderate hazard and corrosive.

2. A teacher-led demonstration is used to reinforce the extent of the catalytic effect of manganese(IV) oxide on hydrogen peroxide – possibly comparing the effect of changing the amount of catalyst used (see Technician's notes).

The way that a catalyst increases the rate of a reaction – by offering a reaction pathway with a lower activation energy – can be seen as an extension. The specificity of catalysts is also worthy of attention.

Technician's notes

Be sure to check the latest safety notes on these resources before proceeding.
The following resources are needed for the class practical C6a.6, per group:

40 cm^3 of 10 volume hydrogen peroxide solution
a range of metal oxides:
copper(II) oxide
iron(III) oxide
manganese(IV) oxide
magnesium oxide
aluminium oxide
test-tubes and test-tube rack
teat pipette
spatula
eye protection

Dilute solutions deteriorate rapidly and may be more dilute than labelled. Check dilute solutions produce the expected result before the lesson.

The following resources are needed for the demonstration of a catalyst on hydrogen peroxide:

up to 100 cm^3 of 20 volume hydrogen peroxide solution – only 20 cm^3 needed for each demonstration
manganese(IV) oxide
conical flask with delivery tube to gas syringe
100 cm^3 measuring cylinder
spatula
eye protection

Learning episode C6a.7 Consolidation and summary

Learning aims

- Review the learning points of the topic summarised in the End of topic checklist.
- Test understanding of the topic content by answering the End of topic questions.

Resources

Student Book End of topic checklist and End of topic questions

Approach

Ask students to work with a partner to make a list of key words from this topic. They could then work together to produce a spider diagram showing how the different concepts are linked. They could compare their list with the list of key terms given in the Student Book. Discuss the checklist and ask questions to see how much of the content students are comfortable with.

Students could make flashcards of the key content and then use the flashcards to quiz each other on the information.

Ask students to work individually through the End of topic questions in the Student Book without looking at the text. As they work, walk around the classroom observing their answers and questioning them as necessary to find out which questions are causing difficulties.

After a set period, ask students to stop working. Discuss any areas of difficulty you observed as you walked round the class.

Students should complete any unanswered questions for homework, but you should stress that they should answer the questions without looking at the text, so that they can see how much they have remembered.

C6b Redox reactions

Introduction

This is a short topic. When you teach the topic should depend on when it is most appropriate to your class. This Teacher's Guide recommends teaching this topic towards the end of the course, and so it may be seen as a consolidation or revision topic. This topic builds on the understanding of ions from Topic C2b *Ions and ionic bonds.* In topic C9a *Metals* the concept of redox reactions links to the extraction of metals. The concept of combustion (oxidation) is covered in a number of topics, including Topic C10a *Chemistry of the environment*, Topic C11b *Fuels* and Topic C11c *Alkanes.* Students may be familiar with the use of oxidation numbers when naming compounds of transition metals.

Links to other topics

Section	Essential background knowledge	Useful links
C2 Atoms, elements and compounds	C2b Ions and ionic bonds	
C4 Electrochemistry		C4a Electrochemistry
C9 Metals		C9a Metals
C10 Chemistry of the environment		C10a Chemistry of the environment
C11 Organic chemistry		C11a Fuels C11b Alkanes

Topic overview

C6b.1	**Orientation** This learning episode recaps on examples of oxidation and reduction encountered so far in the course. The extraction of metals from their oxides provides a suitable context to revise definitions of oxidation and reduction in terms of gain or loss of oxygen. The terminology of oxidation numbers is introduced in the context of the naming of transition metal compounds.
C6b.2	**Consolidation and summary** This learning episode provides an opportunity for a quick recap of the ideas encountered in the topic.

Career links

These are some scientific careers that focus on this area of chemistry but careers in many other fields use the knowledge and skills gained studying science. **Dentists** make use of redox chemistry with the use of hydrogen peroxide to whiten teeth. **Archaeologists** can determine the composition of materials through redox analysis.

Learning episode C6b.1 Orientation

Learning objectives

- Define redox reactions as involving simultaneous oxidation and reduction.
- Define oxidation as gain of oxygen and reduction as loss of oxygen.
- Identify redox reactions as reactions involving gain and loss of oxygen.
- Identify oxidation and reduction in redox reactions. (Oxidation number limited to its use to name ions, e.g., iron(II), iron(III), copper(II).)

Common misconceptions

Some students think that there are reactions, such as combustion, where a substance is oxidised but nothing is reduced.

Resources

Student Book Oxidation, reduction and redox

Worksheet C6b.1 Oxidation and reduction

Approach

This learning episode is likely to be relatively short, but this depends on the extent of students' background knowledge.

Ask students to define the processes of oxidation and reduction. Ask for examples of reactions where these processes take place.

Ask if the processes always go together as in the example of the reaction of copper(II) oxide and zinc in the Student Book.

Define the term 'redox'. Then introduce the concept of an oxidation number, drawing on the examples provided in the Student Book.

Use the worksheet provided to consolidate students' learning of these basic ideas. Question 3 on the combustion of carbon is intended to emphasise the nature of a redox reaction – if something is oxidised, then something must be reduced.

SAFETY INFORMATION
Wear eye protection.

Learning episode C6b.2 Consolidation and summary

Learning aims

- Review the learning points of the topic summarised in the End of topic checklist.
- Test understanding of the topic content by answering the End of topic questions.

Resources

Student Book End of topic checklist and End of topic questions

Approach

Ask students to work with a partner to make a list of key words from this topic. They could then work together to produce a spider diagram showing how the different concepts are linked. They could compare their list with the list of key terms given in the Student Book. Discuss the checklist and ask questions to see how much of the content students are comfortable with.

Students could make flashcards of the key content and then use the flashcards to quiz each other on the information.

Ask students to work individually through the End of topic questions in the Student Book without looking at the text. As they work, walk around the classroom observing their answers and questioning them as necessary to find out which questions are causing difficulties.

After a set period, ask students to stop working. Discuss any areas of difficulty you observed as you walked round the class.

Students should complete any unanswered questions for homework, but you should stress that they should answer the questions without looking at the text, so that they can see how much they have remembered.

C7 Acids, bases and salts

Contents

Teacher Guide	Syllabus reference
C6a Rate of reaction	C6.1 Physical and chemical changes C6.2 Rate of reaction
C6b Redox reactions	C6.3 Redox

Overview of the section

Students should be familiar with some of the common acids such as hydrochloric acid, sulfuric acid and nitric acid but may not be aware of their industrial importance in today's world. Some may be familiar with organic acids such as acetic acid (ethanoic acid) as vinegar and formic acid (methanoic acid) as a result of stings and bites. Initially they will be less familiar with bases and salts (apart from common salt), although once these are introduced they are very likely to have encountered a number of these previously.

The industrial applications are not covered in this section but there are a relatively large number of links with other sections.

Starting points

The Student Book section opener puts the ideas in the section into context and sets the scene. It also allows students to acknowledge and value their prior learning, and provides a benchmark against which future learning can be compared.

The questions provide a structure for introducing the section and can be used in a number of different ways:

- You could ask students to consider the questions as an introductory homework task.
- You could put students into groups to share their own ideas and understanding and then to report back to the whole class.
- Students could be given access to the Internet, preferably with a tight timescale, to find out the information required.

You could then use a spider chart or other form of wall chart to summarise everybody's ideas.

Recording these initial ideas allows you to retain them for reference as the individual topics are developed. In this way, your students' progress in learning can be readily acknowledged.

C7a Acids, bases and salts

Introduction

This topic builds on knowledge of compounds introduced in Topic C2a *Atoms, elements and compounds*. Topic C10a *Chemistry of the environment* considers the production and impact of acid rain.

The topic starts with the use of indicators to identify acids, alkalis and neutral substances. This is likely to be a revision for most students. You will then consider the characteristics of acids, bases and alkalis and then the types of oxides. The topic then focuses on soluble salts, their nature and methods of preparation. You will find opportunities to practise writing and balancing chemical equations.

Supplement There is supplementary work on the preparation of insoluble salts.

Students have good opportunities to develop their practical/investigative skills, including accurate use of the titration technique.

Links to other topics

Section	Essential background knowledge	Useful links
C2 Atoms, elements and compounds	C2a Atoms, elements and compounds	
C3 Stoichiometry		C3a Formulas
C10 Chemistry of the environment		C10a Chemistry of the environment
C12 Experimental techniques and chemical analysis		C12a Experimental design and chromatography

Topic overview

C7a.1	**Orientation** This learning episode seeks to revise and consolidate ideas on acids and alkalis, using indicators to classify them. Several indicators are used to emphasise the benefits of a universal indicator. The pH scale facilitates the classification of solutions as strongly or weakly acidic or strongly or weakly alkaline.
C7a.2	**Acids, bases, alkalis and salts** This learning episode considers the nature of acids, bases, alkalis and salts. The characteristic properties of acids and bases are considered and in this context the importance of controlling acidity in soil. A focus on the types of oxides is supported by a practical activity. Salts are introduced as the substances produced when the acidic hydrogen(s) of an acid is/are replaced by a metal.
C7a.3	**Making soluble salts** This learning episode starts with considering four methods for producing soluble salts, and then focuses on the practical aspects of two methods. The first method is neutralisation, the reaction between an acid and an alkali and the titration technique. The second focuses on the preparation of hydrated magnesium sulfate by the reaction between magnesium carbonate and dilute sulfuric acid. Both preparation techniques provide students with opportunities for developing their practical skills.

C7a.4	Supplement **Making insoluble salts** Students perform a series of test-tube reactions to familiarise themselves with some precipitation reactions. They are also given practice at constructing chemical equations for these reactions.
C7a.5	**Consolidation and summary** This learning episode recaps on the key ideas encountered in the topic. Time can be allocated to answering the End of topic questions in the Student Book.

Career links

These are some scientific careers that focus on this area of chemistry but careers in many other fields use the knowledge and skills gained studying science.

Water quality analysts check the purity and safety of drinking water and ensure that any impurities, such as salts, are below legal limits. **Agronomists** study the factors that affect crop growth including soil acidity, and work with farmers to improve plant yields.

Learning episode C7a.1 Orientation

Learning objectives

- Describe acids in terms of their effect on litmus; methyl orange.
- Describe alkalis in terms of their effect on litmus; methyl orange.

Common misconceptions

Take care how you use the terms ‘strong’ and ‘weak’ in relation to acids and alkalis. The concept of strong and weak acids is not in the syllabus. It will be helpful to use terms such as ‘strongly acidic’ rather than ‘strong acid’ etc. and refer to relative acidity and relative alkalinity. The terms ‘strong’ and ‘weak’ should not be confused with *concentration*, e.g., you can have a concentrated solution of a weak acid.

Resources

Student Book The characteristic properties of acids and bases – Aqueous solutions

Worksheet C7a.1 Using indicators

Resources for class practical (see Technician’s notes, following)

Approach

Your students are very likely to have existing experience and knowledge of acids and alkalis, and will probably have used indicators such as litmus and universal indicator. This learning episode will revise and consolidate their ideas.

Give them an overview of the topic; it provides a more detailed study of the nature of acids and alkalis as well as focusing on the class of compounds called salts.

Ask students to list their current knowledge of acids, alkalis and indicators. Take feedback and summarise on a chart. Explain that they are going to undertake a straightforward practical activity to consolidate their understanding, possibly including some indicators they are not familiar with: methyl orange.

In the context of pH and indicators ensure that students understand the meaning of the terms ‘strongly acidic’, ‘weakly acidic’, ‘neutral’, ‘weakly alkaline’ or ‘strongly alkaline’.

Use Worksheet C7a.1 *Using indicators* to support students in testing some ‘everyday’ chemicals. The advantages of using universal indicator and the pH scale will become obvious – it allows differentiation into strongly or weakly acidic and strongly or weakly alkaline.

SAFETY INFORMATION

Wear eye protection (splash-proof goggles).

Methyl orange is acutely toxic, corrosive, hazardous to the aquatic environment, a health hazard, a moderate hazard and flammable.

Universal indicator is flammable.

Technician's notes

Be sure to check the latest safety notes on these resources before proceeding. The following resources are needed for the class practical C7a.1, per group:

test-tubes, test-tube rack
0.1 mol / dm^3 sodium hydroxide solution
0.1 mol / dm^3 hydrochloric acid
0.1 mol / dm^3 solution of bicarbonate of soda (sodium hydrogen carbonate)
soapy water, milk of magnesia, household 'ammonia' cleaner
grapefruit juice, vinegar, distilled water
solutions of methyl orange, universal indicator and chart
eye protection (splash-proof goggles)

Learning episode C7a.2 Acids, bases, alkalis and salts

Learning objectives

- Describe the characteristic properties of acids in terms of their reactions with metals, bases and carbonates.
- State that bases are oxides or hydroxides of metals and that alkalis are soluble bases.
- Describe the characteristic properties of bases in terms of their reactions with acids.
- Describe how to compare neutrality, relative acidity and relative alkalinity in terms of colour and pH (using universal indicator paper).
- Classify oxides as acidic, including SO_2 and CO_2, or basic, including CuO and CaO, related to metallic and non-metallic character.

Resources

Student Book The characteristic properties of acids and bases; Oxides; Salts Worksheet C7a.2 Investigating oxides

Resources for class practical (see Technician's notes, following)

Approach

There is a large amount of information to cover in this learning episode. It would be helpful to spread it over more than one lesson. Use the section in the Student Book to cover the learning objectives listed above.

After covering the properties of acids and bases, consider the classification of oxides – you may use a practical activity here. Spend some time clarifying the nature of a salt and the names of salts linked to the acids that they are formed from.

The sources of acidity in soil, the reasons for controlling this acidity and then the substances that are used could form the basis of a research task prior to feedback and discussion in the learning episode.

SAFETY INFORMATION
Wear eye protection (splash-proof goggles).
Sodium hydroxide and phosphoric acid are irritants.

Technician's notes

Be sure to check the latest safety notes on these resources before proceeding.

The following resources are needed for the class practical C7a.2, per group:

sodium hydroxide 0.1 mol / dm^3 (labelled 'sodium oxide (aq)')
phosphoric acid 0.1 mol / dm^3 (labelled 'phosphorus oxide (aq)')
sulfuric acid 0.1 mol / dm^3 (labelled 'sulfur oxide (aq)')
samples of magnesium oxide, aluminium oxide and silicon oxide (sand)
test-tubes, universal indicator and pH chart
distilled water, glass stirring rod
eye protection (splash-proof goggles)

Learning episode C7a.3 Making soluble salts

Learning objectives

- Describe the preparation, separation and purification of soluble salts by reaction of an acid with:
- an alkali by titration; excess metal; excess insoluble base; excess insoluble carbonate (students do not need to know general solubility rules for salts).

Resources

Student Book Salts –s Making soluble salts; In the laboratory Worksheet C7a.3(1) Making a soluble salt (1) – using a titration method Worksheet C7a.3(2) Making a soluble salt (2)

Resources for class practical (see Technician's notes, following)

Approach

Explain that the focus in this learning episode will be the preparation of soluble salts.

Look at the four general methods as listed in the Student Book. Explain to students that they will undertake some investigative work focused on Methods 1 and 3 in the Student Book.

1. Acid and alkali

Students can use Worksheet C7a.3(1) to investigate Method 1 (acid + alkali). The focus is on the experimental technique; students do not actually produce a sample of a salt, although the worksheet indicates how this would be done.

Emphasise the importance of working carefully – the titration technique can give an accuracy of one drop. Also show how safely to use the equipment, for example fitting pipette fillers.

Explain that the first set of results should be considered as a trial. The first titration can be done relatively quickly to avoid drop-by-drop addition over a long period of time. Your students are likely to need to perform at least two further titrations to get the required agreement between results.

SAFETY INFORMATION

Wear eye protection (splash-proof goggles).

2. Acid and carbonate

Students can use Worksheet C7a.3(2) to investigate Method 3 (acid + carbonate) by preparing a sample of hydrated magnesium sulfate. The method is the same as in the Student Book and may be applied to any soluble salt preparation involving a solid + acid.

Students may find it difficult to determine the crystallisation point. You might help them out with advice such as 'Don't reduce the volume of the evaporating solution by more than half'.

Students should be able to suggest suitable starting chemicals to make a range of different soluble salts.

SAFETY INFORMATION

Wear eye protection (splash-proof goggles).

Dilute sulfuric acid is an irritant.

Technician's notes

Be sure to check the latest safety notes on these resources before proceeding.

The following resources are needed for the class practical C7.1c(1), per group:

burette and stand, 25 cm^3 pipette and bulb filler, conical flask, white tile, wash bottle containing distilled water
0.04 mol / dm^3 sodium hydroxide solution
0.15 mol / dm^3 hydrochloric acid, methyl orange
eye protection (splash-proof goggles)

The following resources are needed for the class practical C7.1c(2), per group:

250 cm^3 beaker, measuring cylinder, spatula, glass rod, filter funnel and filter paper, evaporating basin
Bunsen burner, tripod, gauze and heatproof mat, crystallising dish, microscope slide
0.5 mol / dm^3 sulfuric acid
magnesium carbonate
access to an electronic balance accurate to 1 decimal place

Supplement Learning episode C7a.4 Making insoluble salts

Learning objective

- Supplement Describe the preparation of insoluble salts by precipitation (candidates do not need to know general solubility rules for salts).

Resources

Student Book Salts – Making insoluble salts

Worksheet C7a.4 Making an insoluble salt

Resources for class practical (see Technician's notes, below)

Approach

Explain that the method for preparing an insoluble salt involves a precipitation reaction. Introduce the learning episode by referring to the equation with state symbols and the practical method described in the Student Book and then use Worksheet C7a.4 to enable students to carry out a number of different precipitation reactions.

Supplement This is a good opportunity for students to practise constructing chemical equations for the precipitation reactions.

Additionally, the students should be able to suggest suitable starting chemicals to make a range of different insoluble salts.

SAFETY INFORMATION
Wear eye protection (splash-proof goggles).
Silver nitrate causes skin and eye irritation.

Technician's notes

Be sure to check the latest safety notes on these resources before proceeding. The following resources are needed for the class practical C7a.4, per group:

test-tubes, test-tube rack, droppers and eye protection (splash-proof goggles)
0.01 mol / dm^3 solutions of sodium carbonate, sodium sulfate
0.05 mol / dm^3 solution of silver nitrate
0.01 mol / dm^3 solutions of calcium nitrate, copper(II) sulfate
0.4 mol / dm^3 solution of sodium chloride
0.01 mol / dm^3 solution of potassium sulfate

Learning episode C7a.5 Consolidation and summary

Learning aims

- Review the learning points of the topic summarised in the End of topic checklist.
- Test understanding of the topic content by answering the End of topic questions.

Resources

Student Book End of topic checklist and End of topic questions

Approach

Ask students to work with a partner to make a list of key words from this topic. They could then work together to produce a spider diagram showing how the different concepts are linked. They could compare their list with the list of key terms given in the Student Book. Discuss the checklist and ask questions to see how much of the content students are comfortable with.

Students could make flashcards of the key content and then use the flashcards to quiz each other on the information.

Refer to the Science in Context box, which provides additional information on some acids and alkalis.

Ask students to work individually through the End of topic questions in the Student Book without looking at the text. As they work, walk around the classroom observing their answers and questioning them as necessary to find out which questions are causing difficulties.

After a set period, ask students to stop working. Discuss any areas of difficulty you observed as you walked round the class.

Students should complete any unanswered questions for homework, but you should stress that they should answer the questions without looking at the text, so that they can see how much they have remembered.

C8 The Periodic Table

Contents

Teacher Guide	Syllabus reference
C8a The Periodic Table	C8.1 Arrangement of elements
C8b Group 1 elements	C8.2 Group I properties
C8c Group 7 elements	C8.3 Group VII properties
C8d Transition metals and noble gases	C8.4 Transition elements C8.5 Noble gases

Overview of the section

The Periodic Table could be considered to be the foundation of inorganic chemistry. Students will already be familiar with the link between electronic structure and the arrangement of elements in the Periodic Table. This section starts with a brief recap on the structure of the Periodic Table before a more detailed study of two series of metals, the Group 1 metals and the transition metals, providing a sharp contrast of the properties of these two groups. Similarly there is a detailed study of two groups of non-metals, the Group VII elements and the noble gases, which again provide a sharp contrast in chemical properties. Unsurprisingly this section links to a relatively large number of other sections.

Starting points

The Student Book section opener puts the ideas in the section into context and sets the scene. It also allows students to acknowledge and value their prior learning, and provides a benchmark against which future learning can be compared.

The questions provide a structure for introducing the section and can be used in a number of different ways:

- You could ask students to consider the questions as an introductory homework task.
- You could put students into groups to share their own ideas and understanding and then to report back to the whole class.
- Students could be given access to the Internet, preferably with a tight timescale, to find out the information required.

You could then use a spider chart or other form of wall chart to summarise everybody's ideas.

Recording these initial ideas allows you to retain them for reference as the individual topics are developed. In this way, your students' progress in learning can be readily acknowledged.

C8a The Periodic Table

Introduction

This topic builds on students' understanding of Topic C2a *Atoms, elements and compounds*, which covered atomic structure and elements as substances that cannot be broken down by chemical means.

The Periodic Table is introduced as a very useful way to arrange the elements and to separate metals from non-metals. This topic provides the basis for studying groups and arrangements in the Periodic Table, including Topic C8b *Group I elements*, Topic C8c *Group VII elements* and Topic C8d *Transition elements and noble gases*. The differences in the reactivity of metals, and about the characteristic properties of metals and how they differ from those of typical non-metals are covered in Topic C9a *Metals*. The topic concentrates on how elements are arranged in the Periodic Table and how their arrangement is linked to their atomic structures. You will consider how reactivity changes within a group of metals and a group of non-metals. Later topics in this section provide more detailed study of this.

Links to other topics

Section	Essential background knowledge	Useful links
C2 Atoms, elements and compounds	C2a Atoms, elements and compounds	C2b Ions and ionic bonds C2c Simple molecules and covalent bonds
C8 The Periodic Table		C8b Group I elements C8c Group VII elements C8d Transition elements and noble gases
C9 Metals		C9a Metals

Topic overview

C8a.1	**Orientation** The purpose of this short learning episode is to recap students' understanding of the nature of elements and atomic structure. You may use demonstrations to set the scene for this topic and some of the subsequent topics on particular groups of elements.
C8a.2	**Arrangement of the elements** This learning episode concentrates on the structure of the Periodic Table, the use of the proton number to order the elements and how they are arranged in groups and periods. It also covers the link between the position of an element in the Periodic Table and its likely metal/non-metal classification and electron arrangement.
C8a.3	**Consolidation and summary** This learning episode provides an opportunity for a quick recap on the ideas encountered in the topic and time for the students to answer the End of topic questions in the Student Book.

Career links

These are some scientific careers that focus on this area of chemistry but careers in many other fields use the knowledge and skills gained studying science.

A knowledge of the periodic table underpins many careers in science. For example, **solar energy research chemists and technicians** use non-metals to improve the efficiency of solar panels; **computational toxicologists** use computer models to assess the risks of different elements and compounds; and **geochemists** investigate the elements present in different rocks and minerals.

Learning episode C8a.1 Orientation

Learning aims

- To revise ideas about the differences between elements, compounds and mixtures.
- To revise ideas on atomic structure and the arrangement of electrons.
- To map out the main purposes of the topic.

Resources

Resources for demonstration (see Technician's notes, following)

Approach

This introductory learning episode can be very short, particularly if students show a good knowledge and understanding of previous work on atoms and atomic structure. You could use a quick quiz of 10 questions to probe their understanding of the nature of an element (as opposed to a compound or a mixture) and the arrangement of protons, neutrons and electrons in an atom.

You can then justify the study of the Periodic Table by a quick demonstration of the nature of three elements: sodium (symbol Na; a shiny, soft element stored under oil), copper (symbol Cu; a hard, shiny metal with high melting point) and chlorine (symbol Cl, a pale yellow-green gas) (see Technician's notes).

Cut a small piece of sodium with a sharp knife on filter paper to show how soft it is. Show students the samples of the other two elements.

These elements are clearly very different from each other. Where are they found in the Periodic Table? Perhaps there is a clear link between the properties of different elements and their position in the Periodic Table? Perhaps the Periodic Table as an arrangement of elements might be a really useful way of simplifying the study of all these elements?

SAFETY INFORMATION

Wear eye protection (splash-proof goggles).

Keep chlorine gas in fume cupboard.

Chlorine is oxidising, toxic and hazardous to the environment.

Small amounts of Cl_2, which can cause respiratory distress in some people, may be produced.

The laboratory must be well ventilated.

Technician's notes

Be sure to check the latest safety notes on these resources before proceeding.

The following resources are needed for the demonstration of sodium, copper and chlorine:

bottle of sodium
copper foil
gas jar of chlorine (should be kept in a fume cupboard)
sharp knife, tongs and filter paper
eye protection (splash-proof goggles)

Learning episode C8a.2 Arrangement of the elements

Learning objectives

- Describe the Periodic Table as an arrangement of elements in periods and groups and in order of increasing proton number / atomic number.
- Describe the change from metallic to non-metallic character across a period.
- State that: (b) the number of outer-shell electrons is equal to the group number in Groups I to VII.
- Supplement Identify trends in groups, given information about the elements.

Resources

Student Book Arrangement of elements in the Periodic Table

Approach

Refer briefly to the historical development of the Periodic Table and the work of Mendeleev, who is generally credited with being the leading originator of the modern Periodic Table. Emphasise the following aspects of the arrangement of the table:

1. The labelling of groups and periods.
2. The relative positions of metals, non-metals and metalloids.

Supplement 3. Trends in reactivity for metals and non-metals. Students need to identify trends in groups from information provided.

Learning episode C8a.3 Consolidation and summary

Learning aims

- Review the learning points of the topic summarised in the End of topic checklist.
- Test understanding of the topic content by answering the End of topic questions.

Resources

Student Book End of topic checklist and End of topic questions

Approach

Ask students to work with a partner to make a list of key words from this topic. They could then work together to produce a spider diagram showing how the different concepts are linked. They could compare their list with the list of key terms given in the Student Book. Discuss the checklist and ask questions to see how much of the content students are comfortable with.

Students could make flashcards of the key content and then use the flashcards to quiz each other on the information.

Ask students to work individually through the End of topic questions in the Student Book without looking at the text. As they work, walk around the classroom observing their answers and questioning them as necessary to find out which questions are causing difficulties.

After a set period, ask students to stop working. Discuss any areas of difficulty you observed as you walked round the class.

Students should complete any unanswered questions for homework, but you should stress that they should answer the questions without looking at the text, so that they can see how much they have remembered.

Supplement Ideas on trends in reactivity will feature strongly in Topic C8b *Group I elements*, Topic C8c *Group VII elements* and Topic C9a *Metals*.

C8b Group I elements

Introduction

This topic follows on naturally from Topic C8a *The Periodic Table*. Students will know that Group I is a group of metals and will probably know that sodium is an unusual metal, as it is stored under oil in the laboratory. As the topic is very short, you could easily cover the first two learning episodes in the same lesson.

The topic centres on the reactions of the Group I elements with water. You will find opportunities to practise writing and balancing chemical equations. Alkali metals have obvious links with Topic C7a *Acids, bases and salts* and the reactivity of metals in Topic C8d *Transition elements and noble gases.* The differences in the reactivity of metals, and about the characteristic properties of metals and how they differ from those of typical non-metals are covered in Topic C9a *Metals*.

Links to other topics

Section	Essential background knowledge	Useful links
C2 Atoms, elements and compounds		C2b Ions and ionic bonds
C3 Stoichiometry		C3a Formulas
C7 Acids, bases and salts		C7a Acids, bases and salts
C8 The Periodic Table	C8a The Periodic Table	C8d Transition elements and noble gases
C9 Metals		C9a Metals

Topic overview

C8b.1	**Orientation** This is a very short learning episode re-emphasising the position of the Group I elements in the Periodic Table and introducing the elements lithium, sodium and potassium. These are very unusual metals, as they are soft and stored under oil in the laboratory. As metals they have very low densities and melting points. The trend in reactivity of the elements is emphasised. Supplement There is an opportunity for students to construct equations for the reaction of the elements with oxygen.
C8b.2	**Reactions of the Group I elements** This learning episode involves demonstrating the reactions of lithium, sodium and potassium with water. It emphasises the formation of hydrogen gas and an alkaline solution. Supplement Practice can be provided for identifying trends in the group using information on melting points and densities. An explanation of the trend in reactivity can be provided in terms of the ease of removal of the outermost electron, although this is beyond the syllabus requirement. Equations for the reactions are developed.

C8b.3	**Consolidation and summary** This is another short learning episode that provides an opportunity for a quick recap on learning points, as well as giving students time to answer the End of topic questions in the Student Book.

Career links

These are some scientific careers that focus on this area of chemistry but careers in many other fields use the knowledge and skills gained studying science. **Soil scientists** conduct research to study the effects of potassium and sodium deficiencies in plants. **Neuroscientists and researchers** study the effect of the Group I ions on brain activity in the drive to find treatments for diseases such Alzheimer's and epilepsy.

Learning episode C8b.1 Orientation

Learning objective

- Describe the Group I alkali metals, lithium, sodium and potassium, as relatively soft metals with general trends down the group, limited to: decreasing melting point; increasing density.

Resources

Student Book Reactivity of Group I elements; Group I properties

Resources for a demonstration (see Technician's notes, below)

SAFETY INFORMATION
Wear eye protection (splash-proof goggles).
Dispense only the quantity of each metal required, and keep the remainder under oil.

Approach

Quickly recap on previous work on the Periodic Table. Ask students what they already know about the Group I elements. Show samples of the elements in bottles. Ask students why the elements are stored in oil. Explain that the melting points of the elements decrease down the group (indicating that the strength of the metallic bonding decreases) and overall, the densities of the metals increase down the group. You can also cut small samples of the metals on a filter paper with a sharp knife. *Make sure that pieces of metal that are not to be used immediately are returned to their bottles and so are kept under oil* (see Technician's notes).

Refer to Fig C8.7 in the Student Book, which shows that the reactivity of the Group I elements increases down the group.

Supplement Construct equations for the reactions of the metals with oxygen (from the air).

Technician's notes

Be sure to check the latest safety notes on these resources before proceeding.

The following resources are needed for the demonstration of the Group I metals:

lithium – one piece, 5 mm cube; sodium – one piece, 4 mm cube; potassium – one piece, 3 mm cube
forceps, filter paper and knife, eye protection (splash-proof goggles or a face shield)

Learning episode C8b.2 Reactions of the Group I elements

Learning objective

- Describe the Group I alkali metals, lithium, sodium and potassium, as relatively soft metals with general trends down the group, limited to: increasing reactivity with water.
- Supplement Predict the properties of other elements in Group I, given information about the elements.

Resources

Student Book Reactivity of Group I elements; Group I properties

Resources for a demonstration (see Technician's notes, below)

Approach

Demonstrate the reaction of lithium, sodium and potassium with water (see Technician's notes). The oil must be wiped off the metal once it has been cut. A 500 cm^3 beaker can be used for lithium. For sodium and potassium, a small trough of water on an overhead projector provides good visibility. Alternatively, you may use a large trough of water.

Use a safety screen and carefully control the size of the metal samples. In a small trough of water, use samples no bigger than 1 mm × 1 mm; in a large trough of water the metal samples must be no bigger than 2 mm × 2 mm. Demonstrate the reactions in turn, using fresh water each time. Add universal indicator to the solution to illustrate how an alkaline solution forms – hence the family name 'alkali metals'.

The increased reactivity down the group should be easy to observe from lithium (which doesn't usually melt) to sodium (which melts into a sphere instantaneously) to potassium (which melts and catches fire on contact with the water).

On the Internet there are many video clips of rubidium reacting with water. This can form a stimulating conclusion to the demonstrations. Ask students to predict what will happen before they watch the clip.

More detailed observations are provided in the Student Book. Introduce equations for the reactions.

SAFETY INFORMATION

Wear eye protection (splash-proof goggles).

Use safety screens. The screens should be positioned close to the trough to avoid the risk of sodium or potassium shooting out of the trough.

In a small trough of water, use samples no bigger than 1mm × 1mm; in a large trough of water the metal samples must be no bigger than 2mm × 2mm.

Technician's notes

Be sure to check the latest safety notes on these resources before proceeding.

The following resources are needed for the demonstration of the reactions of Group I metals:

Lithium
Sodium
Potassium
forceps, filter paper and sharp knife (all dry)
500cm^3 beaker (for lithium); glass trough (for sodium and potassium)
universal indicator paper or solution
overhead projector and 2 safety screens
goggles or face shield

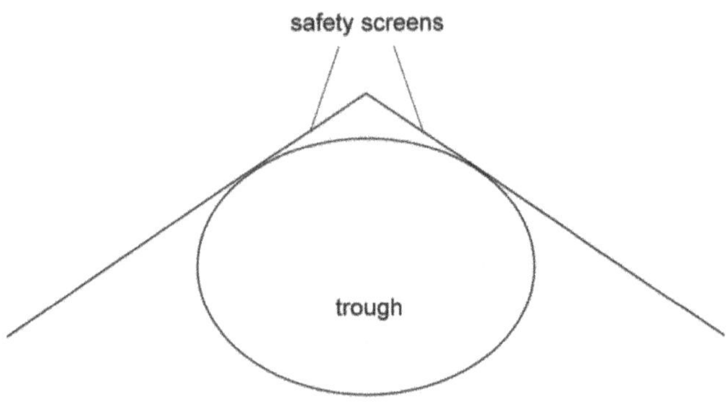

The safety screens should be positioned close to the trough to avoid the risk of sodium or potassium shooting out of the trough, as shown in the diagram above. Take steps to prevent theft.

Learning episode 8b.3 Consolidation and summary

Learning aims

- Review the learning points of the topic summarised in the End of topic checklist.
- Test understanding of the topic content by answering the End of topic questions.

Resources

Student Book End of topic checklist and End of topic questions

Approach

Ask students to work with a partner to make a list of key words from this topic. They could then work together to produce a spider diagram showing how the different concepts are linked. They could compare their list with the list of key terms given in the Student Book. Discuss the checklist and ask questions to see how much of the content students are comfortable with.

Students could make flashcards of the key content and then use the flashcards to quiz each other on the information.

Ask students to work individually through the End of topic questions in the Student Book without looking at the text. As they work, walk around the classroom observing their answers and questioning them as necessary to find out which questions are causing difficulties.

After a set period, ask students to stop working. Discuss any areas of difficulty you observed as you walked round the class.

Students should complete any unanswered questions for homework, but you should stress that they should answer the questions without looking at the text, so that they can see how much they have remembered.

C8c Group VII elements

Introduction

This is a short topic that is likely to follow on from Topic C8a *The Periodic Table* and Topic C8b *Group I elements*. Students should be familiar with the position of non-metals in the Periodic Table, so they should recognise Group VII elements as non-metals.

The topic concentrates on the appearance of the three common halogens (chlorine, bromine and iodine), their relative reactivities and, at supplement level, how these are confirmed by displacement reactions involving the halogen and solutions of their salts. There is an opportunity to practice the redox nature of these reactions if Topic C6b *Redox reactions* has been covered. You will find opportunities to practise writing and balancing chemical equations.

Topic C12b *Identification of ions and gases* covers the test for halide ions. There is an opportunity for some student practical work on displacement reactions.

Links to other topics

Section	Essential background knowledge	Useful links
C2 Atoms, elements and compounds		C2b Ions and ionic bonds
C3 Stoichiometry		C3a Formulas
C6 Chemical reactions		C6b Redox reactions
C8 The Periodic Table	C8a The Periodic Table	C8b Group I elements
C12 Experimental techniques and chemical analysis		C12b Identification of ions and gases

Topic overview

C8c.1	**Orientation** This learning episode very briefly recaps on the structure of the Periodic Table, followed by an introduction to the elements of chlorine, bromine and iodine. Fluorine and astatine are briefly mentioned in the context of predicting their physical states and relative reactivities.
C8c.2	**Supplement Displacement reactions of the halogens** This learning episode focuses on the potential reactions of chlorine, bromine and iodine with solutions of their salts.
C8c.3	**Consolidation and summary** This learning episode allows a quick recap on the ideas encountered in the topic, particularly displacement reactions, and an opportunity for students to answer the End of topic questions in the Student Book.

Career links

These are some scientific careers that focus on this area of chemistry but careers in many other fields use the knowledge and skills gained studying science.

Laboratory technicians and **science teachers** need to have a good understanding of the properties and dangers of the halogens in order to plan practical work that can be done safely by students when using them.

Learning episode C8c.1 Orientation

Learning objectives

- Describe the Group VII halogens, chlorine, bromine and iodine, as diatomic non-metals with general trends down the group, limited to: increasing density; decreasing reactivity.
- State the appearance of the halogens at room temperature and pressure, r.t.p. as: chlorine, a pale yellow-green gas; bromine, a red-brown liquid; iodine, a grey-black solid.
- Supplement Predict the properties of other elements in Group VII, given information about the elements.

Resources

Student Book Reactivity of Group VII elements; Group VII properties

Resources for a demonstration (see Technician's notes, following)

Approach

Ask students to tell you what they already know about the Group VII elements – they should know that they are non-metals, that they all have seven electrons in their outer electron shells and that reactivity (for non-metals) decreases as you descend the group.

Explain that the elements have the family name 'halogens', which means 'salt maker'. In other words, these elements react with metals to form salts (such as chlorides, bromides and iodides).

Show students samples of the three halogens: chlorine is a pale green gas; bromine is a red-brown liquid and iodine is a black solid (see Technician's notes). Explain that because chlorine and bromine are toxic in the laboratory they are often used in solution – chlorine solution is colourless; bromine solution is orange.

Put the three elements in order of reactivity and then put cards for fluorine and astatine in the correct positions. Ask students to predict the appearance and reactivity of these other two elements (fluorine – a pale yellow gas; astatine – a black solid). Refer students to the Science in Context box on fluorine in the Student Book. The content of these Science in Context boxes is beyond the syllabus but provides the opportunity to explore the content more deeply.

SAFETY INFORMATION

Wear eye protection (splash-proof goggles).

Chlorine is oxidising, toxic and hazardous to the environment.

Small amounts of Cl_2, which can cause respiratory distress in some people, may be produced. ***The laboratory must be well ventilated.***

Keep all the samples in a fume cupboard.

Technician's notes

Be sure to check the latest safety notes on these resources before proceeding.

The following resources are needed for the demonstration of the Group VII elements:

a gas jar of chlorine (in a fume cupboard)
bromine liquid in a bottle/gas jar (in a fume cupboard) and iodine solid (0.25 g)
eye protection (splash-proof goggles)

Learning episode C8c.2 Displacement reactions of the halogens

Learning objectives

- Supplement Describe and explain the displacement reactions of halogens with other halide ions.
- Supplement Predict the properties of other elements in Group VII, given information about the elements.

Resources

Student Book Group VII properties – Supplement section

Worksheet C8c.2 Displacement reactions of halogens Resources for demonstration (see Technician's notes, following)

Approach

Supplement Explain displacement reactions in terms of the halogen atoms competing for a metal. Demonstrate the practical shown on Worksheet C8c.2. Students should not use cyclohexane in the classroom, so this must be a teacher demonstration. Students can record observations during the demonstration using the worksheet. Explain that halogens are used in solution to reduce the risks posed by these chemicals. Cyclohexane, a non-polar solvent, is used to distinguish more precisely between solutions containing bromine and those containing iodine. You will need to spend time on developing the equations for the reactions.

The Developing Practical Skills activity in the Student Book deals with displacement reactions. You could use this as a consolidation after the practical activity or leave it for the consolidation and summary learning episode at the End of the topic.

Supplement of the text in the Student Book explains the redox nature of these reactions with an ionic equation and then with half-equations.

SAFETY INFORMATION

Wear eye protection (splash-proof goggles).

The non-polar/ hydrocarbon solvent is highly flammable and harmful, and irritating to skin.

Chlorine solution, bromine solution and iodine solution are all irritants.

Warn asthmatics.

Ensure that there are no naked flames.

Ensure that the laboratory is well ventilated.

Cyclohexane highly flammable and harmful and should not be used by students.

Technician's notes

Be sure to check the latest safety notes on these resources before proceeding.

The following resources are needed for the class practical C8c.2, per group:

chlorine solution (0.1% w/v) – diluted until the colour is just visible
bromine solution (0.3% w/v) – diluted until the colour is just visible
iodine solution (0.5% w/v) – diluted until the colour is just visible
0.1 mol / dm^3 solution sodium chloride
0.1 mol / dm^3 solution sodium bromide
0.1 mol / dm^3 solution sodium iodide
Cyclohexane
test-tubes, droppers, glass rods
eye protection (splash-proof goggles)
beaker for waste chemicals

Learning episode C8c.3 Consolidation and summary

Learning aims

- Review the learning points of the topic summarised in the End of topic checklist.
- Test understanding of the topic content by answering the End of topic questions.

Resources

Student Book End of topic checklist and End of topic questions

Approach

Ask students to work with a partner to make a list of key words from this topic. They could then work together to produce a spider diagram showing how the different concepts are linked. They could compare their list with the list of key terms given in the Student Book. Discuss the checklist and ask questions to see how much of the content students are comfortable with.

Students could make flashcards of the key content and then use the flashcards to quiz each other on the information.

Supplement Revisit displacement reactions and, if not already covered, ask students to answer the questions related to the Developing Practical Skills box in the Student Book.

Ask students to work individually through the End of topic questions in the Student Book without looking at the text. As they work, walk around the classroom observing their answers and questioning them as necessary to find out which questions are causing difficulties.

Supplement Spend some time revising the equations for displacement reactions.

After a set period, ask students to stop working. Discuss any areas of difficulty you observed as you walked round the class.

Students should complete any unanswered questions for homework, but you should stress that they should answer the questions without looking at the text, so that they can see how much they have remembered.

C8d Transition elements and noble gases

Introduction

This topic completes the systematic study of the Periodic Table. The two aspects are combined because they represent two important 'regions' of the Periodic Table, not because they are particularly similar.

There are clear links to Topic C8a *The Periodic Table*, Topic C8b *Group I elements* and Topic C8c *Group VII elements*, and Topic C9a *Metals*. Students should be familiar with the position of metals and non-metals in the Periodic Table. They should know the significance of the electron configuration of noble gases in chemical bonding as explored in Topic C2b *Ions and ionic bonds* and Topic C2c *Simple molecules and covalent bonds*.

The characteristic colours of the transition metal compounds, and specifically their hydroxides, link to Topic C12b *Identification of ions and gases*. There is also a link to Topic C6a *Rate of reaction* through the use of transition metals as catalysts.

After a brief recap on the Periodic Table's structure and organisation, the topic concentrates on the properties of the transition metals. There is a brief look at the noble gases, focusing primarily on their electronic configurations. The topic provides an opportunity for developing the practical/investigative skills of Observing, measuring and recording.

Links to other topics

Section	Essential background knowledge	Useful links
C2 Atoms, elements and compounds		C2b Ions and ionic bonds C2c Molecules and covalent bonds
C6 Chemical reactions		C6a Rate of reaction
C8 The Periodic Table	C8a The Periodic Table	C8b Group I elements C8c Group VII elements
C9 Metals		C9a Metals
C12 Experimental techniques and chemical analysis		C12b Identification of ions and gases

Topic overview

C8d.1	**Orientation** This learning episode very briefly recaps on the structure of the Periodic Table and highlights the positions of the transition metals and the noble gases. The transition metals are contrasted with the Group I metals and the noble gases are contrasted with the Group VII elements.
C8d.2	**The transition elements** This learning episode focuses on the properties of the transition metals (high densities and high melting points), forming coloured compounds and the use of transition metals and their compounds as catalysts. A class practical is used to emphasise the fact that many transition metals form coloured compounds.

C8d.3	**The noble gases** The unreactive nature of the noble gases provides the core of this short learning episode, with links being made to the students' knowledge of ionic and covalent bonding. The monatomic nature of the noble gases is emphasised.
C8d.4	**Consolidation and summary** This learning episode allows a quick recap on the ideas encountered in the topic and an opportunity for students to answer the End of topic questions in the Student Book.

Career links

These are some scientific careers that focus on this area of chemistry but careers in many other fields use the knowledge and skills gained studying science. **Aircraft engineers** work with transition metals such as titanium to design aircrafts which are strong but lightweight. **Anaesthesiologists** use some noble gases such as xenon in general anaesthetics and welders often do their job in an atmosphere of inert argon to keep them safe.

Learning episode C8d.1 Orientation

Learning aims

- Know the positions of the transition metals and noble gases in the Periodic Table.
- Know some of the key differences between the transition metals and the Group I metals.
- Know some of the key differences between the noble gases and the Group VII elements.

Resources

Resources for a demonstration (see Technician's notes, following)

Approach

Ask students to consider the element copper, Cu. What does the position of copper in the Periodic Table indicate about the element? In what ways is copper similar to, and different from, a Group I element such as sodium?

Show students samples of a selection of transition metals and compare with a sample of the element sodium (see Technician's notes).

SAFETY INFORMATION
Wear eye protection (splash-proof goggles).
Warn students who have allergies to heavy metals, such as nickel not to handle it unless they are wearing disposable gloves.

Ask students about the element argon, Ar. What does the position of argon in the Periodic Table indicate about the element? In what ways is argon similar to, and different from, a Group VII element such as chlorine?

Ask students about the electronic structure of the noble gases. What is the significance of the noble gas electron structure when atoms combine to form ionic or covalent bonds?

Technician's notes

Be sure to check the latest safety notes on these resources before proceeding.

The following resources are needed for the demonstration of transition metals:

samples of transition metals, e.g., chromium, iron, copper, nickel, zinc
a bottle of sodium, with the metal stored in oil

Learning episode C8d.2 The transition elements

Learning objectives

- Describe the transition elements as metals that:

  have high densities; have high melting points; form coloured compounds; often act as catalysts as elements and in compounds.

Resources

Student Book Transition elements

Worksheet C8d.2 Transition metal compounds

Resources for a class practical (see Technician's notes, following)

Approach

Ask students to look up the densities and melting points of the following transition metals and to compare with the corresponding data for magnesium and aluminium:

- titanium, vanadium, chromium, iron, nickel, copper, silver, gold, mercury.

Refer to the comparison in the Student Book between the transition metals and the Group I metals. Students could research the use of transition metals as catalysts as a homework task.

Introduce the class practical, which highlights the characteristic colours of some transition metal compounds. It also provides a link to the identification of cations by precipitation reactions. (See Topic C12b *Identification of ions and gases* and Worksheet C8d.2 [using a flame test to identify metal ions] to check that you are happy with the amount of overlap between these practical activities.)

The worksheet suggests that zinc is not a typical transition metal – this could be the focus of a short discussion. Stress the use of Roman numerals to indicate the oxidation number of the metal. Ask students why Roman numerals tend *not* to be used when naming zinc compounds.

Supplement The text in the Student Book describing coloured compounds will reinforce the outcomes of the practical work. The practical also provides an opportunity for those students following the supplementary syllabus to consider ionic equations for the precipitation reactions.

SAFETY INFORMATION

Wear eye protection (splash-proof goggles).

Solutions of copper(II) sulfate, iron(II) ammonium sulfate, iron(III) chloride, zinc sulfate and ammonia are irritants. 0.1M sodium hydroxide solution is an irritant.

If students are allergic to nickel, they should wear disposable gloves.

For example, for the reaction between copper(II) sulfate solution and sodium hydroxide solution, the ionic equation is:

$$Cu^{2+}(aq) + 2OH^{-}(aq) \rightarrow Cu(OH)_2(s)$$

Technician's notes

Be sure to check the latest safety notes on these resources before proceeding.

The following resources are needed for the class practical C8d.2, per group:

0.1 mol / dm^3 solutions of copper(II) sulfate, iron(II) ammonium sulfate, iron(III) chloride, zinc sulfate
0.1 mol / dm^3 sodium hydroxide solution
0.5 mol / dm^3 ammonia solution
test-tubes, droppers, glass rods
eye protection (splash-proof goggles)

Learning episode C8d.3 The noble gases

Learning objective

- Describe the Group VIII noble gases as unreactive, monatomic gases and explain this in terms of electronic configuration.
- State that: Group VIII noble gases have a full outer shell.

Resources

Student Book Noble gases

Approach

Students might be interested to know that there is more argon (0.9%) in the air than there is carbon dioxide (0.04%).

Students should be very familiar with the unreactive nature of the noble gases and should be able to explain this in terms of full outer electron shells (or eight electrons in the outer electron shell).

Ask students to undertake some independent research on the uses of the noble gases and on any reactions that they do undergo. This will make the point that they are not totally unreactive and will react with other elements under certain circumstances, e.g., xenon will react with the highly reactive element fluorine.

Learning episode C8d.4 Consolidation and summary

Learning aims

- Review the learning points of the topic summarised in the End of topic checklist.
- Test understanding of the topic content by answering the End of topic questions.

Resources

Student Book End of topic checklist and End of topic questions

Approach

Ask students to work with a partner to make a list of key words from this topic. They could then work together to produce a spider diagram showing how the different concepts are linked. They could compare their list with the list of key terms given in the Student Book. Discuss the checklist and ask questions to see how much of the content students are comfortable with.

Students could make flashcards of the key content and then use the flashcards to quiz each other on the information.

Ask students to work individually through the End of topic questions in the Student Book without looking at the text. As they work, walk around the classroom observing their answers and questioning them as necessary to find out which questions are causing difficulties.

After a set period, ask students to stop working. Discuss any areas of difficulty you observed as you walked round the class.

Students should complete any unanswered questions for homework, but you should stress that they should answer the questions without looking at the text, so that they can see how much they have remembered.

C9 Metals

Contents

Teacher Guide	Syllabus reference
C9a Metals	C9.1 Properties of metals C9.2 Uses of metals C9.3 Alloys and their properties C9.4 Reactivity series C9.5 Corrosion of metals C9.6 Extraction of metals

Overview of the section

Metals are common everyday materials and students will be very familiar with the common metals. Awareness of the nature of alloys may also be present, although this section will extend this by looking in more detail at the structure of alloys and how this affects their properties. Their properties then link to their uses.

A major part of this section is a study of the reactivity series, which brings together the chemical properties of a range of metals. The corrosion of iron and ways of reducing/preventing this corrosion are studied.

The final part of the section focuses on the extraction of iron and aluminium from their ores with the supplementary study concentrating on the associated chemical equations for the extraction of iron from iron oxide.

Starting points

The Student Book section opener puts the ideas in the section into context and sets the scene. It also allows students to acknowledge and value their prior learning, and provides a benchmark against which future learning can be compared.

The questions provide a structure for introducing the section and can be used in a number of different ways:

- You could ask students to consider the questions as an introductory homework task.
- You could put students into groups to share their own ideas and understanding and then to report back to the whole class.
- Students could be given access to the Internet, preferably with a tight timescale, to find out the information required.

You could then use a spider chart or other form of wall chart to summarise everybody's ideas.

Recording these initial ideas allows you to retain them for reference as the individual topics are developed. In this way, your students' progress in learning can be readily acknowledged.

C9a Metals

Introduction

This topic is extensive and you will need to give it a relatively large amount of time.

Students will be aware of the differences in the reactivity of metals from their work in Topic C8a *The Periodic Table* and Topic C8b *Group I elements*. The reduction of metal oxides to form metals provides a link to Topic C6b *Redox reactions*. There are also links to Topic C7a *Acids, bases and salts* through the reactions of metals with acid to produce hydrogen.

This topic starts with a section on the physical properties of metals, which is extended to include the structure and uses of alloys. The next learning episode focuses on the reactivity series of metals, concentrating on the reaction of metals with water/steam and hydrochloric acid.

The reduction of oxides with carbon leads to a consideration of the extraction of metals, using iron as an example. Reference is made to the extraction of aluminium in this topic. Finally, you will consider the uses of metals such as aluminium and copper.

There are some opportunities for students to develop their practical skills.

Links to other topics

Section	**Essential background knowledge**	**Useful links**
C6 Chemical reactions		C6b Redox reactions
C7 Acids, bases and salts		C7a Acids, bases and salts
C8 The Periodic Table	C8a The Periodic Table C8b Group I elements C8d Transition metals and noble gases	

Topic overview

C9a.1	**Orientation** This learning episode provides an opportunity to revise the reactions of the Group I metals with water, in comparison with the relative lack of reactivity of the transition metals. You might like to demonstrate the reaction of sodium and water and compare this with adding a piece of copper to water. Explain that the topic will not only concentrate on the reactivity series of metals but how this can be used to predict ways of extracting metals from their ores. Some consideration will also be given to alloys, particularly brass and stainless steel.
C9a.2	**Properties of metals** This learning episode focuses on the characteristic physical properties of metals. A straightforward practical activity using Worksheet C9a.2 can be used to emphasise the characteristic properties of metals and how they differ from those of typical non-metals. Supplement The nature and structure of alloys are explored and a simple model is provided to explain why alloys are stronger than the pure metal from which they are made.

C9a.3	**Reactivity series** This learning episode starts with reference to reactions of metals with oxygen and then focuses on the reaction of metals with water (or steam) and hydrochloric acid. The reaction of magnesium with steam can be demonstrated. Worksheet C9a.3(1) can be used as a student practical to investigate the reaction of a range of metals with dilute hydrochloric acid. Displacement reactions involving heating a metal oxide with carbon feature in Worksheet C9a.3(2) can be used to help develop a greater understanding of the reduction of metal oxides using carbon. These reactions can be used to introduce or revise the concept of a redox reaction.
C9a.4	**Extraction of metals** This learning episode focuses on the manufacture of iron in a blast furnace. Supplement This focuses on the reduction of the iron(III) oxide and the associated key equations. Understanding of the processes involved in the blast furnace can be checked using Worksheet C9a.4. The conversion of iron into the alloy steel provides a link to the next learning episode on the uses of metals.
C9a.5	**Uses of metals, and corrosion** The uses of aluminium and copper provide the starting point for this learning episode. The focus then switches to the problems caused by the rusting of iron and steel and the relative merits of the methods used to prevent or limit such corrosion. Worksheet C9a.5 provides students with a relatively straightforward plan and design activity relating to the corrosion of metals.
C9a.6	**Consolidation and summary** This learning episode provides an opportunity for a quick recap on the ideas encountered in the topic and time for the students to answer the End of topic questions in the Student Book.

Career links

These are some scientific careers that focus on this area of chemistry but careers in many other fields use the knowledge and skills gained studying science. Some **environmental chemists** analyse soil and water samples for contamination from heavy metals. **Powder metallurgy engineers** create components and equipment from metal powders to reduce waste.

Learning episode C9a.1 Orientation

Learning aims

- To revise the reactions of Group I metals with water
- To revise that reactivity increases down the group for the Group I metals
- To revise the lack of reactivity of the transition metals.

Common misconceptions

Some students think that alloys are pure metals.

Some students think that alloys are always a mixture of metals (e.g., discounting the use of carbon in steel)

Resources

Student Book N/A

Resources for a demonstration (see Technician's notes, following)

Approach

If appropriate, you may demonstrate the reaction of sodium with water to revise the highly reactive nature of the Group I metals (see Technician's notes). Compare this reaction with the lack of reactivity of copper. Explain that metals have a range of reactivity: sodium is one of the most reactive metals and copper is one of the least reactive.

SAFETY INFORMATION

Wear eye protection (splash-proof goggles).

Use safety screens. The screens should be positioned close to the trough to avoid the risk of the sodium shooting out of the trough.

In a small trough of water, use a sample no bigger than 1 mm × 1 mm; in a large trough of water the metal sample must be no bigger than 2 mm × 2 mm.

Sodium reacts violently with water, releasing flammable hydrogen gas that may ignite spontaneously.

Ask students to name a metal that is more reactive than sodium and a metal that is less reactive than copper (metals like silver and gold are so unreactive they are found in their native state). Explain that chemists have found it useful to produce a reactivity series for the most common metals: the reactivity series indicates the method that is needed to extract the metal from its ore.

Ask students to find out the name of an ore of iron and how the iron is extracted from this ore. (The Student Book is a quick source for this information, or students could use the Internet.)

Ask students to find (stainless) steel and brass (two common 'metals') in the Periodic Table. Why are these substances not included in the Periodic Table? Are they metals? How is a metal different from an alloy? You could set a homework task of a focused piece of research on the nature of alloys, the names of some common alloys and their uses.

Technician's notes

Be sure to check the latest safety notes on these resources before proceeding.

The following resources are needed for the demonstration of the reaction of sodium with water:

Sodium
large glass trough of water
2 safety screens
forceps
filter paper and sharp knife (all dry)
eye protection (splash-proof goggles) (goggles or face shield)

The safety screens should be positioned close to the trough to avoid the risk of sodium shooting out of the trough, as shown in the diagram.

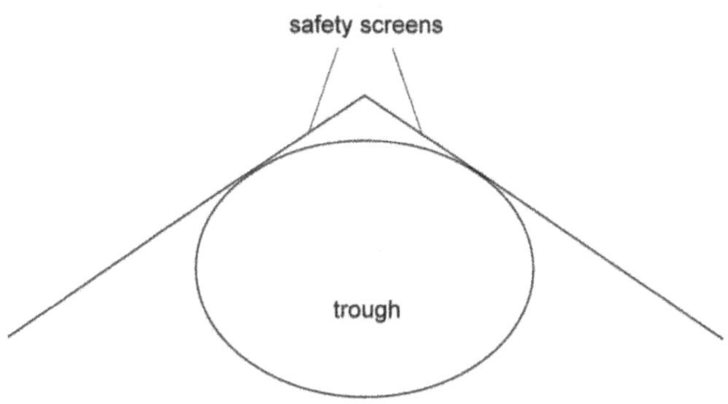

Learning episode C9a.2 Properties of metals

Learning objectives

- Compare the general physical properties of metals and non-metals, including: thermal conductivity; electrical conductivity; malleability; melting points and boiling points.
- Describe the general chemical properties of metals, limited to their reactions with: dilute acids; cold water and steam.
- Describe alloys as mixtures of a metal with other elements, including: brass as a mixture of copper and zinc; stainless steel as a mixture of iron and other elements, such as chromium, nickel and carbon.
- State that alloys can be harder or stronger than the pure metals and are more useful.
- Describe the uses of alloys in terms of their physical properties, including stainless steel in cutlery because of its hardness and resistance to rusting.
- Identify representations of alloys from diagrams of structure.
- Supplement Explain in terms of structure how alloys can be harder or stronger than the pure metals because the different sized atoms in alloys mean the layers can no longer slide over each other.

Common misconceptions

Some students think that all metals are magnetic. There are only three: iron, nickel and cobalt.

Resources

Student Book Properties of metals; Reactivity series; Alloys and their properties

Worksheet C9a.2 Physical properties of metals and non-metals Resources for a class practical (see Technician's notes, below)

Approach

Worksheet C9a.2 gives students an opportunity to explore the physical properties of some elements and identify some of the properties of metals in comparison with those of non-metals.

The worksheet will require you to demonstrate a comparison of the conduction of heat by the different elements (see Technician's notes).

The characteristic physical properties of metals are summarised in Fig C9.2 in the Student Book.

The Student Book sections on 'Alloys and their properties' considers the uses of some common metals and alloys, and include some structural diagrams to explain why alloys are stronger than the metals from which they are formed. If students researched alloys as a homework task, they could present their findings and then the information in the Student Book could be used as a summary.

SAFETY INFORMATION

Wear eye protection (splash-proof goggles).

Wash hands after handling lead.

Lead metal will rub off on skin. Lead metal can be wrapped in plastic wrap.

Technician's notes

Be sure to check the latest safety notes on these resources before proceeding.

The following resources are needed for the class practical C9a.2, per group:

samples (3 cm × 1 cm strips, where possible) of aluminium, carbon, copper, iron, lead, sulfur, zinc; avoid sharp edges
electrical circuit – power supply (6 V)
6 V lamp in holder
Leads
Magnet
eye protection (splash-proof goggles)

The following resources are needed for the demonstration of heat conduction:

Bunsen burner, tripod, gauze (or access to hotplate), heatproof mat
longer strips of aluminium, carbon, copper, iron, zinc (all the same length)
candle wax or petroleum jelly

SAFETY INFORMATION
Ensure that the petroleum jelly is low hazard.

Learning episode C9a.3 Reactivity series

Learning objectives

- State the order of the reactivity series as: potassium, sodium, calcium, magnesium, aluminium, carbon, zinc, iron, hydrogen, copper, silver, gold.
- Describe the reactions, if any, of:
  potassium, sodium and calcium with cold water; magnesium with steam; magnesium, zinc, iron, copper, silver and gold with dilute hydrochloric acid.
  and explain these reactions in terms of the position of the metals in the reactivity series.
- Deduce an order of reactivity from a given set of experimental results.

Resources

Student Book Reactivity series

Worksheet C9a.3(1) The reactivity of metals

Worksheet C9a.3(2) Obtaining metals from metal oxides

Resources for demonstrations and class practicals (see Technician's notes, below)

Approach

This is a long learning episode with three main parts, and is likely to span a number of lessons.

Part 1

Refer to the reactions of metals with oxygen – the Student Book contrasts the reaction of calcium with oxygen, with the fact that there is no reaction of gold with oxygen. Students will have seen the reaction of sodium with water. Demonstrate the reaction of magnesium with steam. This reaction can most easily be demonstrated using a hard-glass test-tube with a plug of water-soaked mineral wool in the bottom, a coil of magnesium ribbon in the middle and with a bung and small jet positioned in the open end of the tube. Do not use magnesium powder (see Technician's notes).

The tube is clamped horizontally and the magnesium is heated strongly with the flame occasionally being directed for a few seconds to the mineral wool. When the reaction starts, hydrogen can be ignited at the jet.

Remind students of the reactions of metals with dilute hydrochloric to produce hydrogen (this was covered in Topic C7a *Acids, bases and salts*). Use Worksheet C9a.3(1) to support a student practical activity. Refer to the equation for the reaction between zinc and hydrochloric acid in the Student Book and then to the reactivity series in Fig C9.12. Point out that the metals at the bottom of the series are very unreactive and are found in nature in their native states.

SAFETY INFORMATION
Wear eye protection (goggles or face shield).
Magnesium is highly flammable. Dilute hydrochloric acid and dilute sulfuric acid are corrosive.

Part 2

This part focuses on the reduction of metal oxides by more reactive metals or by carbon. It provides an opportunity to revise the definitions of oxidation and reduction in terms of oxygen gain or loss.

Introduce Worksheet C9a.3(2). This is a practical activity to reinforce ideas about the reduction of metal oxides, this time using carbon. Explain that carbon is used in these reduction (redox) reactions on an industrial scale rather than using more reactive metals. Ask students why carbon is a suitable reducing agent to be used on an industrial scale as opposed to, for example, magnesium.

SAFETY INFORMATION
Wear eye protection (splash-proof goggles).
Copper(II) oxide is harmful. Zinc oxide is dangerous to the environment.
Good ventilation is important when heating metals and oxides, particularly when lead is used if anyone may be pregnant.

Technician's notes

Be sure to check the latest safety notes on these resources before proceeding.

The following resources are needed for the demonstration (Part 1):

magnesium ribbon
Bunsen burner and heatproof mat
mineral wool, test-tube, bung and jet
eye protection (splash-proof goggles)

The following resources are needed for the class practical C9a.3(1), per group:

test-tubes, test-tube rack, wooden splint and droppers
1 cm strip of magnesium ribbon (pre-cut)
1 cm strip of each of the following metals: aluminium, copper and lead foil
1 spatula of granulated zinc
1 spatula of iron powder
about 50 cm^3 0.5 mol / dm^3 hydrochloric acid
about 50 cm^3 0.5 mol / dm^3 sulfuric acid
eye protection (splash-proof goggles)

The following resources are needed for the class practical C9a.3(2), per group:

about 1 g copper(II) oxide
about 1 g each of magnesium oxide, zinc oxide, iron(III) oxide
about 100 cm^3 limewater, about 1 g carbon powder. Label botte of limewater as a moderate hazard.
eye protection (splash-proof goggles)
7 test-tubes, delivery tube, spatula, test-tube holder, clamp and stand
Bunsen burner and heatproof mat

Learning episode C9a.4 Extraction of metals

Learning objectives

- Describe the ease of obtaining metals from their ores, related to the position of the metal in the reactivity series.
- State that iron from hematite is extracted by reduction of iron(III) oxide in the blast furnace.
- Supplement Describe the extraction of iron from hematite in the blast furnace, limited to: the burning of carbon (coke) to provide heat and produce carbon dioxide
  $C + O_2 \rightarrow CO_2$; the reduction of carbon dioxide to carbon monoxide
  $C + CO_2 \rightarrow 2CO$; the reduction of iron(III) oxide by carbon monoxide
  $Fe_2O_3 + 3CO \rightarrow 2Fe + 3CO_2$.
- State that the main ore of aluminium is bauxite and that aluminium is extracted by electrolysis.

Resources

Student Book Extraction of metals

Worksheet C9a.4 Extracting iron

Approach

Remind students that most metals exist as ores, and when a pure metal is needed, the metal must be extracted from its ore. Refer to Table C9.2 in the Student Book, which relates the reactivity of a metal to its method of extraction. Students should be familiar with the use of carbon as a reducing agent from learning episode C9a.3.

Introduce the industrial process used to extract iron from its ore, hematite (iron(III) oxide).

Consider the diagram of the blast furnace (Fig C9.18) and look at the stages given in the Student Book. Emphasise that the overall process is reduction.

Supplement For students following the supplementary syllabus spend some time on the chemical equations for the main reactions.

Reinforce these ideas using Worksheet C9a.4.

Explain that the iron produced in the blast furnace has limited use – it contains too high a proportion of carbon, which means it is brittle and cannot be easily shaped. Instead, most iron is converted into steel, which is an alloy.

Revise the term 'alloy', a mixture of a metal with another element. By reducing the proportion of carbon in iron obtained from a blast furnace, a much more versatile and useful substance can be made.

Refer also to the Science in Context box in the Student Book. The content of these Science in Context boxes is beyond the syllabus but provides the opportunity to explore the content more deeply.

Emphasise also that the main aluminium ore is bauxite (aluminium oxide) and the extraction of aluminium from bauxite is undertaken by electrolysis.

Learning episode C9a.5 Uses of metals, and corrosion

Learning objectives

- Describe the uses of metals in terms of their physical properties, including:
  aluminium in the manufacture of aircraft because of its low density; aluminium in the manufacture of overhead electrical cables because of its low density and good electrical conductivity; aluminium in food containers because of its resistance to corrosion; copper in electrical wiring because of its good electrical conductivity.
- State the conditions required for the rusting of iron (presence of oxygen and water).
- State some common barrier methods, including painting, greasing and coating with plastic.
- Describe how barrier methods prevent rusting by excluding oxygen or water.

Common misconceptions

Some students think that steel does not corrode (they may be confusing steel with stainless steel).

Resources

Student Book Uses of metals; Corrosion of metals Worksheet C9a.5 Preventing corrosion

Resources for a class practical (see Technician's notes, following)

Approach

Start by considering the uses of aluminium and copper, referring to their properties as listed in the section on 'Uses of metals' and the first part of 'Alloys and their properties' in the Student Book. Then concentrate on the section 'Corrosion of metals', in terms of the corrosion or rusting of iron and steel.

Ask students to say under what conditions iron will rust. Ask why iron might rust more rapidly in seawater. Use the information in the Student Book to look at the different ways of protecting the corrosion or rusting of iron.

Ask students to work in small groups and to plan an investigation to compare the effectiveness of different ways of limiting the rusting of iron.

Worksheet C9a.5 provides a structure for students to plan and then put their plan into practice. The task is very straightforward. You will need to judge whether students need additional prompting about the things that they should consider as they plan.

Students will need to consider the best ways of attaching the magnesium and zinc to the iron nails. It is likely that they will need to record observations after about 48 hours.

SAFETY INFORMATION
Wear eye protection (splash-proof goggles)
Magnesium ribbon is flammable.

Technician's notes

Be sure to check the latest safety notes on these resources before proceeding.

The following resources are needed for the class practical C9a.5, per group:

shiny iron/mild steel nails
sodium chloride and spatula
Grease
Oil
2 cm length of magnesium ribbon
2 cm square of zinc foil – avoid sharp edges
test-tubes and test-tube rack
eye protection (splash-proof goggles)

Learning episode 9a.6 Consolidation and summary

Learning aims

- Review the learning points of the topic summarised in the End of topic checklist.
- Test understanding of the topic content by answering the End of topic questions.

Resources

Student Book End of topic checklist and End of topic questions

Approach

Ask students to work with a partner to make a list of key words from this topic. They could then work together to produce a spider diagram showing how the different concepts are linked. They could compare their list with the list of key terms given in the Student Book. Discuss the checklist and ask questions to see how much of the content students are comfortable with.

Students could make flashcards of the key content and then use the flashcards to quiz each other on the information.

The Science in Context box on the extraction of metals in the Student Book could be referred to if this were not used in learning episode C9a.4. The content of these Science in Context boxes is beyond the syllabus but provides the opportunity to explore the content more deeply.

Ask students to work individually through the End of topic questions in the Student Book without looking at the text. As they work, walk around the classroom observing their answers and questioning them as necessary to find out which questions are causing difficulties.

After a set period, ask students to stop working. Discuss any areas of difficulty you observed as you walked round the class.

Students should complete any unanswered questions for homework, but you should stress that they should answer the questions without looking at the text, so that they can see how much they have remembered.

C10 Chemistry of the environment

Contents

Teacher Guide	Syllabus reference
C10a Chemistry of the environment	C10.1 Water C10.2 Air quality and climate

Overview of the section

This section focuses on the global and local issues that are regularly in the news. The students following this syllabus are going to be a generation which will be aware of the issues possibly more than previous generations. The importance of limiting, or removing, the sources of pollution for the air is one of the central issues of this section. The sources of pollution are now generally very well-known, as are the potential strategies to reduce or eliminate the impact they are having. Action is being taken and time will tell whether the pace of change is rapid enough. This section provides a detailed context and a rationale for acting.

Starting points

The Student Book section opener puts the ideas in the section into context and sets the scene. It also allows students to acknowledge and value their prior learning, and provides a benchmark against which future learning can be compared.

The questions provide a structure for introducing the section and can be used in a number of different ways:

- You could ask students to consider the questions as an introductory homework task.
- You could put students into groups to share their own ideas and understanding and then to report back to the whole class.
- Students could be given access to the Internet, preferably with a tight timescale, to find out the information required.

You could then use a spider chart or other form of wall chart to summarise everybody's ideas.

Recording these initial ideas allows you to retain them for reference as the individual topics are developed. In this way, your students' progress in learning can be readily acknowledged.

C10a Chemistry of the environment

Introduction

This topic focuses essentially on the importance of treating domestic water supplies and the environmental factors governing the quality of our air.

The study of carbon dioxide and pollutants such as sulfur dioxide links to the combustion of fuels in Topic C11a *Fuels* and Topic C11b *Alkanes*.

The topic starts with a section on chemical tests for water and water purification methods. The focus then switches to the air and the impacts of climate change.

For students following the supplementary syllabus there is a more detailed focus on global warming and acid rain, including strategies to reduce their effects.

There are some opportunities for students to develop their practical/investigative skills.

Links to other topics

Section	Essential background knowledge	Useful links
C2 Atoms, elements and compounds	C2a Atoms, elements and compounds	
C11 Organic chemistry		C11a Fuels C11b Alkanes
C12 Experimental techniques and chemical analysis		C12b Identification of ions and gases

Topic overview

C10a.1	**Orientation** As this topic is suggested to come near the beginning of the course, this learning episode focuses on the students' knowledge and prior learning about air and water. It is likely that many students will already have a good knowledge of the composition of the air from previous study in geography, if not in science.
C10a.2	**Water** The learning episode has two parts. In part 1 the focus is the use of anhydrous copper(II) sulfate and anhydrous cobalt(II) chloride to detect the presence of water. In part 2 the focus is on water purification.
C10a.3	**The composition of the air** This learning episode looks at the composition of pure air and provides an opportunity to outline the processes in the carbon cycle that maintain the proportion of carbon dioxide. The nature and impact of pollutants on the composition of the air is covered in the next section.

C10a.4	**Air quality and climate** This learning episode considers the sources and impact of common pollutants such as carbon monoxide, sulfur dioxide, particulates and oxides of nitrogen, in addition to carbon dioxide and methane (greenhouse gases). There is a particular focus on the effects of acid rain and climate change. Supplement This part of the learning episode considers in more detail the greenhouse effect, and strategies to reduce the effects of climate change and acid rain.
C10a.5	**Consolidation and summary** This learning episode provides an opportunity for a quick recap on the ideas encountered in the topic and time for the students to answer the End of topic questions in the Student Book.

Career links

These are some scientific careers that focus on this area of chemistry but careers in many other fields use the knowledge and skills gained studying science.

Some **atmospheric chemists** study the changes in atmospheric pollutants which can be harmful to life. **Analytical technicians** may look for micro-plastic pollution in samples taken from different ecosystems. **Environmental consultants** advise companies on how to reduce the impact of their products on the environment.

Learning episode C10a.1 Orientation

Learning aims

- To review knowledge on the composition of the air.
- To highlight the content areas in the topic.
- To revise that oxygen is a key component of the air and that it reacts with elements and compounds to form oxides.

Common misconceptions

Some students think that oxygen is the major component of the air.

Some students think that the percentage of carbon dioxide in the air is much greater than 0.04%. Some students think there are significant amounts of hydrogen in the air.

Approach

The main purpose of this introductory learning episode is to map out the main areas of study in the topic. Although it may come near the beginning of the course, certain aspects will build on previous learning in science, and it is important to acknowledge this learning.

Ask questions to elicit students' prior knowledge, and prompt links to other topics where appropriate. Key points might include:

- Water
- Water purification (link to chlorination?)
- The air
- The composition of the air (the gases and their proportions? Oxygen the 'active' ingredient in rusting and corrosion, or oxidation reactions more generally?)
- How the composition remains approximately constant (respiration, photosynthesis?)
- Substances that act as air pollutants (sources of air pollution – cars, power stations, factories?)
- Supplement Carbon dioxide as a greenhouse gas (greenhouse effect?)

Learning episode C10a.2 Water

Learning outcomes

- Describe chemical tests for the presence of water using anhydrous cobalt(II) chloride and anhydrous copper(II) sulfate.
- State that distilled water is used in practical chemistry rather than tap water because it contains fewer chemical impurities.
- Describe the treatment of the domestic water supply in terms of:

  sedimentation and filtration to remove solids; use of carbon to remove tastes and odours; chlorination to kill microbes.

Resources

Student Book Water; Treatment of water supplies

Resources for a demonstration (see Technician's notes, below)

Approach

The learning episode has two parts.

Part 1

Demonstrate the action of water on anhydrous copper(II) sulfate and anhydrous cobalt(II) chloride (powder or paper form) (see Technician's notes). Explain that these tests will indicate that water is present but do not prove that the liquid tested is pure water (determining the boiling point is one way of doing this although this information is beyond the scope of this syllabus).

SAFETY INFORMATION
Wear eye protection (splash-proof goggles).
Copper(II) sulfate is harmful. Cobalt(II) chloride is toxic by inhalation and a possible sensitiser by inhalation and skin contact. Avoid handling paper, if possible, wash hands if papers are

Part 2

Use the flowchart in the Student Book to introduce a common method of water treatment.

Emphasise the importance of the filtration through sand before the process of chlorination. Students could research other methods of water treatment (in particular desalination), as well as the sorts of water-borne diseases and the problems caused by drinking 'dirty' or untreated water.

Technician's notes

Be sure to check the latest safety notes on these resources before proceeding.

The following resources are needed for the demonstration of water and anhydrous copper(II) sulfate and anhydrous cobalt(II) chloride:

anhydrous copper(II) sulfate, anhydrous cobalt(II) chloride and cobalt(II) chloride (powder or paper)
test-tubes, eye protection (splash-proof goggles)

Learning episode 10a.3 The composition of the air

Learning objectives

- State the composition of clean, dry air as approximately 78% nitrogen, N_2, 21% oxygen, O_2, and the remainder as a mixture of noble gases and carbon dioxide, CO_2.

Common misconceptions

Some students think that oxygen is the major component of the air.

Some students think that the percentage of carbon dioxide in the air is much greater than 0.04%.

Resources

Student Book Air quality and climate

Approach

The learning episode concentrates on the percentage composition of 'clean, dry' air.

Fig C10.4 in the Student Book provides information about the composition of the air – the proportion of oxygen at 21% (approximately one-fifth) is also emphasised in the Developing Practical Skills box.

The apparently surprising low proportion of carbon dioxide can lead to a general discussion of the processes that maintain this proportion. This provides links to biological knowledge about the carbon cycle, though only the idea of combustion is required in the Chemistry part of the Coordinated syllabus. The effect of carbon dioxide as a greenhouse gas is covered in learning episode C10a.4.

The high proportion of nitrogen in the atmosphere poses the question: 'Is nitrogen an important component of the air?' Emphasise that it is very unreactive (inert) and that some plants (leguminous – 'pod' plants, for example, peas) can 'fix' the nitrogen and form protein.

The Science in Context box in the Student Book can be used to explore the process used in the industrial manufacture of oxygen and nitrogen. Stress the important uses of oxygen and nitrogen and how they are stored as liquids under pressure. The content of these Science in Context boxes is beyond the syllabus but provides the opportunity to explore the content more deeply.

Learning episode C10a.4 Air quality and climate

Learning objectives

- State the source of each of these air pollutants, limited to:
  carbon dioxide from the complete combustion of carbon-containing fuels;
  carbon monoxide and particulates from the incomplete combustion of carbon-containing fuels;
  methane from the decomposition of vegetation and waste gases from digestion in animals;
  oxides of nitrogen from car engines;
  sulfur dioxide from the combustion of fossil fuels which contain sulfur compounds.
- State the adverse effects of these air pollutants, limited to:
  carbon dioxide: higher levels of carbon dioxide leading to increased global warming, which leads to climate change;
  carbon monoxide: toxic gas;
  particulates: increased risk of respiratory problems and cancer;
  methane: higher levels of methane leading to increased global warming, which leads to climate change;
  oxides of nitrogen: acid rain, photochemical smog and respiratory problems;
  sulfur dioxide: acid rain.
- Supplement State and explain strategies to reduce the effects of climate change:
  planting trees;
  reduction in livestock farming;
  decreasing use of fossil fuels;
  increasing use of renewable energy, e.g., wind, solar.
- Supplement State strategies to reduce the effects of acid rain: reducing emissions of sulfur dioxide by using low-sulfur fuels.
- Supplement Describe how the greenhouse gases carbon dioxide and methane cause global warming, limited to:
  the absorption, reflection and emission of thermal energy; reducing thermal energy loss to space.

Common misconceptions

Damage to the ozone layer may be seen to be the greatest cause of global warming.

Resources

Student Book Air quality and climate

Worksheet C10a.4 Air pollution

Approach

The focus of this learning episode includes the main pollutants in the atmosphere. For students following the supplementary syllabus this is extended to a description of global warming as a result of greenhouse gases, and strategies to reduce the effects of acid rain and climate change. Detailed information is provided in the Student Book and other resources can be used to provide additional evidence.

This learning episode lends itself to student enquiry, either individually or in groups. Worksheet C10a.4 provides a template that the students could use. For sulfur dioxide, their answers in the last column will only be required at supplementary level and knowing how to remove other pollutants is not required by the syllabus, however it is probably worth asking all students to complete this as these are important environmental issues.

Learning episode C10a.5 Consolidation and summary

Learning aims

- Review the learning points of the topic summarised in the End of topic checklist.
- Test understanding of the topic content by answering the End of topic questions.

Resources

Student Book End of topic checklist and End of topic questions

Approach

Ask students to work with a partner to make a list of key words from this topic. They could then work together to produce a spider diagram showing how the different concepts are linked. They could compare their list with the list of key terms given in the Student Book. Discuss the checklist and ask questions to see how much of the content students are comfortable with.

Students could make flashcards of the key content and then use the flashcards to quiz each other on the information.

Ask students to work individually through the End of topic questions in the Student Book without looking at the text. As they work, walk around the classroom observing their answers and questioning them as necessary to find out which questions are causing difficulties.

After a set period, ask students to stop working. Discuss any areas of difficulty you observed as you walked round the class.

Students should complete any unanswered questions for homework, but you should stress that they should answer the questions without looking at the text, so that they can see how much they have remembered.

C11 Organic chemistry

Contents

Teacher Guide	Syllabus reference
C11a General introduction to organic chemistry	C11.1 Terminology
C11b Fuels	C11.2 Fuels
C11c Alkanes	C11.3 Alkanes
C11d Alkenes	C11.4 Alkenes
C11e Polymers	C11.5 Polymers

Overview of the section

This section covers the organic chemistry content of the course. It is unlikely that students will have studied organic chemistry before starting this course. You will need to spend time developing their understanding of organic formulas, the general terminology they are likely to meet during the section, the naming of organic compounds and improving their familiarity with terms such as 'saturated', and 'unsaturated'.

Supplement In addition a homologous series and its general characteristics can be introduced and then these ideas can be reinforced during the study of individual topics.

After the general introduction hydrocarbon fuels are the focus before concentrating on the two homologous series: alkanes and alkenes. The study of alkenes links to later work on addition polymerisation in the polymers topic.

All the topics in the section require a good understanding of covalent bonding and the properties of simple covalent compounds. Discussion of the combustion of hydrocarbons links to pollution in Topic C10a *Chemistry of the environment*. In the suggested teaching sequence, all the topics are included in the second year of the course. As with other sections, it is important to use orientation to establish the extent of your students' prior knowledge.

Starting points

The Student Book section opener puts the ideas in the section into context and sets the scene. It also allows students to acknowledge and value their prior learning, and provides a benchmark against which future learning can be compared.

The questions provide a structure for introducing the section and can be used in a number of different ways:

- You could ask students to consider the questions as an introductory homework task.
- You could put students into groups to share their own ideas and understanding and then to report back to the whole class.
- Students could be given access to the Internet, preferably with a tight timescale, to find out the information required.

You could then use a spider chart or other form of wall chart to summarise everybody's ideas.

Recording these initial ideas allows you to retain them for reference as the individual topics are developed. In this way, your students' progress in learning can be readily acknowledged.

C11a General introduction to organic chemistry

Introduction

This topic, as its name implies, is a general introduction. Its contents are replicated in the subsequent topics in this section. For students following the supplementary course the concept of a homologous series is introduced and knowledge that in a homologous series all the compounds will have similar chemical properties.

Links to other topics

Section	Essential background knowledge	Useful links
C2 Atoms, elements and compounds	C2c Simple molecules and covalent bonds	
C11 Organic chemistry		C11b Fuels C11c Alkanes C11d Alkenes C11e Polymers

Topic overview

C11a.1	**Orientation** This learning episode introduces and defines the terminology that will be used throughout the section. There are no further activities and you can move straight to Topic C11b Fuels.

Career links

These are some scientific careers that focus on this area of chemistry but careers in many other fields use the knowledge and skills gained studying science.

Forensic analysts, **environmental chemists** and **food scientists** are just three examples of careers which make use of a wide range of knowledge on organic chemistry.

Pharmaceutical chemists develop active ingredients to improve or create new medicines.

Learning episode C11a.1 Orientation

Learning objectives

- State that a saturated compound has molecules in which all carbon–carbon bonds are single bonds.
- State that an unsaturated compound has molecules in which one or more carbon–carbon bonds are not single bonds.
- Supplement State that homologous series are families of similar compounds with similar chemical properties.
- Supplement Describe the general characteristics of a homologous series as:
  a) having the same general formula (recall of specific general formulas is **not** required)
  b) displaying a trend in physical properties
  c) sharing similar chemical properties.

Resources

Student Book What are alkanes?; What are alkenes?

Approach

Students should be familiar with some organic compounds from their study of Topic C2c *Simple molecules and covalent bonds*. By revising some of the molecules they have encountered already it should be possible to provide a quick introduction to some homologous series, as well as saturated and unsaturated compounds. The time spent on this will depend on your overall strategy for studying organic chemistry. You may decide to introduce these ideas in the context of the other topics in this Section.

C11b Fuels

Introduction

This topic introduces organic chemistry as well as providing a very direct link to Topic C11c *Alkanes* and Topic C11d *Alkenes*. Fractional distillation features in Topic C12a *Experimental design and chromatography*. In Topic C2c *Simple molecules and covalent bonds* students explored the nature of bonding in hydrocarbon molecules and learned the physical properties associated with simple molecular compounds. The problems of the combustion and incomplete combustion of hydrocarbon fuels are covered in Topic C10a *Chemistry of the environment*.

The topic starts with a general introduction to the fossil fuels coal, natural gas and petroleum (crude oil) before looking in some detail at separating petroleum into its components by fractional distillation.

Cracking is covered in Topic C11d *Alkenes* to emphasise that cracking is needed to convert the larger molecules in petroleum fractions into more useful fractions. Finally, the continuing use of fossil fuels is considered.

There is some opportunity for developing practical skills in this topic.

Links to other topics

Sections	Essential background knowledge	Useful links
C2 Atoms, elements and compounds	C2c Simple molecules and covalent bonds	
C10 Chemistry of the environment		C10a Chemistry of the environment
C11 Organic chemistry	C11a General introduction to organic chemistry	C11c Alkanes C11d Alkenes
C12 Experimental techniques and chemical analysis		C12a Experimental design and chromatography

Topic overview

C11b.1	**Orientation** This learning episode provides an opportunity to revise or reinforce ideas on the main fossil fuels and their non-renewable nature. An opportunity is taken to revise the nature of covalent bonding and to draw some dot-and-cross diagrams of hydrocarbons. Reference is also made to fractional distillation as a process of separating liquids with different boiling points.
C11b.2	**The fractional distillation of petroleum** This learning episode focuses on the industrial use of the fractional distillation of petroleum (crude oil). The process is demonstrated and emphasis is placed on the nature and uses of the major fractions obtained from the fractional distillation. **Supplement** This focuses on the different properties of the fractions.

C11b.3	**Consolidation and summary** This learning episode considers the advantages and disadvantages of using fossil fuels before recapping the key ideas encountered in the topic. Time can be allocated to studying the End of topic checklist and then answering the End of topic questions in the Student Book.

Career links

These are some scientific careers that focus on this area of chemistry but careers in many other fields use the knowledge and skills gained studying science. **Policy advisors** for the government use their knowledge and understanding of chemistry to brief politicians on scientific research to inform changes that may be needed to their policies. **Petroleum geologists** find sources of crude oil by analysing structures underground using geological and geophysical methods.

Learning episode C11b.1 Orientation

Learning objectives

- Name the fossil fuels: coal, natural gas and petroleum.
- Name methane as the main constituent of natural gas.
- State that hydrocarbons are compounds that contain hydrogen and carbon only.
- State that petroleum is a mixture of hydrocarbons.
- State that a saturated compound has molecules in which all carbon–carbon bonds are single bonds.
- State that an unsaturated compound has molecules in which one or more carbon–carbon bonds are not single bonds.

Resources

Student Book What are fossil fuels?

Approach

Students should be very familiar with the names of the common fossil fuels – coal, natural gas and petroleum (crude oil). They should also be familiar with the term 'non-renewable'.

Check the level of this existing knowledge with a few questions. Ask students what they understand by the term 'hydrocarbon', which they should have met in the earlier Topic C10a *Chemistry of the environment*.

Ask students to draw a dot-and-cross diagram for the simplest hydrocarbon: methane, CH_4. Revise ideas on covalent bonding from Topic C2c *Simple molecules and covalent bonds* if and as required.

Supplement Ask students to draw a dot-and-cross diagram for ethene, C_2H_4.

Explain the terms saturated and unsaturated in the context of methane and ethene. Explain also that methane belongs to a family of organic compounds called alkanes, whereas ethene belongs to a family of organic compounds called alkenes (both of which will be studied later in the course). Ask what the characteristic properties of simple covalent substances are.

Explain that petroleum is a mixture of a large number of organic compounds, many of which are used as fuels. Ask what process can be used to separate a mixture of liquids – if necessary, remind students about fractional distillation. Explain that the next learning episode will look in more detail at the fractional distillation of petroleum.

Learning episode C11b.2 The fractional distillation of petroleum

Learning objectives

- Describe the separation of petroleum into useful fractions by fractional distillation.
- Name the uses of the fractions as:
  (a) refinery gas fraction for gas used in heating and cooking;
  (b) gasoline / petrol fraction for fuel used in cars;
  (c) naphtha fraction as a chemical feedstock;
  (d) diesel oil / gas fraction for fuel used in diesel engines;
  (e) bitumen fraction for making roads.
- Supplement Describe how the properties of fractions obtained from petroleum change from the bottom to the top of the fractionating column, limited to: decreasing chain length; lower boiling points.

Resources

Student Book Fractional distillation

Worksheet C11b.2 The fractional distillation of petroleum (crude oil)

Resources for a demonstration (see Technician's notes, following)

Approach

Ask students what process can be used to separate a mixture of liquids with different boiling points (fractional distillation). Explain that petroleum (crude oil) can be separated into components using fractional distillation, but rather than separating liquid by liquid (or compound by compound), it is separated into fractions with a range of boiling points. These fractions are therefore likely to contain more than one compound.

Demonstrate the fractional distillation of a petroleum or crude oil substitute (see Technician's notes). Explain that petroleum or crude oil is too dangerous to be allowed in schools and that the technician has made up a safer substitute, though it is still harmful.

Use Worksheet C11b.2 so that students can record the observations and measurements and analyse their findings.

Supplement Link the trend in boiling point to the decreasing chain length, as shown in Fig. C11.5 in the Student Book.

It is important that students know some of the key components of the fractions (refinery gases, gasoline/petrol, naphtha, diesel oil/gas oil and bitumen), where they are produced in the fractionating column and, for students following the supplementary course, the approximate number of carbon atoms they have in their molecules and the trend in boiling points.

SAFETY INFORMATION

Wear eye protection (splash-proof goggles).

Petroleum/crude oil substitute is highly flammable and harmful.

Ventilate room well to remove smoke, and only use small amounts.

Technician's notes

Be sure to check the latest safety notes on these resources before proceeding.

You will need the following resources for the demonstration of fractional distillation:

100 cm^3 petroleum/crude oil substitute made from: • 55 cm^3 liquid paraffin; 20 cm^3 paraffin oil (kerosene) • 11 cm^3 white spirit; 4 cm^3 petroleum ether (100–120 °C) • 4 cm^3 petroleum ether (80–100 °C); 6 cm^3 of petroleum ether (60–80 °C) • black artists' oil paint (to provide colour) about 6 cm^3 is needed for the demonstration
Bunsen burner and heatproof mat
mineral wool
Tongs
side-arm borosilicate glass boiling tube, delivery tube and rubber connection tubing
thermometer (0–360 °C), teat pipette, beaker, watch glass
small test-tubes to collect fractions and test-tube holder
eye protection (splash-proof goggles)
SAFETY INFORMATION *Crude oil must not be used in schools.*

Learning episode C11b.3 Consolidation and summary

Learning aims

- Review the learning points of the topic summarised in the End of topic checklist.
- Test understanding of the topic content by answering the End of topic questions.

Common misconceptions

Some students think that all fossil fuels will run out in the next few years. (Supplies of coal are relatively large.)

Resources

Student Book The fossil fuel dilemma; End of topic checklist and End of topic questions

Approach

Ask students to work with a partner to make a list of key words from this topic. They could then work together to produce a spider diagram showing how the different concepts are linked. They could compare their list with the list of key terms given in the Student Book. Discuss the checklist and ask questions to see how much of the content students are comfortable with.

Students could make flashcards of the key content and then use the flashcards to quiz each other on the information.

Use the Science in Context box, *The Fossil Fuel Dilemma*, in the Student Book to consider the advantages and disadvantages of using fossil fuels to generate energy. Are alternative forms of energy the answer? Should coal be used more extensively? The content of these Science in Context boxes is beyond the syllabus but provides the opportunity to explore the content more deeply.

Ask students to work individually through the End of topic questions in the Student Book without looking at the text. As they work, walk around the classroom observing their answers and questioning them as necessary to find out which questions are causing difficulties.

After a set period, ask students to stop working. Discuss any areas of difficulty you observed as you walked round the class.

Reinforce the point that students will do further work on alkanes and alkenes, and on the concept of saturated and unsaturated hydrocarbons, in topics that follow this one.

Students should complete any unanswered questions for homework, but you should stress that they should answer the questions without looking at the text, so that they can see how much they have remembered.

C11c Alkanes

Introduction

Students should already be familiar with covalent bonding from Topic C2c *Simple molecules and covalent bonds*. There are also links to methane as a greenhouse gas and the incomplete combustion of fuels in Topic C10a *Chemistry of the environment*. This topic builds on the introduction to organic chemistry provided in Topic C11a *General introduction to organic chemistry* and Topic C11b *Fuels*. It is also the basis for the subsequent study of Topic C11d *Alkenes*.

The topic reaffirms terms such as 'hydrocarbon' and 'saturated', followed by the use of the names of alkanes in naming all organic compounds. At supplementary level the nature of a homologous series can be revised. Students will have another opportunity, if needed, to revise dot-and-cross diagrams.

Links to other topics

Sections	Essential background knowledge	Useful links
C2 Atoms, elements and compounds	C2c Simple molecules and covalent bonds	
C10 Chemistry of the environment		C10a Chemistry of the environment
C11 Organic chemistry	C11a General introduction to organic chemistry C11b Fuels	C11d Alkenes

Topic overview

C11c.1	**Orientation** This learning episode starts with a revision of covalent bonding and, for students following the supplementary syllabus, introduces or revises the concept of a homologous series within the organic chemistry branch of chemistry. The names of the first four alkanes are introduced, together with information about their states of matter at room temperature and their uses as fuels. The terms 'hydrocarbon' and 'saturated' are revised. At supplementary level the general characteristics of a homologous series are introduced or revised.
C11c.2	**The chemical properties of alkanes** This learning episode starts with the combustion of alkanes, re-emphasising their general use as fuels. The concept of complete and incomplete combustion is revised (first encountered in Topic C10a *Chemistry of the environment*) and the potential dangers associated with incomplete combustion.
C11c.3	**Consolidation and summary** This learning episode provides an opportunity for a quick recap of the ideas encountered in the topic and time for the students to answer the End of topic questions in the Student Book.

Career links

These are some scientific careers that focus on this area of chemistry but careers in many other fields use the knowledge and skills gained studying science. **Gas engineers** service boilers and fires to make sure that combustion of the fuel is complete and does not produce the toxic gas carbon monoxide, or particulates which can be harmful to human health. **Car designers** work with scientists to design car engines which use the fuel in the most efficient ways.

Learning episode C11c.1 Orientation

Learning objectives

- State that the bonding in alkanes is single covalent and that alkanes are saturated hydrocarbons.
- Supplement State that a homologous series is a family of similar compounds with similar chemical properties.
- Supplement Describe the general characteristics of a homologous series as:
- having the same general formula (recall of the alkane general formula is **not** required); displaying a trend in physical properties; sharing similar chemical properties.

Resources

Student Book What are alkanes?

Worksheet C11c.1 Covalent bonding in alkanes

Resources for demonstration (see Technician's notes, below)

Approach

Explain that alkanes are a 'family' (or for supplementary syllabus, a homologous series) of organic chemicals.

Show samples of alkanes: methane (natural gas/Bunsen burner), propane (camping gas canister) and butane (lighter fuel). Mention that they are all fuels. Emphasise the meaning of the term 'hydrocarbon'. The samples are just shown to illustrate their everyday use – photos could be used instead.

Ask students what they can recall about the nature of covalent bonding. Use molecular models to illustrate the structure of alkanes and Worksheet C11c.1 to recap on dot-and-cross bonding diagrams.

SAFETY INFORMATION
Wear eye protection (splash-proof goggles).

Supplement Revisit (from Topic C11a) the idea that members of a homologous series have similar chemical properties – point out that all the alkanes are used as fuels as they can be combusted. Discuss how the samples show a trend in physical properties – this is another characteristic of a homologous series (refer to the data on boiling points in Table C11.1 in the Student Book). Show how to draw the displayed formulas of each alkane and write their structural formulas. You may also wish to show how the formulas of alkanes match the general formula C_nH_{2n+2}.

Technician's notes

Be sure to check the latest safety notes on these resources before proceeding.

The following resources are needed for the demonstration of alkanes:

a bottle or canister of camping gas (propane)
lighter fuel (butane)

Learning episode C11c.2 The chemical properties of alkanes

Learning objectives

- Describe the properties of alkanes as being generally unreactive, except in terms of combustion.

Resources

Student Book The properties of alkanes; Combustion of alkanes

Approach

Students will be familiar with the combustion of alkane fuels. Explain that hydrocarbons burn in a plentiful supply of air or oxygen to form carbon dioxide and water (as shown by the blue flame on a Bunsen burner).

Ask what happens if the air supply is restricted. (This is demonstrated by the yellow flame on the Bunsen burner.) Emphasise that incomplete combustion can produce carbon and carbon monoxide, and that carbon monoxide is toxic.

Supplement Let students practice constructing equations for the complete and incomplete combustion of simple hydrocarbons. Refer to the equations given in the Student Book.

Learning episode C11c.3 Consolidation and summary

Learning aims

- Review the learning points of the topic summarised in the End of topic checklist.
- Test understanding of the topic content by answering the End of topic questions.

Resources

Student Book Some interesting facts about methane; End of topic checklist and End of topic questions

Approach

Ask students to work with a partner to make a list of key words from this topic. They could then work together to produce a spider diagram showing how the different concepts are linked. They could compare their list with the list of key terms given in the Student Book. Discuss the checklist and ask questions to see how much of the content students are comfortable with.

Students could make flashcards of the key content and then use the flashcards to quiz each other on the information.

Refer to the Science in Context box on methane in the Student Book (if not used in Topic

C10a *Chemistry of the environment*), to highlight that methane is a greenhouse gas (supplementary level) and that it contributes to global warming. The content of these Science in Context boxes is beyond the syllabus but provides the opportunity to explore the content more deeply.

Ask students to work individually through the End of topic questions in the Student Book without looking at the text. As they work, walk around the classroom observing their answers and questioning them as necessary to find out which questions are causing difficulties.

After a set period, ask students to stop working. Discuss any areas of difficulty you observed as you walked round the class.

Students should complete any unanswered questions for homework, but you should stress that they should answer the questions without looking at the text, so that they can see how much they have remembered.

C11d Alkenes

Introduction

This is a short topic that extends understanding of covalent bonding from Topic C2c *Simple molecules and covalent bonds*, Topic C11a *General introduction to organic chemistry* and the introductory work on alkanes in Topic C11c *Alkanes*. It also provides the basis for a more detailed consideration of addition polymerisation in Topic C11f *Polymers*.

The topic focuses on the structures of the alkenes, the nature of alkenes as unsaturated hydrocarbons and the test that can be used to distinguish between saturated and unsaturated hydrocarbons.

Supplement The supplementary syllabus includes the use of catalytic cracking to produce alkenes from larger alkane molecules in petroleum fractions and addition reactions with bromine, hydrogen and steam.

Links to other topics

Section	Essential background knowledge	Useful links
C2 Atoms, elements and compounds	C2c Simple molecules and covalent bonds	
C11 Organic chemistry	C11a General introduction to organic chemistry C11b Fuels C11c Alkanes	C11e Polymers

Topic overview

C11d.1	**Orientation** This learning episode picks up on many of the ideas covered in Topic C11a *General introduction to organic chemistry* and Topic C11b *Fuels*, including the distinction between a saturated and unsaturated hydrocarbon and the covalent bonding associated with a C=C double bond. Supplement The characteristics of a homologous series are revised and the structural and displayed formulas of alkenes are introduced.
C11d.2	**Cracking crude oil fractions** Supplement The reason for the importance of the catalytic cracking of the larger molecules in crude oil fractions is emphasised (converting the less useful longer-chain alkanes into more useful shorter-chain alkenes and hydrogen).
C11d.3	**The chemical properties of alkenes** This learning episode focuses on alkenes as examples of hydrocarbons and the reaction with bromine water. The decolourisation of bromine water is introduced as a way of identifying an alkene as an unsaturated hydrocarbon. Supplement Addition reactions with bromine, hydrogen and steam are introduced.

C11d.4	**Consolidation and summary** This learning episode recaps on the ideas encountered in the topic, as well as considering a Science in Context box, *Saturated and Unsaturated Fats* in the Student Book. Time can be allocated to answering the End of topic questions in the Student Book.

Career links

These are some scientific careers that focus on this area of chemistry but careers in many other fields use the knowledge and skills gained studying science. **Petrochemical engineers** develop ways that crude oil can be broken down into smaller fractions which are used to make plastics and fuels.

Learning episode C11d.1 Orientation

Learning objectives

- State that the bonding in alkenes includes a double carbon–carbon covalent bond and that alkenes are unsaturated hydrocarbons.
- Supplement State that a homologous series is a family of similar compounds with similar chemical properties.
- Supplement Describe the general characteristics of a homologous series as:
  (a) having the same general formula (recall of the alkene general formula is **not** required)
  (b) displaying a trend in physical properties

Resources

Student Book What are alkenes?

Approach

Ask students to name the first two members of the alkane family (for supplementary syllabus, a homologous series) and to draw the displayed formulas for these.

Illustrate that in contrast to the alkanes, the alkenes contain a C=C double bond, that is, the alkenes are unsaturated hydrocarbons. Ask students to draw the displayed formula for ethene (and then for supplementary, propene and butene).

Supplement Ask students if they can remember the characteristics of a homologous series (general formula, gradual trend in physical properties and common chemical properties). You may also wish to show how the formulas of alkenes match the general formula C_nH_{2n}.

Supplement Use Table C11. in the Student Book to look at the trend in boiling points, and the subsequent states of matter at room temperature and pressure, of the alkene series.

Learning episode C11d.2 Cracking crude oil fractions

Learning objectives

- Supplement Describe the manufacture of alkenes and hydrogen by the cracking of larger alkane molecules using a high temperature and a catalyst.
- Describe the test to distinguish between saturated and unsaturated hydrocarbons by their reaction with aqueous bromine.

Resources

Student Book What are alkenes?; Cracking

Worksheet C11d.2 Cracking liquid paraffin

Resources for a class practical (see Technician's notes, below)

Approach

Supplement Refer to Table C11.4 in the Student Book, which shows the typical percentages of the different fractions produced by the fractional distillation of crude oil. Explain that the fractions most easily used are the fractions, such as gasoline (13%) containing short-chain hydrocarbon molecules, and so the fractions containing longer chain hydrocarbons must be converted into smaller hydrocarbon molecules, which are more useful. Emphasise that one of the products of cracking is always an alkene. Explain that bromine water is used to identify unsaturated hydrocarbons such as alkenes.

Tell students that alkenes are much more reactive than alkanes, so they can be converted (by addition reactions) into a much wider range of useful chemicals, as will be described in Topic C11e *Polymers*.

Worksheet C11d.2 provides details of a class practical involving the cracking of liquid paraffin. **Warn students to be careful that a 'suck back' does not take place**. This process also features in the Developing Practical Skills box in the Student Book.

You will need to decide whether to refer to both and, if so, in what order you will introduce them.

The test with a lighted spill should confirm the presence of hydrogen in the gas present. Explain that hydrogen may be formed during some cracking reactions, depending on the starting molecule and the conditions.

SAFETY INFORMATION

Wear eye protection (splash-proof goggles).

Aqueous bromine (bromine water) is harmful.

Ensure that the laboratory is well ventilated.

Technician's notes

Be sure to check the latest safety notes on these resources before proceeding. The following resources are needed for the class practical C11.4b, per group:

medicinal paraffin (sometimes called liquid paraffin)
aqueous bromine (bromine water (0.02 M)) – diluted until the colour is barely visible – 5 cm^3
pot fragments
Bunsen burner and heatproof mat
stand and clamp, boiling tube, bung and delivery tube with Bunsen valve (see below for details on how to make a Bunsen valve)
mineral wool and tongs
water basin, test-tubes, 2 dropping pipettes
eye protection (splash-proof goggles)

It is recommended that a Bunsen valve is fitted to the end of the delivery tube to reduce the risk of suck- back. Bunsen valves can be made easily as follows.

You will need a piece of clean, unused soft rubber tubing, about 3 cm long and a piece of glass rod, 1–2 cm long. Attach the rubber tubing to the end of the delivery tube and then attach the glass rod to the other end of the tubing, as shown in the diagram. Make a slit along one side of the rubber tubing, about 1 cm in length.

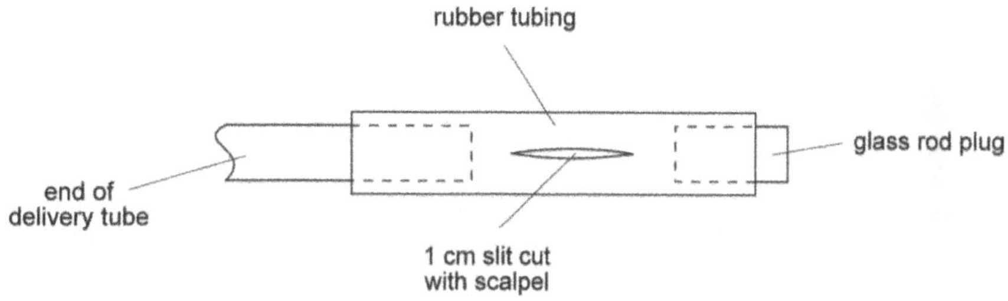

Learning episode C11d.3 The chemical properties of alkenes

Learning objectives

- Describe the test to distinguish between saturated and unsaturated hydrocarbons by their reaction with aqueous bromine.
- Supplement Describe the properties of alkenes in terms of addition reactions with:
- bromine or aqueous bromine; hydrogen in the presence of a nickel catalyst; steam in the presence of an acid catalyst.

Resources

Student Book Addition reactions of alkenes

Resources for demonstration (see Technician's notes following)

Approach

Demonstrate the reaction of an alkene (hexene or cyclohexene) with aqueous bromine (CARE). Students are not allowed to use cyclohexene in the classroom, so this must be a teacher demonstration (see Technician's notes). The bromine water is decolourised.

Explain that this reaction is used to identify alkenes (or more accurately any unsaturated hydrocarbon, including alkynes, which contain a C≡C triple bond).

Supplement Use Fig. C11.14 in the Student Book to illustrate the displayed formula of the product of the reaction and the reason for describing the reaction as 'addition'.

Supplement Describe the addition reactions of ethene with bromine, hydrogen and steam. In each case, ask students to write the displayed formula of the product formed.

SAFETY INFORMATION

Wear eye protection (splash-proof goggles).

Aqueous bromine is harmful.

Hexene and cyclohexene are highly flammable.

Ensure that the laboratory is well ventilated.

Technician's notes

Be sure to check the latest safety notes on these resources before proceeding.

The following resources are needed for the demonstration of an alkene reacting with bromine:

an alkene (such as hexene or cyclohexene)
0.005 mol / dm^3 concentration of aqueous bromine
test-tube, droppers, test-tube holder
eye protection (splash-proof goggles)

Learning episode C11d.4 Consolidation and summary

Learning aims

- Review the learning points of the topic summarised in the End of topic checklist.
- Test understanding of the topic content by answering the End of topic questions.

Resources

Student Book Saturated and unsaturated fats; End of topic checklist; End of topic questions

Approach

Ask students to work with a partner to make a list of key words from this topic. They could then work together to produce a spider diagram showing how the different concepts are linked. They could compare their list with the list of key terms given in the Student Book. Discuss the checklist and ask questions to see how much of the content students are comfortable with.

Students could make flashcards of the key content and then use the flashcards to quiz each other on the information.

Use the Science in Context box on saturated and unsaturated fats in the Student Book to emphasise the importance of recognising this distinction in a healthy diet. The content of these Science in Context boxes is beyond the syllabus but provides the opportunity to explore the content more deeply.

Ask students to work individually through the End of topic questions in the Student Book without looking at the text. As they work, walk around the classroom observing their answers and questioning them as necessary to find out which questions are causing difficulties.

After a set period, ask students to stop working. Discuss any areas of difficulty you observed as you walked round the class.

Students should complete any unanswered questions for homework, but you should stress that they should answer the questions without looking at the text, so that they can see how much they have remembered.

C11e Polymers

Introduction

As with all the organic chemistry topics, this topic builds on knowledge of covalent bonding covered in Topic C2c *Simple molecules and covalent bonds*. In addition, there are links to Topic C11d *Alkenes* in relation to addition polymerisation.

The topic starts with a recap on the family (for supplementary, homologous series) of alkenes. After this revision, the topic focuses on addition polymerisation. There are opportunities to consider some of the environmental aspects of the use of synthetic polymers, particularly recycling and disposal.

There are few opportunities for students to develop their investigative skills.

Links to other topics

Section	**Essential background knowledge**	**Useful links**
C2 Atoms, elements and compounds	C2c Simple molecules and covalent bonds	
C11 Organic chemistry	C11a General introduction to organic chemistry C11c Alkenes	

Topic overview

C11e.1	**Orientation** This learning episode provides an opportunity to revise the organic chemistry of alkenes, focusing on the chemical name ending in –ene and their displayed formulas. The terms 'monomer', 'polymer' and 'polymerisation' are introduced.
C11e.2	**Synthetic polymers** This learning episode is in two parts. **Part 1 Addition polymerisation** This part focuses on how many monomers can combine to form addition polymer molecules, using the example of the formation of poly(ethene) from ethene monomers. Examples are given of common addition polymers and their uses.
	Part 2 The problems of plastics This is beyond the scope of the syllabus but using the Science in Context box in the Student Book provides a focus for discussion of the problems posed by the non-biodegradability of many synthetic polymers.
C11e.3	**Consolidation and summary** This learning episode is an opportunity to recap on the key ideas of the topic. Time can be allocated to studying the End of topic checklist and then answering the End of topic questions in the Student Book.

Career links

These are some scientific careers that focus on this area of chemistry but careers in many other fields use the knowledge and skills gained studying science. **Plastics engineers** research ways in which plastics can be made to be more biodegradable as well as the best ways to recycle them.

Learning episode C11e.1 Orientation

Learning objective

- Define polymers as large molecules built up from many smaller molecules called monomers.

Resources

Student Book Addition polymers

Approach

Ask students to draw the displayed formula for ethene. Check that they can remember the name of this compound.

Introduce the terms 'monomer', 'polymer' and 'polymerisation'. Explain that the topic will consider some synthetic polymers, together with polymers that are often referred to as plastics. Ask students, in groups, to make a list of as many things as they know about polymers – allow only a few minutes to do this. Review the lists. Disposal of plastics and recycling will be common emerging themes.

Explain that the topic focuses on one way of making synthetic polymers. This method involves alkenes and makes use of their reactivity and ability to undergo addition reactions.

Learning episode C11e.2 Synthetic polymers

Learning objectives

- State that the formation of poly(ethene) is an example of addition polymerisation using ethene (an alkene) monomers.

Resources

Student Book Addition polymers

Approach

Part 1 Addition polymerisation

Introduce the definition of an addition polymer. Use displayed formulas to illustrate how ethene molecules can undergo addition reactions to form a polymer chain called poly(ethene). Students do not need to recall details of this diagram.

Describe some of the addition polymers in common use and their uses, such as poly(propene) – carpets; poly(chloroethene) or PVC – plastic sheets; and poly(phenylethene) or polystyrene – yoghurt cartons, packaging.

Part 2 The problems of plastics

Rehearse some of the familiar problems of using plastics, both in terms of their disposal and recycling. Emphasise and explain the non-biodegradability of many addition polymers.

Refer to the 'three Rs' (Reduce, Reuse, Recycle) and ask the students how these three Rs are being used, or could be used more effectively, to prevent waste.

Use the Science in Context box *The Challenges of Recycling Plastics* in the Student Book to explain that recycling is complicated, not just because different plastic materials need to be separated. The content of these Science in Context boxes is beyond the syllabus but provides the opportunity to explore the content more deeply.

Learning episode C11e.3 Consolidation and summary

Learning aims

- Review the learning points of the topic summarised in the End of topic checklist.
- Test understanding of the topic content by answering the End of topic questions.

Resources

Student Book End of topic checklist and End of topic questions

Approach

Ask students to work with a partner to make a list of key words from this topic. They could then work together to produce a spider diagram showing how the different concepts are linked. They could compare their list with the list of key terms given in the Student Book. Discuss the checklist and ask questions to see how much of the content students are comfortable with.

Students could make flashcards of the key content and then use the flashcards to quiz each other on the information.

Take the opportunity to revise aspects of the chemistry of alkenes, particularly if students had limited recall of these at the start of this topic.

Ask students to work individually through the End of topic questions in the Student Book without looking at the text. As they work, walk around the classroom observing their answers and questioning them as necessary to find out which questions are causing difficulties.

After a set period, ask students to stop working. Discuss any areas of difficulty you observed as you walked round the class.

Students should complete any unanswered questions for homework, but you should stress that they should answer the questions without looking at the text, so that they can see how much they have remembered.

C12 Experimental techniques and chemical analysis

Contents

Teacher Guide	Syllabus reference
C12a Experimental design and chromatography	C12.1 Experimental design C12.2 Chromatography C12.3 Separation and purification
C12b Identification of ions and gases	C12.4 Identification of ions and gases

Overview of the section

As the title of this section indicates the focus is very much on practical activities that the students can actively take part in and consider the related risk assessments. Some of the practical work may have featured in earlier study but, nevertheless, a revision or recap could prove beneficial. You will need to decide whether you approach the whole section at one time or you decide to split it so that it links more closely with some of the other sections.

Chemical analysis focuses on simple identification tests that can readily be performed in a school/college laboratory. Essentially it involves the identification of common anions, cations and gases.

Supplement The work on chromatography includes R_f values at supplementary level.

Starting points

The Student Book section opener puts the ideas in the section into context and sets the scene. It also allows students to acknowledge and value their prior learning, and provides a benchmark against which future learning can be compared.

The questions provide a structure for introducing the section and can be used in a number of different ways:

- You could ask students to consider the questions as an introductory homework task.
- You could put students into groups to share their own ideas and understanding and then to report back to the whole class.
- Students could be given access to the Internet, preferably with a tight timescale, to find out the information required.

You could then use a spider chart or other form of wall chart to summarise everybody's ideas.

Recording these initial ideas allows you to retain them for reference as the individual topics are developed. In this way, your students' progress in learning can be readily acknowledged.

C12a Experimental techniques

Introduction

You may find that this topic overlaps significantly with what students have already learned. After the orientation learning episode, you will find it easier to judge how much teaching time to devote to the topic.

There are some useful links to other topics: the use of filtration and crystallisation in the preparation of soluble salts in Topic C7a *Acids, bases and salts*, and the use of distillation and fractional distillation in Topic C11b *Fuels*.

This topic focuses first on measurement, then on the criteria for purity and finally on some familiar methods of purification, including filtration, distillation and fractional distillation.

Links to other topics

Section	Essential background knowledge	Useful links
C7 Acids, bases and salts		C7a Acids, bases and salts
C11 Organic chemistry		C11b Fuels

Topic overview

C12a.1	**Orientation** Students will be familiar with a range of laboratory measuring equipment and are also likely to have some prior knowledge of ways of separating mixtures. The purpose of this learning episode is to assess their prior knowledge, identify any misconceptions and then decide how much teaching time will be needed on the rest of the topic.
C12a.2	**Chromatography** This learning episode considers purity, initially using chromatography to separate pure substances from a mixture. Supplement The use of R_f values is introduced to identify the components separated in chromatography.
C12a.3	**Separation and purification** This learning episode is intended to make sure that the students are familiar with the following separation processes: simple distillation, fractional distillation, filtration, crystallisation and chromatography. These aspects may be very familiar to students and so the learning episode may be quite short.
C12a.4	**Consolidation and summary** This learning episode provides a quick recap of the key ideas of the topic.

Career links

These are some scientific careers that focus on this area of chemistry but careers in many other fields use the knowledge and skills gained studying science. Experimental skills also underpin all careers in chemistry and not only specifically in the lab. **Analytical chemists** develop analytical and purification methods to purify compounds such as medicines before they are used in biological tests. **Water purification chemists** analyse drinking water to make sure that it is suitable for human use and consumption.

Learning episode C12a.1 Orientation

Learning objectives

- Name appropriate apparatus for the measurement of time, temperature, mass and volume, including:
- stopwatches; thermometers; balances; burettes; volumetric pipettes; measuring cylinders; gas syringes.
- Describe a:
  (a) solvent as a substance that dissolves a solute;
  (b) solute as a substance that is dissolved in a solvent;
  (c) solution as a mixture of one or more solutes dissolved in a solvent;
  (d) saturated solution as a solution containing the maximum concentration of a solute dissolved in the solvent at a specified temperature;
  (e) residue as a solid substance that remains after evaporation, distillation, filtration or any similar process;
  (f) filtrate as a liquid or solution that has passed through a filter.

Resources

Student Book Measurement

Resources for demonstration (see Technician's notes, below)

Approach

Start with a quick quiz in which students have to name some common laboratory measuring equipment. This could include a stopwatch, thermometer, electronic scales or balance, burettes, dropping pipettes, volumetric pipettes and measuring cylinders.

Set up demonstrations of distillation, fractional distillation, filtration, crystallisation and chromatography (see Technician's notes). As students observe each process, ask them to identify which states of matter are being separated in each process. You will need to judge whether students have enough prior knowledge to omit learning episode C12a.3.

SAFETY INFORMATION
Wear eye protection (splash-proof goggles).
Ethanol is highly flammable.

Technician's notes

Be sure to check the latest safety notes on these resources before proceeding.

The following resources are needed for a demonstration of methods of purification:

stopwatch, thermometer, electronic balance, burette, dropping pipette and measuring cylinder
distillation apparatus, with dilute blue washable ink, for example
fractional distillation apparatus, with ethanol and water, for example
filtration, with sand and water, for example
crystallisation apparatus, with salt solution, for example
chromatography apparatus, with black soluble ink, for example

Learning episode C12a.2 Chromatography

Learning objectives

- Describe how paper chromatography is used to separate mixtures of soluble coloured substances, using a suitable solvent.
- Interpret simple chromatograms to identify:
  unknown substances by comparison with known substances; pure and impure substances.
- Supplement State and use the equation for R_f:

$$R_f = \frac{\text{distance travelled by substance)}}{\text{distance travelled by solvent}}$$

Resources

Student Book Chromatography

Worksheet C12a.2 Paper chromatography

Resources for a class practical (see Technician's notes, following)

Approach

From the Orientation learning episode, you will be able to judge whether it is appropriate to provide some practical investigation of the separation of mixtures. If so, chromatography lends itself to class practical work.

Worksheet C12a.2 can be used to support this work. Instructions are given on the worksheet.

SAFETY INFORMATION
Wear eye protection (splash-proof goggles).

Technician's notes

Be sure to check the latest safety notes on these resources before proceeding.

The following resources are needed for the class practical C12a.2, per group:

paper chromatography apparatus
250 cm^3 beaker, strips of paper, clip to hold paper, pencil (to sit across top of beaker), sticky tape
4 samples of black (water soluble) ink labelled A, B, C and D – two of the inks should be the same, so 3 different inks are needed

Learning episode C12a.3 Separation and purification

Learning objectives

- Describe and explain methods of separation and purification using:
  a suitable solvent; filtration; crystallisation; simple distillation; fractional distillation.
- Suggest suitable separation and purification techniques, given information about the substances involved.

Common misconceptions

Some students may think that these methods of purification involve chemical changes.

Resources

Student Book Separation and purification

Worksheet C12a.3 Methods of purification

Resources for class practical in worksheet C12a.3 (see below)

Approach

Review the following processes, making use of the relevant sections in the Student Book: dissolving, filtration, crystallisation, simple distillation and fractional distillation.

You will know from learning episode C12a.1 whether it is appropriate to give students some practical opportunities to investigate purification/separation of mixtures. If so, then a simple distillation lends itself to class practical work.

Worksheet C12a.3 can be used to support this work.

SAFETY INFORMATION
Wear eye protection (splash-proof goggles).

Technician's notes

Be sure to check the latest safety notes on these resources before proceeding.

The following resources are needed for the class practical C12a.3, per group:

a dilute solution of washable blue ink
eye protection (splash-proof goggles)
simple distillation apparatus as shown on Worksheet C12a.3: boiling tube and test-tube, bung with delivery tube, porcelain pieces, 250 cm^3 beaker, Bunsen burner, heatproof mat, clamp and stand

Learning episode C12a.4 Consolidation and summary

Learning aims

- Review the learning points of the topic summarised in the End of topic checklist.
- Test understanding of the topic content by answering the End of topic questions.

Resources

Student Book End of topic checklist and End of topic questions

Approach

Ask students to work with a partner to make a list of key words from this topic. They could then work together to produce a spider diagram showing how the different concepts are linked. They could compare their list with the list of key terms given in the Student Book. Discuss the checklist and ask questions to see how much of the content students are comfortable with.

Students could make flashcards of the key content and then use the flashcards to quiz each other on the information.

Ask students to work individually through the End of topic questions in the Student Book without looking at the text. As they work, walk around the classroom observing their answers and questioning them as necessary to find out which questions are causing difficulties.

After a set period, ask students to stop working. Discuss any areas of difficulty you observed as you walked round the class.

Students should complete any unanswered questions for homework, but you should stress that they should answer the questions without looking at the text, so that they can see how much they have remembered.

C12b Identification of ions and gases

Introduction

This may be approached as a stand-alone topic, although there is a strong link, for students following the supplementary syllabus, between the precipitation reactions used to identify metal ions in solution and the formation of insoluble compounds as described in Topic C7a *Acids, bases and salts*. In this context there are also links to the formation of coloured precipitates as described in Topic C8d *Transition elements and noble gases*. Students will probably have already encountered the tests for carbon dioxide and hydrogen in Topic C9a *Metals*. The idea of ions in ionic compounds and use of the terms 'cation' and 'anion' is described in Topic C2b *Ions and ionic bonds*. The use of the terms cation and anion also link to Topic C4a *Electrochemistry*.

The topic focuses on identification of cations by flame tests and then tests for cations in solutions and includes the test for the ammonium ion. It then considers tests for anions, which are essentially solution tests with the possible exception of the test for the carbonate ion. The topic concludes with the tests for gases.

There are plenty of opportunities for students to practise and develop their practical skills through qualitative analysis and identification activities.

Links to other topics

Section	Essential background knowledge	Useful links
C2 Atoms, elements and compounds	C2b Ions and ionic bonds	
C3 Stoichiometry		C3a Formulas
C4 Electrochemistry		C4a Electrochemistry
C7 Acids, bases and salts		C7a Acids, bases and salts
C8 The Periodic Table		C8a The Periodic Table C8c Group VII elements C8d Transition elements and noble gases
C9 Metals		C9a Metals

Topic overview

C12b.1	**Orientation** This short introductory learning episode can be used to revise ideas on ionic bonding and the charges on ions. In addition, you may revise some of the tests for gases already encountered. You may introduce the terms 'cation' and 'anion', particularly if students are unfamiliar with them.

C12b.2	**Identifying cations** This learning episode has three themes: the identification of metal ions using a flame test, the identification of metal ions in solution using sodium hydroxide solution and ammonia solution, and the identification of the ammonium ion. Worksheets C12b.2(1) and C12b.2(2) support student practical work.
C12b.3	**Identifying anions** This learning episode has three themes: the identification of carbonates by reaction with dilute acid, the identification of the halide ions (chloride, bromide and iodide) by acidifying silver nitrate solution with nitric acid, and the use of acidifying with nitric acid barium nitrate solution to identify the sulfate ion in solution Worksheet C12b.3 supports the identification of the halide and sulfate ions in solution.
C12b.4	**Identifying gases** This learning episode brings together information about individual confirmatory tests for gases, including tests for hydrogen, oxygen, carbon dioxide, chlorine and ammonia. Most of these tests should be familiar to students. Worksheet C12b.4 provides a key that supports a systematic approach to identifying the five gases.
C12b.5	**Consolidation and summary** This learning episode provides an opportunity for a quick recap of the ideas encountered in the topic and time for the students to answer the End of topic questions in the Student Book.

Career links

These are some scientific careers that focus on this area of chemistry but careers in many other fields use the knowledge and skills gained studying science.

Astrochemists design probes used to study the atmosphere and surface composition of other plants in the solar system through gas and ion analysis, among other techniques.

Learning episode C12b.1 Orientation

Learning aims

- To recap on existing knowledge of ions and identification tests for gases.
- Know that metals form positive ions which are called cations.
- Know that non-metals form negative ions which are called anions.
- Know the identification tests for hydrogen, oxygen and carbon dioxide.

Approach

This is a short learning episode. Ask students to summarise in three or four points what they know about ionic bonding and how the Periodic Table can help them to work out the charges on ions.

Explain that the topic has three main sections. The first looks at some tests that can be used to identify positive ions (also called cations), mainly metal ions but also the ammonium ion. The second section deals with tests for non-metal ions (also called anions). These tests together will identify the composition of some simple compounds. The third looks at the test for gases.

Recap on the terms 'cation' and 'anion'. Relate these to how ions behave in electrolysis.

Ask students to write down any gas identification tests they already know about. Explain that in this topic the following gases will be considered: hydrogen, oxygen, carbon dioxide, chlorine and ammonia. Students will probably not have encountered the test for ammonia.

Learning episode C12b.2 Identifying cations

Learning objectives

- Describe tests using aqueous sodium hydroxide and aqueous ammonia to identify the aqueous cations:
  (a) ammonium, NH_4^+
  (b) calcium, Ca^{2+}
  (c) copper(II), Cu^{2+}
  (d) iron(II), Fe^{2+}
  (e) iron(III), Fe^{3+}
  (f) zinc, Zn^{2+}.
- Describe the use of a flame test to identify the cations:
  (a) lithium, Li^+
  (b) sodium, Na^+
  (c) potassium, K^+
  (d) copper(II), Cu^{2+}.

Resources

Student Book Identifying metal ions (cations); Identifying ammonium ions, NH_4^+

Worksheet C12b.2(1) Using a flame test to identify metal ions Worksheet C12b.2(2) Identifying metal ions in solution

Resources for a demonstration and a class practical (see Technician's notes, following)

Approach

There are three parts to this learning episode:

1. Identifying metal ions using a flame test

This part is supported by Worksheet C12b.2(1). Demonstrate the flame test technique, emphasising the importance of 'cleaning' the nichrome wire between tests (see Technician's notes). Allow students to test the solids provided, including unknown solids A and B.

SAFETY INFORMATION

Wear eye protection (splash-proof goggles).

Copper(II) nitrate is corrosive, a moderate hazard and hazardous to the aquatic environment.

Copper(II) chloride is a moderate hazard and hazardous to the aquatic environment.

Lithium chloride is a moderate hazard.

Some of the solids are hazardous to the aquatic environment and so great care is needed with disposal.

2. Identifying metal ions in solution

This part is supported by Worksheet C12b.2(2). The basis of the test is the precipitation of insoluble metal hydroxides. The hydroxide of zinc is amphoteric and so will dissolve in excess sodium hydroxide solution. In addition, if ammonia solution is used copper(II) hydroxide will dissolve in excess ammonia forming a deep blue solution of a complex ion. (The formulas of complex ions are not required by the syllabus.)

Students have to use their results to identify the metal ions present in two unknown solutions.

Supplement It is worth spending time checking that the students are able to construct equations for these precipitation reactions.

SAFETY INFORMATION
Wear eye protection (splash-proof goggles).
Sodium hydroxide solution is caustic and ammonia solution is an irritant and releases toxic vapour even at low concentrations.
Ensure that the laboratory is well ventilated.

3. Identifying the ammonium ion

Demonstrate the test using sodium hydroxide solution as described in the Student Book (see Technician's notes). Make the point that the ammonia gas is the only common alkaline gas and use of this fact will be made later in the topic when identifying gases.

SAFETY INFORMATION
Wear eye protection (splash-proof goggles).

Technician's notes

Be sure to check the latest safety notes on these resources before proceeding.

The following resources are needed for the class practical C12b.2(1), per group:

nichrome wire in holder (cork or hard glass)
Bunsen burner and heatproof mat
watch glasses, spatula
1 mol / dm^3 hydrochloric acid
copper(II) chloride, lithium chloride
potassium chloride, sodium chloride
potassium sulfate (or other potassium salt) labelled as 'substance A'
copper(II) nitrate (or other copper salt) labelled as 'substance B'
eye protection (splash-proof goggles)

The following resources are needed for the class practical C12b.2(2), per group:

beakers, droppers, test-tubes, test-tube rack
0.1 mol / dm^3 sodium hydroxide solution, 0.1 mol / dm^3 ammonia solution
0.1 mol / dm^3 solutions of calcium nitrate, copper(II) sulfate, iron(III) chloride, zinc nitrate
0.1 mol / dm^3 solutions of iron(II) ammonium sulfate (labelled iron(II) sulfate)
0.1 mol / dm^3 solutions of copper(II) chloride (labelled as 'solution A')
0.1 mol / dm^3 iron(II) ammonium sulfate (labelled as 'solution B')
eye protection (splash-proof goggles)

SAFETY INFORMATION
Some of the solids are hazardous to the aquatic environment and so great care is needed with disposal.
Copper(II) nitrate is corrosive, a moderate hazard and hazardous to the aquatic environment.
Copper(II) chloride is a moderate hazard and hazardous to the aquatic environment.
Lithium chloride is a moderate hazard.

The following resources are needed for the demonstration of identifying the ammonium ion:

test-tube, dropper
Bunsen burner and heatproof mat
universal indicator paper and chart
0.1 mol / dm^3 ammonium chloride solution
0.1 mol / dm^3 sodium hydroxide

Note: iron(II) ammonium sulfate is used because the solution of iron(II) ions with ammonium ions is more stable than iron(II) sulfate

Learning episode C12b.3 Identifying anions

Learning objective

- Describe tests to identify the anions:
- carbonate, CO_3^{2-}, by reaction with dilute acid and then testing for carbon dioxide gas
- chloride, Cl^-, bromide, Br^-, and iodide I^-, by acidifying with dilute nitric acid then adding aqueous silver nitrate
- sulfate, SO_4^{2-}, by acidifying with dilute nitric acid then adding aqueous barium nitrate

Resources

Student Book Identifying anions

Worksheet C12b.3 Identifying non-metal ions in solution

Resources for a demonstration and a class practical (see Technician's notes, below)

Approach

There are three parts to the learning episode:

1. Identifying the carbonate ion

Students should be familiar with the test to identify carbon dioxide. Demonstrate the reaction of a carbonate, such as calcium carbonate, with dilute hydrochloric acid and bubble the gas through limewater (see Technician's notes). You may use apparatus similar to that shown in the Student Book.

Supplement Give students time to construct equations for this carbonate + acid reaction. Refer also to the equations given in the Student Book.

2. Identifying halide ions

Use the first part of Worksheet C12b.3. In this part students test four unknown solutions A–D, record their observations and then identify the halide ion present using the information given on colours of halide precipitates.

SAFETY INFORMATION

Wear eye protection (splash-proof goggles).

Silver nitrate solution is harmful.

Nitric acid causes skin and eye irritation.

Sodium nitrate is oxidising and a moderate hazard.

Mention that the precipitates are light sensitive and so may darken if left to stand.

3. Identifying sulfate ions

Use the second part of Worksheet C12b.3. Using the information provided on what constitutes a positive sulfate ion test, students test four unknown solutions E–H, record their results and then identify which solutions contain sulfate ions.

Supplement Again, it is worth spending some time on writing chemical equations for these reactions.

SAFETY INFORMATION

Wear eye protection (splash-proof goggles).

Barium nitrate solution is harmful.

Nitric acid causes skin and eye irritation.

Technician's notes

Be sure to check the latest safety notes on these resources before proceeding.

The following resources are needed for a demonstration (parts 1, 4 and 5):

test-tube, bung and delivery tube, test-tube
0.1 mol / dm^3 hydrochloric acid, calcium carbonate, limewater. Limewater is a moderate hazard
0.1 mol / dm^3 sodium hydroxide solution, sodium nitrate
boiling tube, spatula, filter

The following resources are needed for the class practical C12b.3 (parts 2 and 3), per group:

test-tubes, test-tube rack, droppers
0.05 mol / dm^3 silver nitrate solution, 0.1 mol / dm^3 nitric acid
0.1 mol / dm^3 potassium iodide (labelled as 'solution A'), 0.1 mol / dm^3 sodium chloride (labelled as 'solution B')
0.1 mol / dm^3 sodium bromide (labelled as 'solution C'), 0.1 mol / dm^3 sodium iodide (labelled as 'solution D')
0.1 mol / dm^3 sodium chloride (labelled as 'solution E'), 0.1 mol / dm^3 sodium sulfate (labelled as 'solution F')
0.1 mol / dm^3 potassium sulfate (labelled as 'solution G'), 0.1 mol / dm^3 sodium chloride (labelled as 'solution H')
0.1 mol / dm^3 nitric acid
eye protection (splash-proof goggles)

SAFETY INFORMATION
Wear eye protection (splash-proof goggles).
Barium nitrate solution is harmful.
Nitric acid causes skin and eye irritation.

Learning episode C12b.4 Identifying gases

Learning objective

- Describe tests to identify the gases:
- ammonia, NH_3, using damp red litmus paper
- carbon dioxide, CO_2, using limewater
- chlorine, Cl_2, using damp litmus paper
- hydrogen, H_2, using a lighted splint
- oxygen, O_2, using a glowing splint

Resources

Student Book Identifying gases

Worksheet C12b.4 Identifying gases

Resources for a class practical (see Technician's notes, below)

Approach

This learning episode reinforces or revises the tests for the gases in this topic. Students should be familiar with these. Worksheet C12b.4 provides a key for identifying some samples of unknown gases.

The flow diagram uses a glowing splint to distinguish between hydrogen and oxygen. The syllabus requires students to know that a glowing splint is used to detect oxygen.

SAFETY INFORMATION

Wear eye protection (splash-proof goggles).

Chlorine is oxidising, toxic and hazardous to the environment.

Small amounts of Cl_2, can cause respiratory distress in some people. ***The laboratory must be well ventilated.***

Ammonia gas is flammable and toxic by inhalation, and should be used in a fume cupboard.

Ensure that the laboratory is well ventilated. If you are asthmatic, take care when handling these chemicals.

Technician's notes

Be sure to check the latest safety notes on these resources before proceeding.

The following resources are needed for the class practical C12b.4, per group:

universal indicator paper and chart, splints
Bunsen burner and heatproof mat
Limewater is a moderate hazard.
samples of gases (labelled A–E) in stoppered test-tubes: A – hydrogen, B – carbon dioxide, C – ammonia, D – chlorine, E – oxygen
eye protection (splash-proof goggles), water to dampen litmus paper

Learning episode C12b.5 Consolidation and summary

Learning aims

- Review the learning points of the topic summarised in the End of topic checklist.
- Test understanding of the topic content by answering the End of topic questions.

Resources

Student Book End of topic checklist and End of topic questions

Approach

Supplement Students may need practice at writing the chemical equations.

For these students there are also a number of ionic equations that can be used to describe some general reactions (that is, with the spectator ions removed).

Ask students to work with a partner to make a list of key words from this topic. They could then work together to produce a spider diagram showing how the different concepts are linked. They could compare their list with the list of key terms given in the Student Book. Discuss the checklist and ask questions to see how much of the content students are comfortable with.

Students could make flashcards of the key content and then use the flashcards to quiz each other on the information.

Ask students to work individually through the End of topic questions in the Student Book without looking at the text. As they work, walk around the classroom observing their answers and questioning them as necessary to find out which questions are causing difficulties.

The topic lends itself to a '20 questions' approach to revision or consolidation. Use questions such as:

- What is the test for hydrogen?
- What observation indicates a positive result in the identification of sulfate ions?

After a set period, ask students to stop working. Discuss any areas of difficulty you observed as you walked round the class.

Students should complete any unanswered questions for homework, but you should stress that they should answer the questions without looking at the text, so that they can see how much they have remembered.

Chemistry Teaching sequence

The course can be taught in syllabus order and it will work well to do so. If you want to change the teaching sequence for the course, you should keep in mind the considerations below. An example alternative teaching sequence has been provided.

- There should be a logical progression in the development of concepts and knowledge in order to aid students' understanding.
- The course should have a 'storyline' so that wherever possible the sequence can be easily explained to students.
- Consider carefully how to introduce the more difficult concepts. For example, ensure that they are not introduced too early in the course, which may demotivate students, or put in a sequence where they all come one after another.
- Maintain the emphasis on practical and investigative work throughout the course.
- You will need to allocate different amounts of teaching time to different topics, depending on levels of difficulty or extent of the content.
- It is important to retain some flexibility so that you can revisit difficult ideas or apply new ideas in different contexts.
- You will need time for revision at the end of the course. It is also recommended that you encourage students to revisit material for revision throughout the course. This could be done using the flashcards that they create at the end of each topic.

Alternative teaching sequence

This example alternative teaching sequence is based on a two-year course taught in six terms. It is mapped over five terms on the assumption that much of the sixth term is spent on revision. For different numbers of terms or semesters, where the course is started early, or where the time allocated each week is either generous or limited, you will need to adjust this sequence.

First year, 1st term

C1a Solids, liquids and gases
C12a Experimental design and chromatography
C2a Materials, atomic structure and the Periodic Table
C2b Ions and ionic bonds
C2c Simple molecules and covalent bonds
C7a Acids, bases and salts
C10a Chemistry of the environment

First year, 2nd term

C3a Formulas
C8a Arrangement of elements
C8b Group I elements
C8c Group VII elements
C8d Transition metals and noble gases
C6a Rate of reaction

First year, 3rd term

C9a Metals
C4a Electrolysis
C12b Identification of ions and gases

Second year, 1st term

C5a Exothermic and endothermic reactions
C11a Fuels
C11b Alkanes

Second year, 2nd term

C6b Redox reactions

C11c Alkenes

C11d Polymers

Chemistry Scheme of work

Learning episode	**Learning objectives and learning aims** Learning objectives are given which link to the syllabus. Learning aims reflect the intended outcomes of the consolidation and summary learning episodes.	**Syllabus reference**	**Student's Book heading/ subheading**
C1a.1 Orientation	Learning objectives • State the distinguishing properties of solids, liquids and gases.	C1.1	Solids, liquids and gases
C1a.2 Solids, liquids and gases	Learning objectives • Describe the structures of solids, liquids and gases in terms of particle separation, arrangement and motion. • Describe changes of state in terms of melting, boiling, evaporating, freezing and condensing. • Describe the effects of temperature and pressure on the volume of a gas. • Supplement Explain changes of state in terms of kinetic particle theory.	C1.1	Solids, liquids and gases
C1a.3 Consolidation and summary	Learning aims • Review the learning points of the topic summarised in the End of topic checklist. • Test understanding of the topic content by answering the End of topic questions.	C1	End of topic checklist and End of topic questions
C2a.1 Orientation	Learning objectives • Describe the differences between elements, compounds and mixtures.	C2.1	Elements, compounds and mixtures
C2a.2 Sub-atomic particles	Learning objectives • Describe the structure of the atom as a central nucleus containing neutrons and protons surrounded by electrons in shells. • State the relative charges and relative masses of a proton, a neutron and an electron. • Define proton number / atomic number as the number of protons in the nucleus of an atom. • Define mass number / nucleon number as the total number of protons and neutrons in the nucleus of an atom.	C2.2	Atomic structure and the Periodic Table
C2a.3 The arrangement of electrons in atoms	Learning objectives • Determine the electronic configuration of elements, and their ions with proton number 1 to 20, e.g., 2,8,3. • State that: (a) Group VIII noble gases have a full outer shell (b) The number of outer shell electrons is equal to the group number in Groups I to VII (c) The number of occupied electron shells is equal to the period number.	C2.2	Electronic configuration: The first 20 elements of the Periodic Table

Learning episode	**Learning objectives and learning aims** Learning objectives are given which link to the syllabus. Learning aims reflect the intended outcomes of the consolidation and summary learning episodes.	**Syllabus reference**	**Student's Book heading/ subheading**
C2a.4 Consolidation and summary	Learning aims • Review the learning points of the topic summarised in the End of topic checklist. • Test understanding of the topic content by answering the End of topic questions.	C2	End of topic checklist and End of topic questions
C2b.1 Orientation	Learning aims • To revise ideas on atomic structure. • To map out the main purposes of the topic. • Know the structure of an atom by interpreting proton and nucleon numbers. • Know the arrangement of electrons in shells for the first 20 elements.	C2	Sub-atomic particles; Proton number and mass number
C2b.2 The formation of ions	Learning objectives • Describe the formation of: (a) positive ions, known as cations, by the loss of electrons (b) negative ions, known as anions, by the gain of electrons • State that an ionic bond is a strong electrostatic attraction between oppositely charged ions. • Describe the formation of ionic bonds between elements from Group I and Group VII, including the use of dot-and-cross diagrams. • Supplement Describe the formation of ionic bonds between ions of metallic and non-metallic elements, including the use of dot-and-cross diagrams.	C2.3	The formation of ions; Electron transfer in ionic bonding
C2b.3 Properties of ionic compounds	Learning objectives • Describe the properties of ionic compounds: (a) high melting points and boiling points (b) good electrical conductivity when aqueous or molten and poor when solid (c) generally soluble in water. • Supplement Describe the giant lattice structure of ionic compounds as a regular arrangement of alternating positive and negative ions, exemplified by sodium chloride.	C2.3	Properties of ionic compounds
C2b.4 Consolidation and summary	Learning aims • Review the learning points of the topic summarised in the End of topic checklist. • Test understanding of the topic content by answering the End of topic questions.	C2	End of topic checklist and End of topic questions

Learning episode	**Learning objectives and learning aims** Learning objectives are given which link to the syllabus. Learning aims reflect the intended outcomes of the consolidation and summary learning episodes.	**Syllabus reference**	**Student's Book heading/ subheading**
C2c.1 Orientation	Learning aims • To revise ideas on atomic structure and ionic bonding. • To map out the main purposes of the topic. • Know the structure of an atom from its proton and nucleon numbers. • Know the arrangement of electrons in shells for the first 20 elements.	n/a	Sub-atomic particles; Proton number and mass number
C2c.2 The formation of covalent molecules	Learning objectives • State that a covalent bond is formed when a pair of electrons is shared between two atoms leading to noble gas electronic configurations. • Describe the formation of covalent bonds in simple molecules, including H_2, Cl_2, H_2O, CH_4, NH_3 and HCl. Use dot-and-cross diagrams to show the electronic configurations in these and similar molecules. • Supplement Describe the formation of covalent bonds in simple molecules, including CH_3OH, C_2H_4, O_2, CO_2 and N_2. Use dot-and-cross diagrams to show the electronic configurations in these and similar molecules.	C2.4	Electron sharing in covalent bonds
C2c.3 Properties of simple covalent molecules	Learning objectives • Describe the properties of simple molecular compounds: ○ low melting points and boiling points ○ poor electrical conductivity.	C2.4	Properties of covalent compounds
C2c.4 Consolidation and summary	Learning aims • Review the learning points of the topic summarised in the End of topic checklist. • Test understanding of the topic content by answering the End of topic questions.	C2	End of topic checklist and End of topic questions
C3a.1 Orientation	Learning objectives • State the formulas of the elements and compounds named in the subject content. • Define the molecular formula of a compound as the number and type of different atoms in one molecule. • Deduce the formula of a simple compound from the relative number of atoms present in a model or diagrammatic representation. • Supplement Deduce the formula of an ionic compound from the relative numbers of the ions present in a model or a diagrammatic representation or from the charges on the ions.	C3.1	Formulas

Learning episode	**Learning objectives and learning aims** Learning objectives are given which link to the syllabus. Learning aims reflect the intended outcomes of the consolidation and summary learning episodes.	**Syllabus reference**	**Student's Book heading/ subheading**
C3a.2 Writing chemical equations	Learning objectives • Construct word equations to show how reactants form products, including state symbols. • Balance and interpret simple symbol equations, including state symbols. • Supplement Construct symbol equations with state symbols, including ionic equations.	C3.1	Writing chemical equations
C3a.3 Consolidation and summary	Learning aims • Review the learning points of the topic summarised in the End of topic checklist. • Test understanding of the topic content by answering the End of topic questions.	C3	End of topic checklist and End of topic questions
C4a.1 Orientation	Learning objectives • Define electrolysis as the decomposition of an ionic compound, when molten or in aqueous solution, by the passage of an electric current. • Identify in simple electrolytic cells (d) the anode as the positive electrode (e) the cathode as the negative electrode (f) the electrolyte as the molten or aqueous substance that undergoes electrolysis.	C4.1	The process of electrolysis
C4a.2 Electrolysis of molten electrolytes	Learning objectives • Identify the products formed at the electrodes and describe the observations made during the electrolysis of molten lead(II) bromide using inert electrodes made of platinum or carbon / graphite. • Supplement State that metals or hydrogen are formed at the cathode and that non-metals (other than hydrogen) are formed at the anode. • Supplement Predict the identity of the products at each electrode for the electrolysis of a binary compound in the molten state.	C4.1	Electrolysis of molten lead(II) bromide; Predicting the products of electrolysis
C4a.3 Electrolysis of aqueous electrolytes	Learning objectives • Identify the products formed at the electrodes and describe the observations made during the electrolysis of: ○ concentrated aqueous sodium chloride ○ dilute sulfuric acid. using inert electrodes made of platinum or carbon / graphite. • Supplement State that metals or hydrogen are formed at the cathode and that non-metals (other than hydrogen) are formed at the anode.	C4.1	The electrolysis of dilute sulfuric acid, Electrolysis of concentrated sodium chloride solution; Predicting the products of electrolysis

Learning episode	**Learning objectives and learning aims** Learning objectives are given which link to the syllabus. Learning aims reflect the intended outcomes of the consolidation and summary learning episodes.	**Syllabus reference**	**Student's Book heading/ subheading**
C4a.4 Consolidation and summary	Learning aims • Review the learning points of the topic summarised in the End of topic checklist. • Test understanding of the topic content by answering the End of topic questions.	C4	End of topic checklist and End of topic questions
C5a.1 Orientation	Learning objectives • State that an exothermic reaction transfers thermal energy to the surroundings leading to an increase in the temperature of the surroundings. • State that an endothermic reaction takes in thermal energy from the surroundings leading to a decrease in the temperature of the surroundings.	C5.1	Thermal energy transfer in chemical reactions
C5a.2 Measuring energy transfers (1)	Learning objectives • To investigate 'cold packs'. • To measure temperature changes in a reaction using an insulated cup as a simple calorimeter. • Describe how an insulated cup can be used to carry out simple calorimetric experiments to determine thermal energy changes (such as in dissolving solids or neutralisation reactions).	n/a	Thermal energy transfer in chemical reactions
C5a.3 Measuring energy transfers (2)	Learning objectives • Supplement Interpret reaction pathway diagrams showing exothermic and endothermic reactions. • Supplement Define activation energy, E_a, as the minimum energy that colliding particles must have in order to react. • Supplement Draw and label reaction pathway diagrams for exothermic and endothermic reactions using information provided, to include: reactants; products; overall energy change of reaction; activation energy, E_a.	C5.1	Thermal energy transfer in chemical reactions
C5a.4 Rection pathway diagrams, bond breaking and bond forming	Learning objectives • Supplement Interpret reaction pathway diagrams showing exothermic and endothermic reactions. • Supplement Draw and label reaction pathway diagrams for exothermic and endothermic reactions using information provided, to include: reactants; products; overall energy change of reaction; activation energy, E_a. • Supplement State that bond breaking is an endothermic process and bond making is an exothermic process.	C5.1	Reaction pathway diagrams; Activation energy; Where does the energy come from
C5a.5 Consolidation and summary	Learning aims • Review the learning points of the topic summarised in the End of topic checklist. • Test understanding of the topic content by answering the End of topic questions.	C5	End of topic checklist and End of topic questions

Learning episode	Learning objectives and learning aims Learning objectives are given which link to the syllabus. Learning aims reflect the intended outcomes of the consolidation and summary learning episodes.	Syllabus reference	Student's Book heading/ subheading
C6a.1 Orientation	Learning objectives • Identify physical and chemical changes and understand the differences between them.	C6.1	Physical and chemical changes; Rate of reaction, Collision theory
C6a.2 Monitoring the rate of reaction	Learning objectives • Describe practical methods for investigating the rate of a reaction including change in mass of a reactant or product and the formation of a gas. • Interpret data, including graphs, from rate of reaction experiments. • Supplement Describe collision theory in terms of: number of particles per unit volume; frequency of collisions between particles; kinetic energy of particles; activation energy, E_a.	C6.2	Collision theory; Monitoring the rate of a reaction; Interpreting graphs from rate of reaction experiments
C6a.3 Concentration of solutions	Learning objectives • Describe the effect on the rate of a reaction of changing the concentration of solutions. • Supplement Describe and explain the effect on the rate of reaction of changing the concentration of solutions using collision theory. • Describe practical methods for investigating the rate of a reaction including change in mass of a reactant or product and the formation of a gas. • Interpret data, including graphs, from rate of reaction experiments.	C6.2	What can change the rate of a reaction?; Concentration
C6a.4 Temperature	Learning objectives • Describe the effect on the rate of reaction of changing the temperature. • Interpret data, including graphs, from rate of reaction experiments. • Describe practical methods for investigating the rate of a reaction including change in mass of a reactant or product and the formation of a gas. • Supplement Describe and explain the effect on the rate of reaction of changing the temperature using collision theory.	C6.2	What can change the rate of a reaction?; Temperature
C6a.5 Surface area of solids	Learning objectives • Describe the effect on the rate of reaction of changing the surface area of solids. • Describe practical methods for investigating the rate of a reaction including the formation of a gas. • Interpret data, including graphs, from rate of reaction experiments. • Supplement Describe and explain the effect on the rate of reaction of changing the surface area of solids using collision theory.	C6.2	What can change the rate of a reaction?; Surface area

Learning episode	**Learning objectives and learning aims** Learning objectives are given which link to the syllabus. Learning aims reflect the intended outcomes of the consolidation and summary learning episodes.	**Syllabus reference**	**Student's Book heading/ subheading**
C6a.6 Use of a catalyst	Learning objectives • Describe the effect on the rate of reaction of: adding or removing a catalyst. • State that a catalyst increases the rate of a reaction and is unchanged at the end of a reaction. • Supplement Describe and explain the effect on the rate of reaction of adding or removing a catalyst using collision theory. • Interpret data, including graphs, from rate of reaction experiments.	C6.2	What can change the rate of a reaction?; Catalysts
C6a.7 Consolidation and summary	Learning aims • Review the learning points of the topic summarised in the End of topic checklist. • Test understanding of the topic content by answering the End of topic questions.	C6.1, C6.2	End of topic checklist and End of topic questions
C6b.1 Orientation	Learning objectives • Define redox reactions as involving simultaneous oxidation and reduction. • Define oxidation as gain of oxygen and reduction as loss of oxygen. • Identify redox reactions as reactions involving gain and loss of oxygen. • Identify oxidation and reduction in redox reactions. (Oxidation number limited to its use to name ions, e.g., iron(II), iron(III), copper(II).)	C6.3	Oxidation, reduction and redox
C6b.2 Consolidation and summary	Learning aims • Review the learning points of the topic summarised in the End of topic checklist. • Test understanding of the topic content by answering the End of topic questions.	C6.3	End of topic checklist and End of topic questions
C7a.1 Orientation	Learning objectives • Describe acids in terms of their effect on: litmus; methyl orange. • Describe alkalis in terms of their effect on: litmus; methyl orange.	C7.1	The characteristic properties of acids and bases – Aqueous solutions
C7a.2 Acids, bases, alkalis and salts	Learning objectives • Describe the characteristic properties of acids in terms of their reactions with metals, bases, and carbonates. • State that bases are oxides or hydroxides of metals and that alkalis are soluble bases. • Describe the characteristic properties of bases in terms of their reactions with acids. • Describe how to compare neutrality, relative acidity and relative alkalinity in terms of colour and pH (using universal indicator paper). • Classify oxides as acidic, including SO_2 and CO_2, or basic, including CuO and CaO, related to metallic and non-metallic character.	C7.1, C7.2	The characteristic properties of acids and bases; Oxides; Salts

Learning episode	**Learning objectives and learning aims** Learning objectives are given which link to the syllabus. Learning aims reflect the intended outcomes of the consolidation and summary learning episodes.	**Syllabus reference**	**Student's Book heading/ subheading**
C7a.3 Making soluble salts	Learning objectives • Describe the preparation, separation and purification of soluble salts by reaction of an acid with: an alkali by titration; excess metal; excess insoluble base; excess insoluble carbonate (students do not need to know general solubility rules for salts).	C7.3	Salts - Making soluble salts; In the laboratory
C7a.4 Making insoluble salts	Learning objectives • Supplement Describe the preparation of insoluble salts by precipitation (candidates do not need to know general solubility rules for salts).	C7.3	Salts – Making insoluble salts
C7a.5 Consolidation and summary	Learning aims • Review the learning points of the topic summarised in the End of topic checklist. • Test understanding of the topic content by answering the End of topic questions.	C7	End of topic checklist and End of topic questions
C8a.1 Orientation	Learning aims • To revise ideas about the differences between elements, compounds and mixtures. • To revise ideas on atomic structure and the arrangement of electrons. • To map out the main purposes of the topic.	C8	n/a
C8a.2 Arrangement of the elements	Learning objectives • Describe the Periodic Table as an arrangement of elements in periods and groups and in order of increasing proton number/atomic number. • Describe the change from metallic to non-metallic character across a period. • State that the number of outer-shell electrons is equal to the group number in Groups I to VII. • Supplement Identify trends in groups, given information about the elements.	C8.1	Arrangement of elements in the Periodic Table
C8a.3 Consolidation and summary	Learning aims • Review the learning points of the topic summarised in the End of topic checklist. • Test understanding of the topic content by answering the End of topic questions.	C8.1	End of topic checklist and End of topic questions
C8b.1 Orientation	Learning objectives • Describe the Group I alkali metals, lithium, sodium and potassium, as relatively soft metals with general trends down the group, limited to: decreasing melting point; increasing density.	C8.2	Reactivity of Group I elements; Group I properties

Learning episode	**Learning objectives and learning aims** Learning objectives are given which link to the syllabus. Learning aims reflect the intended outcomes of the consolidation and summary learning episodes.	**Syllabus reference**	**Student's Book heading/ subheading**
C8b.2 Reactions of the Group I elements	Learning objectives • Describe the Group I alkali metals, lithium, sodium and potassium, as relatively soft metals with general trends down the group, limited to: increasing reactivity with water. • Supplement Predict the properties of other elements in Group I, given information about the elements.	C8.2	Reactivity of Group I elements; Group I properties
C8b.3 Consolidation and summary	Learning aims • Review the learning points of the topic summarised in the End of topic checklist. • Test understanding of the topic content by answering the End of topic questions.	C8.2	End of topic checklist and End of topic questions
C8c.1 Orientation	Learning objectives • Describe the Group VII halogens, chlorine, bromine and iodine, as diatomic non-metals with general trends down the group, limited to: increasing density; decreasing reactivity. • State the appearance of the halogens at room temperature and pressure, r.t.p. as: chlorine, a pale yellow-green gas; bromine, a red-brown liquid; iodine, a grey-black solid. • Supplement Predict the properties of other elements in Group VII, given information about the elements.	C8.3	Reactivity of Group VII elements; Group VII properties
C8c.2 Displacement reactions of the halogens	Learning objectives • Supplement Describe and explain the displacement reactions of halogens with other halide ions. • Supplement Predict the properties of other elements in Group VII, given information about the elements.	C8.3	Group VII properties – Supplement section
C8c.3 Consolidation and summary	Learning aims • Review the learning points of the topic summarised in the End of topic checklist. • Test understanding of the topic content by answering the End of topic questions.	C8.3	End of topic checklist and End of topic questions
C8d.1 Orientation	Learning aims • Know the positions of the transition metals and noble gases in the Periodic Table. • Know some of the key differences between the transition metals and the Group I metals. • Know some of the key differences between the noble gases and the Group VII elements.	C8	n/a
C8d.2 The transition elements	Learning objectives • Describe the transition elements as metals that: have high densities; have high melting points; form coloured compounds; often act as catalysts as elements and in compounds.	C8.4	Transition elements

Learning episode	**Learning objectives and learning aims** Learning objectives are given which link to the syllabus. Learning aims reflect the intended outcomes of the consolidation and summary learning episodes.	**Syllabus reference**	**Student's Book heading/ subheading**
C8d.3 The noble gases	Learning objectives • Describe the Group VIII noble gases as unreactive, monatomic gases and explain this in terms of electronic configuration. • State that: Group VIII noble gases have a full outer shell.	C8.5	Noble gases
C8d.4 Consolidation and summary	Learning aims • Review the learning points of the topic summarised in the End of topic checklist. • Test understanding of the topic content by answering the End of topic questions.	C8.4, C8.5	End of topic checklist and End of topic questions
C9a.1 Orientation	Learning aims • To revise the reactions of Group I metals with water. • To revise that reactivity increases down the group for the Group I metals. • To revise the lack of reactivity of the transition metals.	C8	N/A
C9a.2 Properties of metals	Learning objectives • Compare the general physical properties of metals and non-metals, including: thermal conductivity; electrical conductivity; malleability; melting points and boiling points. • Describe the general chemical properties of metals, limited to their reactions with: dilute acids; cold water and steam. • Describe alloys as mixtures of a metal with other elements, including: brass as a mixture of copper and zinc; stainless steel as a mixture of iron and other elements, such as chromium, nickel and carbon. • State that alloys can be harder or stronger than the pure metals and are more useful. • Describe the uses of alloys in terms of their physical properties, including stainless steel in cutlery because of its hardness and resistance to rusting. • Identify representations of alloys from diagrams of structure. • Supplement Explain in terms of structure how alloys can be harder or stronger than the pure metals because the different sized atoms in alloys mean the layers can no longer slide over each other.	C9.1, C9.3	Properties of metals; Reactivity series; Alloys and their properties
C9a.3 Reactivity series	Learning objectives • State the order of the reactivity series as: potassium, sodium, calcium, magnesium, aluminium, carbon, zinc, iron, hydrogen, copper, silver, gold. • Describe the reactions, if any, of:potassium, sodium and calcium with cold water; magnesium with steam; magnesium, zinc, iron, copper, silver and gold with dilute hydrochloric acid. • and explain these reactions in terms of the position of the metals in the reactivity series. • Deduce an order of reactivity from a given set of experimental results.	C9.4	Reactivity series

Learning episode	**Learning objectives and learning aims** Learning objectives are given which link to the syllabus. Learning aims reflect the intended outcomes of the consolidation and summary learning episodes.	**Syllabus reference**	**Student's Book heading/ subheading**
C9a.4 Extraction of metals	Learning objectives • Describe the ease of obtaining metals from their ores, related to the position of the metal in the reactivity series. • State that iron from hematite is extracted by reduction of iron(III) oxide in the blast furnace. • Supplement Describe the extraction of iron from hematite in the blast furnace, limited to: the burning of carbon (coke) to provide heat and produce carbon dioxide $C + O_2 \rightarrow CO_2$; the reduction of carbon dioxide to carbon monoxide $C + CO_2 \rightarrow 2CO$; the reduction of iron(III) oxide by carbon monoxide $Fe_2O_3 + 3CO \rightarrow 2Fe + 3CO_2$. • State that the main ore of aluminium is bauxite and that aluminium is extracted by electrolysis. • State that the main ore of aluminium is bauxite and that aluminium is extracted by electrolysis.	C9.6	Extraction of metals
C9a.5 Uses of metals, and corrosion	Learning objectives • Describe the uses of metals in terms of their physical properties, including:aluminium in the manufacture of aircraft because of its low density; aluminium in the manufacture of overhead electrical cables because of its low density and good electrical conductivity; aluminium in food containers because of its resistance to corrosion; copper in electrical wiring because of its good electrical conductivity. • State the conditions required for the rusting of iron (presence of oxygen and water). • State some common barrier methods, including painting, greasing and coating with plastic. • Describe how barrier methods prevent rusting by excluding oxygen or water.	C9.2, C9.5	Uses of metals; Corrosion of metals
C9a.6 Consolidation and summary	Learning aims • Review the learning points of the topic summarised in the End of topic checklist. • Test understanding of the topic content by answering the End of topic questions.	C9	End of topic checklist and End of topic questions
C10a.1 Orientation	Learning aims • To review knowledge on the composition of the air. • To highlight the content areas in the topic. • To revise that oxygen is a key component of the air and that it reacts with elements and compounds to form oxides.	C10	n/a

Learning episode	**Learning objectives and learning aims** Learning objectives are given which link to the syllabus. Learning aims reflect the intended outcomes of the consolidation and summary learning episodes.	**Syllabus reference**	**Student's Book heading/ subheading**
C10a.2 Water	Learning objectives • Describe chemical tests for the presence of water using anhydrous cobalt(II) chloride and anhydrous copper(II) sulfate. • State that distilled water is used in practical chemistry rather than tap water because it contains fewer chemical impurities. • Describe the treatment of the domestic water supply in terms of: sedimentation and filtration to remove solids; use of carbon to remove tastes and odours; chlorination to kill microbes.	C10.1	Water, Treatment of water supplies
C10a.3 The composition of the air	Learning objectives • State the composition of clean, dry air as approximately 78% nitrogen, N_2, 21% oxygen, O_2, and the remainder as a mixture of noble gases and carbon dioxide, CO_2.	C10.2	Air quality and climate

Learning episode	**Learning objectives and learning aims** Learning objectives are given which link to the syllabus. Learning aims reflect the intended outcomes of the consolidation and summary learning episodes.	**Syllabus reference**	**Student's Book heading/ subheading**
C10a.4 Air quality and climate	Learning objectives • State the source of each of these air pollutants, limited to: (a) carbon dioxide from the complete combustion of carbon-containing fuels; (b) carbon monoxide and particulates from the incomplete combustion of carbon-containing fuels; (c) methane from the decomposition of vegetation and waste gases from digestion in animals; (d) oxides of nitrogen from car engines; (e) sulfur dioxide from the combustion of fossil fuels which contain sulfur compounds. • State the adverse effects of these air pollutants, limited to: (a) carbon dioxide: higher levels of carbon dioxide leading to increased global warming, which leads to climate change; (b) carbon monoxide: toxic gas; (c) particulates: increased risk of respiratory problems and cancer; (d) methane: higher levels of methane leading to increased global warming, which leads to climate change; (e) oxides of nitrogen: acid rain, photochemical smog and respiratory problems; (f) sulfur dioxide: acid rain. • Supplement State and explain strategies to reduce the effects of climate change: (a) planting trees; (b) reduction in livestock farming; (c) decreasing use of fossil fuels; (d) increasing use of renewable energy, e.g., wind, solar. • Supplement State strategies to reduce the effects of acid rain: reducing emissions of sulfur dioxide by using low-sulfur fuels. • Supplement Describe how the greenhouse gases carbon dioxide and methane cause global warming, limited to: the absorption, reflection and emission of thermal energy; reducing thermal energy loss to space.	C10.2	Air quality and climate
C10a.5 Consolidation and summary	Learning aims • Review the learning points of the topic summarised in the End of topic checklist. • Test understanding of the topic content by answering the End of topic questions.	C10	End of topic checklist and End of topic questions

Learning episode	**Learning objectives and learning aims** Learning objectives are given which link to the syllabus. Learning aims reflect the intended outcomes of the consolidation and summary learning episodes.	**Syllabus reference**	**Student's Book heading/ subheading**
C11a.1 Orientation	Learning objectives • State that a saturated compound has molecules in which all carbon–carbon bonds are single bonds. • State that an unsaturated compound has molecules in which one or more carbon–carbon bonds are not single bonds. • Supplement State that a homologous series is a family of similar compounds with similar chemical properties. • Supplement Describe the general characteristics of a homologous series as: (a) having the same general formula (recall of specific general formulas is not required) (b) displaying a trend in physical properties (c) sharing similar chemical properties.	C11.1	What are alkanes?
C11b.1 Orientation	Learning objectives • Name the fossil fuels: coal, natural gas and petroleum. • Name methane as the main constituent of natural gas. • State that hydrocarbons are compounds that contain hydrogen and carbon only. • State that petroleum is a mixture of hydrocarbons. • State that a saturated compound has molecules in which all carbon–carbon bonds are single bonds. • State that an unsaturated compound has molecules in which one or more carbon–carbon bonds are not single bonds.	C11.2, C11.1	What are fossil fuels?
C11b.2 The fractional distillation of petroleum	Learning objectives • Describe the separation of petroleum into useful fractions by fractional distillation. • Name the uses of the fractions as: (a) refinery gas fraction for gas used in heating and cooking; (b) gasoline / petrol fraction for fuel used in cars; (c) naphtha fraction as a chemical feedstock; (d) diesel oil / gas fraction for fuel used in diesel engines; (e) bitumen fraction for making roads. • Supplement Describe how the properties of fractions obtained from petroleum change from the bottom to the top of the fractionating column, limited to: decreasing chain length; lower boiling points.	C11.2	Fractional distillation
C11b.3 Consolidation and summary	Learning aims • Review the learning points of the topic summarised in the End of topic checklist. • Test understanding of the topic content by answering the End of topic questions.	C11.1, C11.2	End of topic checklist and End of topic questions

Learning episode	**Learning objectives and learning aims** Learning objectives are given which link to the syllabus. Learning aims reflect the intended outcomes of the consolidation and summary learning episodes.	**Syllabus reference**	**Student's Book heading/ subheading**
C11c.1 Orientation	Learning objectives • Name and draw the displayed formulas of methane and ethane. • State that the bonding in alkanes is single covalent and that alkanes are saturated hydrocarbons. • Supplement State that a homologous series is a family of similar compounds with similar chemical properties • Supplement Describe the general characteristics of a homologous series as: having the same general formula (recall of the alkane general formula is **not** required); displaying a trend in physical properties; sharing similar chemical properties.	C11.3, C11.1	What are alkanes?
C11c.2 The chemical properties of alkanes	Learning objectives • Describe the properties of alkanes as being generally unreactive, except in terms of combustion.	C11.3	The properties of alkanes; Combustion of alkanes
C11c.3 Consolidation and summary	Learning aims • Review the learning points of the topic summarised in the End of topic checklist. • Test understanding of the topic content by answering the End of topic questions.	C11.3	End of topic checklist and End of topic questions
C11d.1 Orientation	Learning objectives • Name and draw the displayed formula of ethene. • State that the bonding in alkenes includes a double carbon–carbon covalent bond and that alkenes are unsaturated hydrocarbons. • Supplement State that a homologous series is a family of similar compounds with similar chemical properties. • Supplement Describe the general characteristics of a homologous series as: having the same general formula (recall of the alkene general formula is **not** required); displaying a trend in physical properties	C11.1, C11.4	What are alkenes?
C11d.2 Cracking crude oil fractions	Learning objectives • Supplement Describe the manufacture of alkenes and hydrogen by the cracking of larger alkane molecules using a high temperature and a catalyst. • Describe the test to distinguish between saturated and unsaturated hydrocarbons by their reaction with aqueous bromine.	C11.4	What are alkenes?; Cracking

Learning episode	**Learning objectives and learning aims** Learning objectives are given which link to the syllabus. Learning aims reflect the intended outcomes of the consolidation and summary learning episodes.	**Syllabus reference**	**Student's Book heading/ subheading**
C11d.3 The chemical properties of alkenes	Learning objectives • Describe the test to distinguish between saturated and unsaturated hydrocarbons by their reaction with aqueous bromine. • Supplement Describe the properties of alkenes in terms of addition reactions with: bromine or aqueous bromine; hydrogen in the presence of a nickel catalyst; steam in the presence of an acid catalyst.	C11.4	Addition reactions of alkenes
C11d.4 Consolidation and summary	Learning aims • Review the learning points of the topic summarised in the End of topic checklist. • Test understanding of the topic content by answering the End of topic questions.	C11.4	End of topic checklist and End of topic questions
C11e.1 Orientation	Learning aims • Define polymers as large molecules built up from many smaller molecules called monomers.	C11.5	Addition polymers
C11e.2 Synthetic polymers	Learning objectives • State that the formation of poly(ethene) is an example of addition polymerisation using ethene (an alkene) monomers.	C11.5	Addition polymers
C11e.3 Consolidation and summary	Learning aims • Review the learning points of the topic summarised in the End of topic checklist. • Test understanding of the topic content by answering the End of topic questions.	C11.5	End of topic checklist and End of topic questions
C12a.1 Orientation	Learning objectives • Name appropriate apparatus for the measurement of time, temperature, mass and volume, including: stopwatches; thermometers; balances; burettes; volumetric pipettes; measuring cylinders; gas syringes. • Describe a: (a) solvent as a substance that dissolves a solute; (b) solute as a substance that is dissolved in a solvent; (c) solution as a mixture of one or more solutes dissolved in a solvent; (d) saturated solution as a solution containing the maximum concentration of a solute dissolved in the solvent at a specified temperature; (e) residue as a substance that remains after evaporation, distillation, filtration or any similar process; (f) filtrate as a liquid or solution that has passed through a filter.	C12.1	Measurement

Learning episode	**Learning objectives and learning aims** Learning objectives are given which link to the syllabus. Learning aims reflect the intended outcomes of the consolidation and summary learning episodes.	**Syllabus reference**	**Student's Book heading/ subheading**
C12a.2 Chromato-graphy	Learning objectives • Describe how paper chromatography is used to separate mixtures of soluble coloured substances, using a suitable solvent. • Interpret simple chromatograms to identify: • unknown substances by comparison with known substances; pure and impure substances. • Supplement State and use the equation for R_f: $R_f = \frac{\text{distance travelled by substance}}{\text{distance travelled by solvent}}$	C12.2	Chromatography
C12a.3 Separation and purification	Learning objectives • Describe and explain methods of separation and purification using: a suitable solvent; filtration; crystallisation; simple distillation; fractional distillation. • Suggest suitable separation and purification techniques, given information about the substances involved.	C12.3	Separation and purification
C12a.4 Consolidation and summary	Learning aims • Review the learning points of the topic summarised in the End of topic checklist. • Test understanding of the topic content by answering the End of topic questions.	C12.1, C12.2, C12.3	End of topic checklist and End of topic questions
C12b.1 Orientation	Learning aims • To recap on existing knowledge of ions and identification tests for gases. • Know that metals form positive ions which are called cations. • Know that non-metals form negative ions which are called anions. • Know the identification tests for hydrogen, oxygen and carbon dioxide.	C2.3	n/a

Learning episode	**Learning objectives and learning aims** Learning objectives are given which link to the syllabus. Learning aims reflect the intended outcomes of the consolidation and summary learning episodes.	**Syllabus reference**	**Student's Book heading/ subheading**
C12b.2 Identifying cations	Learning objectives • Describe tests using aqueous sodium hydroxide and aqueous ammonia to identify the aqueous cations: (a) ammonium NH_4^+ (b) calcium, Ca^{2+} (c) copper(II), Cu^{2+} (d) iron(II), Fe^{2+} (e) iron(III), Fe^{3+} (f) zinc, Zn^{2+}. • Describe the use of a flame test to identify the cations: (a) lithium, Li^+ (b) sodium, Na^+ (c) potassium, K^+ (d) copper(II), Cu^{2+}.	C12.4	Identifying metal ions (cations); Identifying ammonium ions, NH_4^+
C12b.3 Identifying anions	Learning objectives • Describe tests to identify the anions: (a) carbonate, CO_3^{2-}, by reaction with dilute acid and then testing for carbon dioxide gas (b) chloride, Cl^-, bromide, Br^-, and iodide I^-, by acidifying with dilute nitric acid then adding aqueous silver nitrate (c) sulfate, SO_4^{2-}, by acidifying with dilute nitric acid then adding aqueous barium nitrate.	C12.4	Identifying anions
C12b.4 Identifying gases	Learning objectives • Describe tests to identify the gases: (a) ammonia, NH_3, using damp red litmus paper (b) carbon dioxide, CO_2, using limewater (c) chlorine, Cl_2, using damp litmus paper (d) hydrogen, H_2, using a lighted splint (e) oxygen, O_2, using a glowing splint.	C12.4	Identifying gases
C12b.5 Consolidation and summary	Learning aims • Review the learning points of the topic summarised in the End of topic checklist. • Test understanding of the topic content by answering the End of topic questions.	C12.4	End of topic checklist and End of topic questions

P1 Motion, forces and energy

Contents

Overview of the section

This section includes the basics needed for a physics course at this level: a brief consideration of physical quantities and their measurement; motion; mass and weight; density; forces; energy; work and power; and finally, pressure.

Students will learn how to interpret graphs describing motion, and students following the supplement syllabus also learn how acceleration is defined. Students will then learn about the distinction between mass and weight before considering density and how it can be determined for various objects.

Topic P1e considers the effects of forces on motion. Topic P1.f *Energy, work and power* discusses the equations for kinetic and potential energy, and the principle of conservation of energy. There is also a discussion of energy resources and efficiency. Work is described in terms of a relationship between force and distance moved. Similarly, power is described in terms of a relationship between work done and time taken.

The final topic considers how pressure is related to force and area.

Starting points

The Student Book section opener puts the ideas in the section into context and sets the scene. It also allows students to acknowledge and value their prior learning, and provides a benchmark against which future learning can be compared.

The questions provide a structure for introducing the section and can be used in a number of different ways:

- You could ask students to consider the questions as an introductory homework task.
- You could put students into groups to share their own ideas and understanding and then to report back to the whole class.
- Students could be given access to the Internet, preferably with a tight timescale, to find out the information required.

You could then use a spider chart or other form of wall chart to summarise everybody's ideas.

Recording these initial ideas allows you to retain them for reference as the individual topics are developed. In this way, your students' progress in learning can be readily acknowledged.

P1a Physical quantities and measurement techniques

Introduction

In this topic students are introduced to two physical quantities: length and time. They learn about the importance of using measuring instruments accurately. This topic is therefore the basis for most of the practical work during the course, and should be taught at the start of the course. This topic provides essential knowledge for papers 5 and 6 in the practical assessment. The idea of measuring time intervals is fundamental to the measurement of light-years in Topic P6 *Stars and the universe*.

Links to other topics

Sections	**Essential background knowledge**	**Useful links**
P1 Motion, forces and energy		P1b Motion P1c Mass and weight P1d Density P1e Forces P1f Energy, work and power P1g Pressure
P2 Thermal physics		P2b Thermal properties and temperature
P3 Waves		P3a General properties of waves
P5 Space physics		P5b Stars and the Universe

Topic overview

P1a.1	**Length and time** In this learning episode students are taught about the basic techniques of measurement required for the course.
P1a.2	**Consolidation and summary** This learning episode provides an opportunity for a quick recap on the ideas encountered in the topic, as well as time for the students to answer the End of topic questions in the Student Book.

Career links

Many careers linked to physics require knowledge and use of careful measurement techniques. The study of measurement is called Metrology. **Metrologists** work to develop new measurement techniques or instruments, and improve methods used to make measurements.

Learning episode P1a.1 Length and time

Learning objectives

- Describe the use of rulers and measuring cylinders to find a length or a volume.
- Describe how to measure a variety of time intervals using clocks and digital timers.
- Determine an average value for a small distance and for a short interval of time by measuring multiples (including the period of oscillation of a pendulum).

Resources

Student Book Making measurements

Worksheet P1a.1(1) Measuring 1

Worksheet P1a.1(2) Measuring 2

Resources for class practicals (see Technician's notes, following)

Approach

Introduce the learning episode

Ask students to work in pairs to discuss the problems that would occur in building a house if the instructions were vague, such as 'big' and 'small'. Take feedback and discuss the main points that come out.

Develop the learning episode

Introduce the use of the stop clock (or other timer). If possible, ask students to work in small groups to time various activities, such as jumping on the spot ten times. If you can take students outside, ask them to time activities such as running round a given area, etc.

Ask the students how they would measure the thickness of a single sheet of paper. Take answers from the class and discuss the feasibility of the answers. You should explain that measuring the thickness of a single sheet of paper is difficult with just a ruler.

Now ask students to measure the thickness of a pile of paper and count how many sheets of paper there are. They should then divide the thickness of the pile by the number of pages to obtain the thickness of a single sheet of paper. Students could use the inside pages of the Student Book for this, which would mean that they would not have to count individual sheets but can use the page numbers to work out the number of sheets of paper.

Next, ask each group to read how much liquid is in a given measuring cylinder. Make sure that students know that they need to have the meniscus of the liquid at eye level in order to take an accurate reading.

Worksheet P1a.1(1) supports this activity.

SAFETY INFORMATION: Ensure that students do not have any health issues that may prevent them from taking part in physical activity. Discuss with the PE department.

Students should then be shown how to measure small intervals of time using a pendulum. If time allows, they could then try this themselves in pairs or small groups.

Worksheet P1a.1(2) supports this activity.

SAFETY INFORMATION: Check that the mass is secured tightly onto the string, and that the pendulum is secured tightly onto the boss, to prevent the mass falling on someone's foot.

Finish the learning episode

Ask students to work in pairs to produce a summary of what they have learned in this learning episode. Allow a set period of time for them to do this, and then bring the class back together to compare notes. Allow students to add to their summaries in this discussion.

Technician's notes

Be sure to check the latest safety notes on these resources before proceeding.

You will need the following resources for class practical P1a.1(1), per group:

ruler (30 cm marked in mm)
ream of paper or similar
stop clock or other timer
measuring cylinders containing different volumes of liquid

You will need the following resources for class practical P1a.1(2), per group:

pendulum made from piece of string and 100 g mass suspended from a boss on a clamp stand
clamp stand and boss
stop clock or other timer

Learning episode P1a.2 Consolidation and summary

Learning aims

- Review the learning points of the topic summarised in the End of topic checklist
- Test understanding of the topic content by answering the End of topic questions

Resources

Student Book End of topic checklist and End of topic questions

Approach

Introduce the learning episode

Ask students to work with a partner to make a list of key words from this topic. They could then work together to produce a spider diagram showing how the different concepts are linked. They could compare their list with the list of key terms given in the Student Book. Discuss the checklist in the Student Book and ask questions to see how much of the content students are comfortable with.

Students could make flashcards of the key content and then use the flashcards to quiz each other on the information.

Develop the learning episode

Ask students to work individually through the End of topic questions in the Student Book without looking at the text. As they work, walk around the classroom observing their answers and questioning them as necessary to find out which questions are causing difficulties.

Bring the learning episode together

After a set period, ask students to stop working. Discuss any areas of difficulty you observed as you walked round the class. Students should complete any unanswered questions for homework, but you should stress that they should answer the questions without looking at the text, so that they can see how much they have remembered.

P1b Motion

Introduction

In this topic students define speed and velocity, and calculate speed using measurements of distance and time. The concept of speed is used in Section P3 where students learn about wave motion, and, for students studying the supplementary syllabus, in Section P6 for calculating orbital speed. They plot and interpret graphs of motion (speed–time and distance–time graphs). Students studying the supplement topics learn about acceleration as rate of change of velocity, and how to determine if an object is moving with constant or changing acceleration. These students will also learn about the acceleration of free fall near to the Earth. Acceleration is an essential concept for understanding the effect of a resultant force in topic P1e *Forces*.

Links to other topics

Sections	Essential background knowledge	Useful links
P1 Motion, forces and energy	P1a Physical quantities and measurement techniques	P1c Mass and weight P1d Density P1e Forces P1f Energy, work and power P1g Pressure
P3 Waves		P3a General properties of waves
P5 Space physics		P5b Stars and the Universe

Topic overview

P1b.1	**Speed and graphs of motion** In this learning episode students learn how to calculate average speed from total distance travelled and the total time taken. They learn about using distance–time graphs to describe motion, and carry out a practical activity to draw a distance–time graph. Supplement Students will also use a speed–time graph to determine the distance an object has travelled.
Supplement **P1b.2**	**Acceleration** In this learning episode students learn how to define acceleration. They learn how to find acceleration using a speed–time graph. They carry out a practical activity to measure acceleration. They also learn the value of acceleration of free fall, *g*.
P1b.3	**Consolidation and summary** This learning episode provides an opportunity for a quick recap on the ideas encountered in the topic as well as time for the students to answer the End of topic questions in the Student Book.

Career links

Accurate measurement and the prediction of object motion are important in a range of different careers. One area where this is used is in sports and exercise science, where the motion of sports people is analysed to improve their performance. A very different career which also requires understanding of object motion is that of **air traffic controller**, who ensure the safe movement of airplanes in the sky.

Learning episode P1b.1 Speed and graphs of motion

Learning objectives

- Define speed as distance travelled per unit time; recall and use the equation $v = st$.
- Recall and use the equation average speed = total distance travelledtotal time taken.
- Sketch, plot and interpret distance-time and speed-time graphs.
- Know that an object moving with increasing speed is accelerating, and that an object moving with decreasing speed is decelerating.
- Determine, qualitatively from the shape of a distance−time graph when an object is: (a) at rest (b) moving with constant speed.
- Determine, qualitatively from the shape of a speed−time graph when an object is: at rest, moving with constant speed, accelerating, decelerating.
- Calculate speed from the gradient of a straight-line section of a distance–time graph.
- Supplement Calculate the area under a speed−time graph to work out the distance travelled for motion with (a) constant speed (b) constant acceleration.

Common misconceptions

Some students may need to be shown how to follow the line of a distance–time graph from left to right in real time, being aware of the variable plotted on the vertical axis.

Resources

Student Book Calculating speed; Using graphs to study motion; Speed–time graph

Worksheet P1b.1(1) Distance–time data

Worksheet P1b.1(2) Distance–time graphs

Worksheet P1b.1(3) Distance–time graphs and speed–time graphs

Resources for class practicals (see Technician's notes, following)

Approach

Introduce the learning episode

Ask the students what they understand by the word *speed*. Where do they come across this concept in everyday life? Draw a large set of labelled distance–time axes on the board. You do not need to put scales on the axes. Ask a student to describe their journey from home to school that morning. Sketch a graph on the axes to summarise the journey.

Develop the learning episode

Introduce the equation relating speed, distance travelled and time taken. Study the worked examples in the Student Book by setting students the problem to attempt in pairs (without the solution). After a given time, bring the class back together to compare students' solutions with the given answer.

Move on to discuss how distance–time graphs can be used to describe motion. The practical on Worksheet P1b.1(1) could be carried out at this point. Students will line up along a long corridor, playing field, quadrangle or hall.

Each pair of students will have a stop clock and will work out their distance from the start of the line.

At a signal from you, all students start their stop clocks together.

They note the times at which you are in front of them.

You should start from 0 m and walk–run–stand along the line for about a minute.

After the practical is complete, a computer and data projector will simplify data sharing so that students can draw the distance–time graph. They can plot the points for their own graph as they are entered – they do not need to wait until all of the data have been entered.

You should now move onto discuss how speed–time graphs can also be used to describe motion. You can repeat the earlier activity of asking a student to describe their journey to school but this time use axes for a speed–time graph.

Ask students to describe the motion of the bicycle in the example graph in the Student Book. Help them to identify constant speed and when speed is increasing (acceleration). You could use Table P1.2 in the Student Book at this point, or ask students to create their own table.

Use the worked example in the Student Book to show students how to calculate the distance travelled by an object using a speed–time graph.

Students should attempt Worksheet P1b.1(2). Note that the tasks on the worksheet become progressively more demanding. In task 3, students may need some guidance about how to calculate the speed from the graph.

Finish the learning episode

Sketch a distance–time graph on the board and challenge a student to act out the journey that it represents. Allow their peers to criticise/approve the student's interpretation. Repeat as time permits.

Ask students to compare and contrast distance–time and speed–time graphs using Worksheet P1b.1(3) or a similar structure. They should identify what information each graph provides, and what that information can be used for, e.g., a non-zero horizontal line on a distance–time graph represents a stationary object, but a non-zero horizontal line on a speed–time graph represents an object moving at a constant speed.

Students should answer the questions in the Student Book.

Technician's notes

Be sure to check the latest safety notes on these resources before proceeding.

The following resources are needed for the class practical P1b.1(1), per group:

The activity will need to take place in a long corridor, hall, playing field or quadrangle.

stop clock (or other timer) for each pair of students
metre rule for each pair
data projector
computer running a spreadsheet program (e.g., Excel)
graph paper

Supplement Learning episode P1b.2 Acceleration

Learning objectives

- Supplement Define acceleration for an object moving in a straight line as change in speed per unit time; recall and use the equation
  $a = \Delta v \Delta t$.
- Supplement Determine from given data or the shape of a speed–time graph when an object is moving with constant acceleration.
- Supplement Calculate acceleration from the gradient of a straight-line section of a speed–time graph.
- Supplement Know that deceleration is a negative acceleration and use this in calculations.
- Supplement Know that the acceleration of free fall *g* for an object near to the surface of the Earth is approximately constant and is approximately 9.8 m/s^2.

Common misconceptions

Students may need reminding that the beam of infrared light passing through a light gate is invisible.

Resources

Student Book Acceleration; Falling objects

Worksheet P1b.2(1) Finding the speed of a moving object

Worksheet P1b.2(2) Acceleration calculations

Resources for demonstration and class practical (see Technician's notes, following)

Approach

Introduce the learning episode

Talk about a car's speedometer reading 50 km/h and then a few seconds later it displays 70 km/h, and explain that this means the car has *accelerated*. Demonstrate the use of a light gate to measure the speed of a small object passing through it, such as a book (see Technician's notes).

Ask students to calculate the speed of the book from its length and the time recorded by the software. If a data projector is used, you could also teach the class how to use the software. Do not show them how to use the software to calculate speed.

Develop the learning episode

Discuss the text under the heading 'Acceleration' in the Student Book.

Students should try the worked example in the Student Book, without the solution, and then compare their answers to the given solution and discuss any differences.

Students can then carry out the practical on Worksheet P1b.2(1) in small groups. Although the light gate software can calculate a value for *v* directly, it is good for the students to do this themselves.

SAFETY INFORMATION: If this experiment is carried out on a bench or desk, beware of a trolley running off the end of the slope onto the floor or someone's foot!

The next part of the learning episode focuses on speed–time graphs. Discuss the section on speed–time graphs in the Student Book before students answer the questions. Make sure that they understand how to find distance from a velocity–time graph and then ask them to complete Worksheet P1b.2(2).

The questions in the Developing Practical Skills box could be done by students working in pairs. Discussion of the questions is a useful learning tool.

Finish the learning episode

Drop a small ball towards the floor. Ask students to sketch speed–time graphs for the ball as it falls. Some students will show the speed as constant. Point out that it has no speed when it is released. Discuss that the ball's weight makes it accelerate (the relationship between force, mass and acceleration will be covered in Topic P1e). Introduce the acceleration of free fall as the acceleration of an object when the only force acting on it is its weight, due to gravity. If possible, show a time-lapse photo of a ball being dropped. Explain that in the absence of air resistance, all falling objects would have the same acceleration. Near to the Earth, this is 9.8 m/s^2.

Run a short test to see if students can recall the meanings and units of the symbols *u*, *v*, *a* and *t* and the acceleration equation itself.

Technician's notes

Be sure to check the latest safety notes on these resources before proceeding.

The following resources are needed for the acceleration demonstration:

light gate, computer running associated software (to record time object takes to pass through the light gate) and data projector (if available)
book
small ball e.g., golf ball

The following resources are needed for the class practical P1b.2(1), per group:

track and trolley
clamp, sticky tape, card, 30 cm ruler and stop clock (or other timer)
light gate and block
computer

Learning episode P1b.3 Consolidation and summary

Learning aims

- Review the learning points of the topic summarised in the End of topic checklist
- Test understanding of the topic content by answering the End of topic questions

Resources

Student Book End of topic checklist and End of topic questions

Approach

Introduce the learning episode

Ask students to work with a partner to make a list of key words from this topic. They could then work together to produce a spider diagram showing how the different concepts are linked. They could compare their list with the list of key terms given in the Student Book. Discuss the checklist in the Student Book and ask questions to find out how much of the content students are comfortable with.

Students could make flashcards of the key content and then use the flashcards to quiz each other on the information.

Develop the learning episode

Ask students to work individually through the End of topic questions in the Student Book without looking at the text. As they work, walk around the classroom observing their answers and questioning them as necessary to find out which questions are causing difficulties.

In question 7 c), students need to work out the total distance by adding the 200 m and the 160 m together. Work out the total time by adding up the time for each section – they may forget to include the time waiting at the junction.

Bring the learning episode together

After a set period, ask the students to stop working and discuss any areas of difficulty you observed as you walked round the class. Students should complete any unanswered questions for homework, but you should stress that they should attempt the questions without looking at the text, so that they can see how much they have remembered.

P1c Mass and weight

Introduction

In this topic students are introduced to the concept of mass and how it differs from weight. Mass is an essential concept for understanding Newton's second law in the supplement content in Topic P1e *Forces* and when considering potential and kinetic energy in the supplement content in Topic P1f *Energy, work and power*.

Links to other topics

Sections	Essential background knowledge	Useful links
P1 Motion, forces and energy	P1a Physical quantities and measurement techniques P1b Motion	P1d Density P1e Forces P1f Energy, work and power P1g Pressure

Topic overview

P1c.1	**Mass and weight** In this learning episode students are introduced to the concepts of mass and weight and the difference between them. Supplement Students apply the idea of the effect of a gravitational field on a mass to explore weight and the link between gravitational field strength and the acceleration of free fall.
P1c.2	**Consolidation and summary** This learning episode provides an opportunity for a quick recap on the ideas encountered in the topic as well as time for the students to answer the End of topic questions in the Student Book.

Career links

An understanding of the difference between mass and weight is important for careers in the satellite and space industry. For example, the **mechanical engineers** responsible for designing the rovers that are sent to the Moon or Mars have to consider the different gravitational field strengths that the rovers will experience when they leave Earth.

Learning episode P1c.1 Mass and weight

Learning objectives

- Know that mass is a measure of the quantity of matter in an object.
- Know that weight is the gravitational force on an object that has mass.
- Define gravitational field strength, *g*, as the gravitational force per unit mass; recall and use the equation $g = Wm$ and know that near the surface of the Earth g is approximately 9.8 N/kg.
- Supplement Describe, and use the concept of, weight as the effect of a gravitational field on a mass.
- Supplement Know gravitational field strength is equivalent to the acceleration of free fall.

Common misconceptions

The main difficulty is that the term *weight* in physics does not mean the same as the everyday usage.

Students often confuse mass with weight. Use the word *weight* only for forces due to gravity.

Resources

Student Book Mass and weight; The Earth's gravitational field

Worksheet P1.3c.1 Weighing objects

Resources for class practical (see Technician's notes, following)

Approach

Introduce the learning episode

Ask students to discuss in pairs what they understand by *mass* and *weight*. Take feedback. They are likely to define *weight* in its everyday usage: 'weighing' something and getting a quantity in g or kg.

Develop the learning episode

Introduce the meaning of the term *mass* to describe how much material is present in a body. Now explain that *weight* is the force on the object due to gravity.

Use the text 'mass and weight' in the Student Book to introduce gravitational field strength and how to calculate weight. Give students some example values of mass for different objects and ask them to calculate the weight of each mass on Earth.

Discuss how objects are weighed. Use Worksheet P1c.1 to help with this. If you have spring balances you could give students an opportunity to use them to weigh objects around the classroom.

SAFETY INFORMATION: Make sure that students attach any object they weigh securely to the spring balance so that it does not fall on someone's foot. Remind students to not overload spring balances.

Supplement Students should go on to consider the effect of different gravitational fields on the weight of an object. Point out that gravitational field strength is equivalent to the acceleration of free fall. This is because the gravitational force on a unit mass produces an acceleration (from Newton's second law), and in free fall (no other forces acting) acceleration equals the gravitational force on a unit mass.

Finish the learning episode

Ask students to work in pairs to write definitions of mass and weight that could be included in a physics dictionary. After a short time, take feedback and compare ideas.

Technician's notes

Be sure to check the latest safety notes on these resources before proceeding.

The following resources are needed for the class practical P1c.1, per group:

spring balance with a range 0–10 N
selection of objects to weigh (at least five different objects), of different masses up to about 1 kg

Learning episode P1c.2 Consolidation and summary

Learning aims

- Review the learning points of the topic summarised in the End of topic checklist.
- Test understanding of the topic content by answering the End of topic questions.

Resources

Student Book End of topic checklist and End of topic questions

Approach

Introduce the learning episode

Ask students to work with a partner to make a list of key words from this topic, and compare their list with the list of key terms in the Student Book. Discuss the checklist and ask questions to see how much of the content students are comfortable with.

Students could make flashcards of the key content and then use the flashcards to quiz each other on the information.

Develop the learning episode

Ask students to work individually through the End of topic questions in the Student Book without looking at the text. As they work, walk around the classroom observing their answers and questioning them as necessary to find out which questions are causing difficulties.

Bring the learning episode together

After a set period, ask students to stop working and discuss any areas of difficulty you observed as you walked round the class. They should complete any unanswered questions for homework, but you should stress that they should attempt the questions without looking at the text, so that they can see how much they have remembered.

P1d Density

Introduction

In this topic students learn about density and how to determine the density of an object. They need to understand density in order to understand convection in Topic P2c *Transfer of thermal energy*.

Links to other topics

Sections	Essential background knowledge	Useful links
P1 Motion, forces and energy	P1a Physical quantities and measurement techniques P1b Motion P1c Mass and weight	P1e Forces P1f Energy, work and power P1g Pressure
P2 Thermal physics		P2c Transfer of thermal energy

Topic overview

P1d.1	**Density** In this learning episode students learn about density. They learn how to determine the density of a regular object. They learn how to find the density of a liquid and of an irregularly shaped object, and how to use density data to determine whether an object floats or sinks.
P1d.2	**Consolidation and summary** This learning episode provides an opportunity for a quick recap on the ideas encountered in the topic as well as time for the students to answer the End of topic questions in the Student Book.

Career Links

An understanding of density, particularly of liquids, is needed in careers such as chemical engineering and food technologists. As part of process design for creating new products, chemical engineers and food technologists both have to understand how food and liquids move through pipes and equipment. They also need to know what happens when different liquids are mixed together.

Learning episode P1d.1 Density

Learning objectives

- Define density as mass per unit volume; recall and use the equation $\rho = mV$
- Describe how to determine the density of a liquid, of a regularly shaped solid and of an irregularly shaped solid which sinks in a liquid (volume by displacement), including appropriate calculations
- Determine whether an object floats or sinks based on density data

Resources

Student Book Density

Worksheet P1d.1 Finding the density of an irregularly shaped object

Resources for demonstration and class practical (see Technician's notes, following)

Approach

Introduce the learning episode

Ask students which is heavier: a tonne of feathers or a tonne of iron? This is, of course, a trick question. Discuss why the feathers would take up more space if both quantities were loaded onto a truck. Lead them to the need for ideas about density.

As a possible extension, demonstrate to students how to pour a density tower (see Technician's notes). In pairs, ask them to suggest why the liquids are in layers, encourage them to use the terms density, more dense and less dense in their explanations.

Develop the learning episode

Help students to realise that density is defined as mass per unit volume. Students should tackle the worked example in the Student Book without the given solution. After a given time, the given solution should be compared with students' solutions and the differences discussed.

The Developing Practical Skills box in the Student Book introduces students to a method of determining the density of regular objects. They should work through the questions and after a set period of time, bring the class back together to discuss the answers to the questions. You could demonstrate how to find the density of a liquid using the method described in the Student Book. You should also discuss the idea that an object which is less dense than a liquid it is placed in will float, but an object that is more dense than a liquid it is placed in will sink. Students should use the data in the table in the Student Book to determine if gold, aluminium, plastic, and wood will float on water.

Students could then discuss the Science in Context box in the Student Book. At this point, students should tackle the questions individually.

Students should use the method described in the Student Book to determine the density of an irregularly shaped object. Instructions are also given on Worksheet P1d.1. If you do not have sufficient equipment for students to do this themselves, then do it as a demonstration (see Technician's notes) but involve students at every stage using suitable questions. Students should then tackle the worked example in the Student Book using the technique described above for worked examples.

SAFETY INFORMATION: Water should be used for the investigation.

Finish the learning episode

Ask students to list three things they have learned in the learning episode and one thing that they feel they need to get more information about or need further practice with. Take feedback and use this in your planning.

Technician's notes

Be sure to check the latest safety notes on these resources before proceeding.

The following resources are needed for the measuring density of a liquid demonstration:

Tall glass or measuring cylinder
Liquids of different density* e.g., vegetable oil, honey, washing up liquid.
*If liquids of different density are not available, you can create your own using different amounts of salt or sugar dissolved in water. Use food colouring in each different density solution to enable students to see the difference between the layers.

Starting with the most dense liquid, carefully pour the liquids into the glass.

You might find it helpful to pour each liquid slowly over the back of a spoon into the glass to avoid mixing of the layers.

You may need to practice pouring the column to obtain clear layers.

The following resources are needed for the measuring density of a liquid demonstration:

balance
measuring cylinders
liquids e.g., vegetable oil, glycerol
small objects that will fit into the measuring cylinder and that will either float or sink in the liquid (e.g., cork, stone)

The following resources are needed for the class practical P1d.1, per group:

irregularly shaped object
balance
measuring cylinder wide and deep enough to hold object

Learning episode P1d.2 Consolidation and summary

Learning aims

- Review the learning points of the topic summarised in the End of topic checklist
- Test understanding of the topic content by answering the End of topic questions

Resources

Student Book End of topic checklist and End of topic questions

Approach

Introduce the learning episode

Ask students to work with a partner to make a list of key words from this topic. They could then work together to produce a spider diagram showing how the different concepts are linked. They could compare their list with the list of key terms given in the Student Book. Discuss the checklist and ask questions to see how much of the content students are comfortable with.

Students could make flashcards of the key content and then use the flashcards to quiz each other on the information.

Develop the learning episode

Ask students to work individually through the End of topic questions in the Student Book without looking at the text. As they work, walk around the classroom observing their answers and questioning them as necessary to find out which questions are causing difficulties.

Bring the learning episode together

After a set period, ask the students to stop working and discuss any areas of difficulty you observed as you walked round the class. Students should complete any unanswered questions for homework, but you should stress that they should attempt the questions without looking at the text, so that they can see how much they have remembered.

P1e Forces

Introduction

In this topic students are introduced to forces and their effects, balanced and unbalanced forces and the effects of friction. Students learn how to calculate the resultant of two or more forces acting along the same straight line and describe the effect on the motion of a body. Students following the supplementary syllabus explore the link between force and acceleration.

Links to other topics

Sections	Essential background knowledge	Useful links
P1 Motion, forces and energy	P1a Physical quantities and measurement techniques P1b Motion P1c Mass and weight P1d Density	P1f Energy, work and power P1g Pressure

Topic overview

P1e.1	**Introducing forces** In this learning episode students are introduced to the effects of forces. They learn about the types of force and about friction.
P1e.2	**Resultant forces** In this learning episode students learn how to calculate the resultant of two or more forces acting along the same straight line, and identify their effect on the motion of an object.
Supplement **P1e.3**	**Forces and acceleration** In this learning episode students learn about the relationship between force, mass and acceleration.
P1e.4	**Consolidation and summary** This learning episode provides an opportunity for a quick recap on the ideas encountered in the topic as well as time for the students to answer the End of topic questions in the Student Book.

Career links

An understanding of forces underpins careers in many different sectors. For example, an understanding of resultant forces is used by aeronautical engineers and vehicle designers to increase the efficiency of their vehicles by minimising drag forces.

Learning episode P1e.1 Introducing forces

Learning objectives

- Know that forces may produce changes in the size, shape and motion of an object
- Describe friction as the force between two surfaces that may impede motion and produce heating
- Know that friction (drag) acts on an object moving through a liquid
- Know that friction (drag) acts on an object moving through a gas (e.g., air resistance)
- Know that an object either remains at rest or continues in a straight line at constant speed unless there is a resultant force on the object

Common misconceptions

Students are introduced to the concept of vectors acting along the same straight line. The idea that if forces and velocities in one direction are positive, then forces in the opposite direction are negative does not come easily to many students.

Students often don't recognise the effect of friction and air resistance when thinking about the motion of an object.

Resources

Student Book What are forces?

Student Book What is friction?

Worksheet P1e.1 Investigating forces

Resources for demonstration and class practical (see Technician's notes, following)

Approach

Note to teacher

To give students a full picture of the effect of forces, and to allow for a consideration of friction, students should be told that forces can change the size, shape or motion of an object. Fuller consideration of resultant forces will be covered in learning episode P1e.2

Introduce the learning episode

Give pairs of students a small ball of modelling clay. Using Fig.1.23 in the Student Book, ask them to apply the forces to their modelling clay and describe the changes to the shape, size and motion.

SAFETY INFORMATION: Ensure students do not throw the modelling clay, and collect it all back in at the end of the learning episode.

Develop the learning episode

Using the Student Book, discuss the nature of friction and how it is sometimes useful and sometimes something which is a disadvantage. Ask students to work in pairs to give examples of situations where solid friction is experienced and examples where drag is experienced. Students could also discuss the questions in the Student Book in pairs.

Next students can carry out the practical activity on Worksheet P1e.1. Introduce a force meter as a device which uses the length of a spring to measure a force. Show them how to set the force meter to zero and take readings from it. If you do not have enough equipment, this could be done as a demonstration (see Technician's notes), but make sure that students are fully involved by asking questions and involving them in the activity.

The practical gives students some practice in thinking about some of the forces acting on an object on a ramp (friction, weight and pull). The investigation can expand or contract to fit the time available and students can repeat their measurements using a different surface on the ramp if there is time.

Students may be interested to find that the frictional force is proportional to the mass of the object when the ramp is horizontal. There is no such easy pattern as the slope of the ramp is increased; the pulling force must become the weight of the sled when the ramp is vertical.

Finish the learning episode

Discuss the results of the practical. Ask students to tell you two things that they thought worked well and one thing that they would change if they had a chance to repeat the practical.

Ask students to make a summary of the main points of the learning episode.

Technician's notes

Be sure to check the latest safety notes on these resources before proceeding.

The following resources are needed for the demonstration in the introduction:

Modelling clay: sufficient for a small ball (about 2 cm diameter) for each pair of students in the class.

The following resources are needed for the class practical P1e.1.1, per group:

force meter: 0–20 N range
ramp, blocks of different sizes to raise one end of ramp to different levels
sled
five 100 g slotted masses
metre ruler
string loop
balance capable of measuring up to 2 kg
(optional) Different surface coverings to add to the ramp e.g., sandpaper, carpet, plastic

A sled which has a raised lip at the back will stop masses sliding off when the ramp is steep.

Beware of the sled falling off the end of the slope onto anyone's feet.

Learning episode P1e.2 Resultant forces

Learning objectives

- Determine the resultant of two or more forces acting along the same straight line
- Know that an object either remains at rest or continues in a straight line at constant speed unless there is a resultant force on the object

Common misconceptions

Many students think that a force is needed to keep an object moving with a constant speed. Students often do not recognise friction as a force acting on an object.

Resources

Student Book Adding forces; Balanced forces; Unbalanced forces

Worksheet P1e.2(1) Resultant forces

Worksheet P1e.2(2) Forces and motion

Approach

Introduce the learning episode

Introduce students to the concept of vectors and that forces are vectors. You could demonstrate applying a force to an object in different directions to show them that the direction of the force, as well as the magnitude, is important.

Develop the learning episode

Discuss the section 'Adding forces' in the Student Book. Go through the worked examples. Give more simple examples of pairs of forces acting on an object for students to solve. Students should then work through the questions on Worksheet P1e.2(1) to practise finding the resultant force.

Now discuss with students how the motion of an object will change if there is a resultant force acting on it e.g., speeding up, slowing down or changing direction.

If possible, demonstrate or show examples of the motion of an object when there is (almost) no friction (see Technician's notes). This could be done using a video of an ice-hockey puck moving along the ice, with a toy hovercraft or hover ball. Explain to students that if there is no resultant force an object will continue moving with a steady speed. You could use the phrase 'some students think that a force is needed to keep an object moving at a steady speed, but they are not correct'.

Use Worksheet P1e.2(2) to explore the idea of the motion of objects when there is no resultant force.

Students could answer the questions in the Student Book.

Finish the learning episode

Ask students to summarise the main points of this learning episode in a form of their choice.

Supplement Learning episode P1e.3 Forces and acceleration

Learning objective

- Supplement Recall and use the equation $F = ma$ and know that the resultant force and the acceleration are in the same direction

Common misconceptions

Students often confuse mass with weight. Use the word *weight* only for forces due to gravity.

Students have a lot of difficulty in accepting that a resultant force of zero does not automatically result in a velocity of zero.

Resources

Student Book How are mass, force and acceleration related?

Supplement Worksheet P1e.3 Force and acceleration

Resources for demonstrations and class practical (see Technician's notes, following)

Approach

Introduce the learning episode

Release a small ball from above your head. Ask your students why it *accelerates* as it moves towards the floor and *decelerates* on its way up again. They may suggest that there is a downward force (gravity) acting on the ball when it is moving both downwards and upwards.

Ask them to suggest what happens to the acceleration as the force on the ball and its mass are changed (but don't try to demonstrate it).

Develop the learning episode

Show students how to use a light gate and cardboard comb to measure acceleration before they do the practical in Worksheet P1e.3. They will need to operate as a team in groups of four to get reasonable results, so you may prefer to run this as a whole class demonstration (see Technician's notes).

SAFETY INFORMATION: Students need to be aware of the hazard of masses and/or the trolley coming off the end of the track and landing on someone's feet.

There will probably not be enough time to obtain more than one change in each value of mass and force, so agreement with the accepted formulation of Newton's second law will be only approximate.

Discuss the section 'How are mass, force and acceleration related?' in the Student Book. The worked examples should be worked out as described previously.

Use the equation $a = Fm$ to reinforce students understanding that a resultant force is required to acceleration (or change the motion) of an object. Therefore, if there is no resultant force, the object will continue moving at a steady speed. Remind them that friction forces act to slow down the motion of objects.

Give students the following table and ask them to work out the missing quantities. Encourage them to write down all the steps of their working so that they can find out where they go wrong. Some students may wonder about the vertical forces acting on the objects considered in the practical and the questions. Reassure them that they do not need to worry about these at this stage.

Question	Driving force / N	Friction / N	Mass / kg	Resultant force / N	Acceleration / m/s^2
a	1 500		500		+2.5
b	500	200	600		
c	3 000	500			+1.25
d	400	1000	800		

Students can then answer the questions in the Student Book.

Finish the learning episode

Use the board to present students with a variety of forms of transport – cars, bikes, lorries, trains and planes. For each, give suggested values for mass and times to get to 60 km/h from a standing start – students might be able to help with the latter. Challenge students to rank the forms of transport by size of resultant force.

Hand out whiteboards and marker pens (or students can use pen and paper). Set them a number of questions similar to the ones that they answered in the main part of the learning episode. This should allow you to assess their progress in using $F = ma$.

Be sure to check the latest safety notes on these resources before proceeding.

The following resources are needed for the demonstration in the introduction:

small ball such as a golf ball and hard surface such as the ground

The following resources are needed for the class practical P1e.3, per group:

If you do not have light gates and data-logging equipment, students can use a stop clock instead

strong thread, about 1 m length; track
pulley wheel, about 50 mm diameter; 6 × 10 g slotted masses
light gate, cardboard comb for light gate and sticky tape to hold comb in place
ruler to measure comb teeth length
computer running appropriate software
access to scales; force meter (0–1 N)

The following resources are needed for finishing the activity, per group:

mini-whiteboard and marker pen OR sheets of paper

Learning episode P1e.4 Consolidation and summary

Learning aims

- Review the learning points of the topic summarised in the End of topic checklist
- Test understanding of the topic content by answering the End of topic questions

Resources

Student Book End of topic checklist and End of topic questions

Approach

Introduce the learning episode

Ask students to work with a partner to make a list of key words from this topic. They could then work together to produce a spider diagram showing how the different concepts are linked. They could compare their list with the list of key terms given in the Student Book. Discuss the checklist in the Student Book and ask questions to see how much of the content students are comfortable with.

Students could make flashcards of the key content and then use the flashcards to quiz each other on the information.

Develop the learning episode

Ask students to work individually to work through the End of topic questions in the Student Book without looking at the text. As they work, walk around the classroom observing their answers and questioning them as necessary to find out which questions are causing difficulties.

Bring the learning episode together

After a set period, ask the students to stop working and discuss any areas of difficulty you observed as you walked round the class. Students should complete any unanswered questions for homework, but you should stress that they should attempt the questions without the text, so that they can see how much they have remembered.

P1f Energy, work and power

Introduction

In this topic students learn about energy stores and transfers. They learn that energy is conserved. A variety of energy resources are considered, including how electrical power may be generated from these resources. Students learn more about the generation of electricity in Section P4e *Electromagnetic effects*. They consider efficiency and learn the definitions of 'work' and 'power'.

Links to other topics

Sections	Essential background knowledge	Useful links
P1 Motion, forces and energy	P1a Physical quantities and measurement techniques P1b Motion P1c Mass and weight P1d Density P1e Forces	P1g Pressure
P2 Thermal physics		P2c Transfer of thermal energy

Topic overview

P1f.1	**Energy stores and energy transfers** In this learning episode students are introduced to energy stores and consider how energy transfers between stores. They learn about the principle of conservation of energy. Supplement Students learn the equations for calculating kinetic energy and change in gravitational potential energy.
P1f.2	**Energy resources** In this learning episode students learn about renewable and non-renewable energy resources. Supplement Students learn that energy released from the Sun is caused by nuclear fusion reactions, and that energy released in nuclear reactors is caused by nuclear fission.
P1f.3	**Work** In this learning episode students learn about the definition of work in physics.
P1f.4	**Power** In this learning episode students learn about the definition of power in physics.
P1f.5	**Efficiency** In this learning episode students learn about efficiency of energy transfer. Supplement Students learn how to calculate the efficiency of an energy transfer.
P1f.6	**Consolidation and summary** This learning episode provides an opportunity for a quick recap on the ideas encountered in the topic as well as time for the students to answer the End of topic questions in the Student Book.

Career links

Careers in the energy and utilities sector require an understanding of the concepts in this topic. There are jobs in both renewable and non-renewable energy sectors. Some possible careers include nuclear engineers and nuclear operating technicians who work in designing and running nuclear power plants; wind turbine service technicians who optimise the output capacity and safe working of wind turbines; and solar design engineers plan how best to develop infrastructure for efficient inclusion of solar panels into existing electricity grids.

Learning episode P1f.1 Energy stores and energy transfers

Learning objectives

- State that energy may be stored as kinetic, gravitational potential, chemical, elastic (strain), nuclear, electrostatic and internal (thermal)
- Describe how energy is transferred between stores during events and processes, including examples of transfer by forces (mechanical work done), electrical currents (electrical work done), heating, and by electromagnetic, sound and other waves
- Know the principle of the conservation of energy and apply this principle to simple examples including the interpretation of simple flow diagrams (Sankey diagrams are **not** required)
- Supplement Recall and use the equation for kinetic energy $E_k = \frac{1}{2}mv^2$
- Supplement Recall and use the equation for the change in gravitational potential energy $\Delta E_p = mg\Delta h$

Common misconceptions

Students may have developed an understanding of energy from the common uses of the term, which does not match to how the term is used in physics. There is no easy definition of energy which can be used with students, so you will need to develop the concept carefully. Emphasise that energy does not explain 'why' things happen. Rather, energy is a method of calculating if activities are possible. Each energy store introduced in this learning episode has an associated calculation. Students will be introduced to equations for two of the energy stores. If they study physics beyond age 16 then they will meet more equations linked to energy stores.

The Institute of Physics has produced teacher resources which explore misconceptions in energy in more detail.

Resources

Student Book Energy stores; Energy transfers

Worksheet P1f.1 Pendulum swing

Resources for demonstrations and class practical (see Technician's notes, following)

Approach

Introduce the learning episode

In pairs, ask students to suggest how we could answer the following questions:

"If we mix a cold liquid with a hot liquid what will the final temperature be?"

"How many sweets do I need to climb a mountain?"

"How much electricity can renewable energy provide?"

"How many tonnes of plankton does a megamouth shark need to eat to survive?"

Obtain feedback from the students. Point out that to answer these questions we need to do energy calculations. Explain that energy isn't a physical quantity but is a concept that helps us to do calculations in the real world.

Develop the learning episode

Using Table P1.4 'Energy stores' in the Student Book, introduce the idea of different energy stores. For each store, ask students to write down another example (some may be easier than others to think of, so you may need to provide suggestions).

Introduce the idea of energy transfers as a way for energy to move between different stores.

Demonstrate the four different pathways that transfer energy: mechanical work done, electrical work done; heating; and waves. Suggestions are given in the Technician's notes. In each case, choose a clear start and end point and identify the energy store depending on what you might like to calculate in each case. Note that only students studying the supplement material will need to do calculations, but it is useful for students to know that is why we are interested in energy transfers. Do not use 'chains of energy stores' in your explanation as we do not need to calculate the intermediate values. During your demonstration point out the mechanism that is causing the change (e.g., application of a force in mechanical working).

Supplement Introduce the equations for gravitational potential energy and kinetic energy using the text in the Student Book. Students should do the worked examples without the solutions, and then compare their answers with the given solutions. Discuss any differences. You may wish to provide students with further practice questions to answer.

Now introduce the principle of conservation of energy.

Demonstrate a swinging pendulum to introduce the energy transfers between kinetic and gravitational potential stores (see Technician's notes). You could ask students to copy Fig. 1.33 in the Student Book and annotate their picture with a description of the transfers between energy stores. Help them to identify the pathway as mechanical working and that the force due to gravity is the mechanism by which the transfer takes place.

Point out to students that the bottle does not go higher than the initial release point. The energy is conserved.

Supplement Use the worked example in the Student Book to introduce calculations using gravitational potential energy and kinetic energy. Students could carry out the practical on Worksheet P1f.1 at this point.

Use the text in the Student Book on the Energy flow diagram to show how energy transferred between stores can be represented. Ask students to draw their own energy flow diagram for the swinging pendulum.

Note: When considering everyday energy transfers, some energy almost always appears to go missing when the initial and final energy was calculated. Tracking down this 'missing energy' was cutting-edge science 200 years ago. It has been shown over and over again that when calculations of conservation of energy don't balance, energy has been transferred to internal (thermal) stores, the random kinetic energy of atoms and molecules. In mechanical systems, this is often due to friction.

Finish the learning episode

A lot of content has been covered in this learning episode. It would be helpful if students worked with a partner to make a summary of the main points. Bring the class back together after a given period and discuss summaries. Allow students to add to their summaries during the discussion.

Technician's notes

Be sure to check the latest safety notes on these resources before proceeding.

The following resources are needed for the demonstration on energy transfers:

A number of situations that can be used to show energy transfers, e.g.:

Mechanical: ball/toy rolling down a wooden slope [GPE to kinetic OR GPE to thermal].
Electrical: circuit including battery and lamp [chemical to internal (thermal)]
Heating: candle (or other heat source) warming a container of water [chemical to internal (thermal)]
Waves: infrared heater [chemical (in battery or power station) to internal (thermal)]

The following resources are needed for the demonstration on conservation of energy:

A simple pendulum should be made.
One way is to use the following equipment:
string
filled bottle (a 0.5-litre soft drinks bottle filled with sand or water is suitable)
G-Clamp

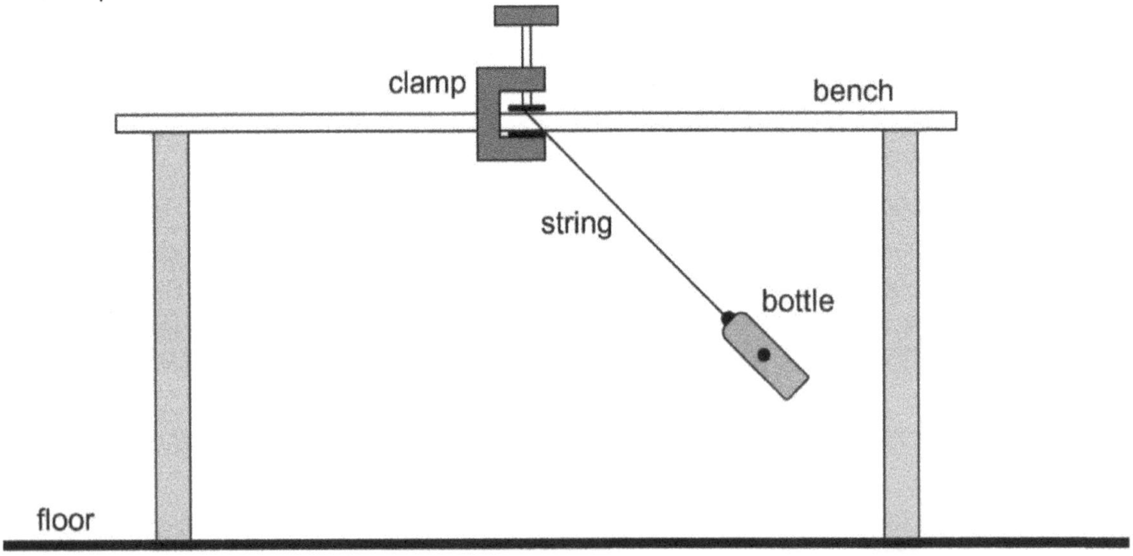

The following resources are needed for the class practical P1f.1 per group:

string
filled bottle (a 0.5-litre soft drinks bottle filled with sand or water is suitable)
G-clamp
light gate
computer
metre rule and set square
scales (should be capable of measuring up to 1 kg) and force meter (should have a range up to 10 N)

Learning episode P1f.2 Energy resources

Learning objectives

- Describe how useful energy may be obtained, or electrical power generated, from: (a) fossil fuels; (b) biofuels; (c) water, including waves, tides, and hydroelectric dams; (d) geothermal resources; (e) nuclear fission; (f) light from the Sun (solar cells); (g) infrared and other electromagnetic waves from the Sun to heat water (solar thermal collectors); (h) wind (wind turbines) including references to a boiler, turbine and generator where they are used
- Supplement Know that radiation from the Sun in the main source of energy for all our energy resources except geothermal, nuclear and tidal
- Supplement Know that energy is released by nuclear fusion in the Sun (detailed knowledge of the process of fusion is **not** required)
- Supplement Know that energy is released by nuclear fission in nuclear reactors (detailed knowledge of the process of fission is **not** required)

Common misconceptions

Some students may think of energy stores as being the same thing as energy resources (fuels). Make sure that you distinguish between the fuel (energy resources) and the energy store associated with it. Although beyond the syllabus, it is worth remembering that the chemical store of energy is filled by chemical reactions such as combustion of a biofuel with oxygen to produce carbon dioxide and water, rather than the energy resource on its own. Although we talk about renewable energy resources, there is no such thing as an energy resource that will last forever. However, in practice, energy resources that rely on the Sun or Earth's core will last at least a billion years, much longer than those that don't.

Resources

Student Book Energy resources

Resources for demonstration and class practical (see Technician's notes, following)

Approach

Introduce the learning episode

Ask students to explain why coal-fired power stations will not be used in the future but wood-fired ones will.

Use a range of items (see list in Technician's notes below) to explain the meaning of *renewable* in the context of energy sources – track each one back to energy transferred from the Sun.

Develop the learning episode

Discuss the section on fossil fuels in the Student Book. Use Fig. 1.36 to explain how power stations use a boiler, turbines and generate to generate electrical power. You should also describe how the fuel used in the boiler can be renewable if it takes the form of biofuels such as wood and ethanol.

Discuss why renewable resources are becoming increasingly important.

Demonstrate a simple wind turbine made out of a motor and a propeller, and show how its output depends upon how fast it spins (see Technician's notes).

Students should then use the Developing practical skills box in the Student Book to plan an investigation into how the output of a simple wind turbine varies with wind speed. If time allows, students could work in groups to carry out their investigation.

SAFETY INFORMATION: Make sure that you check students' plans before they carry out the investigation. Check that students have identified any risks and how they will control them in their plans.

Move on to discuss other renewable resources using of the text in the Student Book. Discuss how renewable resources are used in your country and why there may be opposition to some schemes (hydroelectric power schemes and their effect on local wildlife, for example).

Supplement Discuss with students that nuclear fission releases energy.

Supplement Use the text in the Student Book to introduce the link to the Sun for the majority of energy resources on Earth. Discuss with students how nuclear fusion releases energy in the Sun.

Finish the learning episode

Discuss the questions in the Student Book to check understanding.

Technician's notes

Be sure to check the latest safety notes on these resources before proceeding.

The following resources are needed for the introduction to the activity:

calculator with solar panel
yeast fermenting sugar solution
spirit burner containing ethanol
wooden splint and green plant

SAFETY INFORMATION: Ethanol is HIGHLY FLAMMABLE

The following resources are needed for the demonstration of a simple wind turbine (or per group if carried out as a class practical):

desk top fan
small electric motor to act as dynamo and plastic propeller to fit onto motor shaft
voltmeter 0–50 mV (preferably moving coil) and anemometer
leads and crocodile clips to connect motor to meter
clamp, stand and boss to support motor; metre rule

Desk top fans can be shared between several groups.

Learning episode P1f.3 Work

Learning objectives

- Understand that mechanical or electrical work done is equal to the energy transferred
- Recall and use the equation for mechanical working $W = Fd = \Delta E$

Common misconceptions

Work has a special and precise meaning in physics. Students need to realise that for work to be done, the distance moved must be in the same direction as the force, and that both force and distance are required for an energy transfer to take place.

Resources

Student Book Work

Worksheet P1f.3 Investigating work done

Resources for demonstration and class practical (see Technician's notes, following)

Approach

Introduce the learning episode

Invite a student to stand against a wall and push very hard against it. Explain to the class that although the student is getting tired, they aren't doing any work because the wall is not moving.

Ask for a volunteer. Hold one end of a long length of rubber tubing and invite the student to take hold of the other end and extend it as far as they can. Ask the class if an energy transfer has taken place. Explain that the student has applied a force and moved against it, so work has been done. Be careful to release the tubing safely.

SAFETY INFORMATION: Both you and the volunteer, and all observers, need to wear eye protection. Check tubing for any cracks.

Develop the learning episode

Discuss *work* and explain that it has one specific meaning in physics and a wide range of different meanings in everyday life. Use the section 'Work' in the Student Book to help with this. Students should do the worked examples in the Student Book without the solutions, and then compare their answers with the given solutions. Discuss any differences.

At this point, if time allows, students could carry out the practical given on Worksheet P1f.3. The practical work is a useful test of the idea of energy conservation. Within the random and systematic errors prevalent in practical work at this level, the work done pulling the trolley along the ramp and straight up into the air should be the same, as there should be almost no friction. The string is necessary because the force meter only works correctly when it is vertical.

Note that in Worksheet P1f.3 the calculation of gravitational potential energy is only required by students following the supplement material, and so the worksheet does not expect students to recall the equation. You may wish to provide the values for g.p.e directly to your students if that is appropriate.

Finish the learning episode

In pairs ask students to give examples of everyday situations where work is being done, and also where work is not being done. Use this to reinforce their understanding of the physics meaning of the term. Students should then complete the questions in the Student Book.

Technician's notes

Be sure to check the latest safety notes on these resources before proceeding.

The following resources are needed for the demonstrations in the introduction and at the end:

eye protection for each person in the class
0.5 metre length of rubber tubing, e.g., about 5 mm diameter; heavy weight (for student to hold up)

Supplement **The following resources are needed for practical P1f.3, per group:**

dynamics trolley and trolley track
clamp and stand to raise one end of track
metre rule and force meter (0–10 N)
pulley wheel on end of track
strong thread or string

Students will measure the force required to pull the trolley along the track when one end has been raised above the bench and compare this with the force required to pull the trolley directly up into the air.

Learning episode P1f.4 Power

Learning objectives

- Define power as work done per unit time and also as energy transferred per unit time
- Recall and use the equations:

  (a) $P = Wt$ (b) $P = E\ t$

Common misconceptions

Like *work*, *power* has a precise meaning in science and a much looser meaning in everyday life. Not only is it used to describe things that have nothing to do with energy transfer, but it is also often used to describe a force.

Resources

Student Book Power

Worksheet P1f.4(1) Personal power

Worksheet P1f.4(2) Power calculations

Resources for demonstration and class practical (see Technician's notes, following)

Approach

Introduce the learning episode

Demonstrate the use of an electric motor to raise a mass up from the floor to bench height. Introduce the idea of power and make a quick measurement of its value for the motor (see Technician's notes).

Ask for a volunteer who can lift weights. Get them to show what they can do with dumbbells. Discuss the distinction between their power and the force they can exert.

Develop the learning episode

Using the section 'Power' in the Student Book, introduce the students to power.

Students can try the worked examples in the Student Book, as before, they should attempt to work them out without solutions first. Students need to be familiar with the idea of calculating work done followed by power in order to understand the following practical work.

If time allows, they can then carry out the practical on Worksheet P1f.4(1). Consider doing this activity in the school gym, and contact the PE department for support. Some students can be very competitive when measuring their power. It is important that they don't damage themselves by over-exertion. Some will be unwilling to take part in physical activity – allow them to borrow results from another student.

SAFETY INFORMATION: Take care that masses are not dropped on feet, and that students don't try to lift too heavy a mass.

Students can then do the questions on Worksheet P1f.4(2). For question 2, encourage students to show all the steps in their working. Each question involves two or three steps, so it is easy to get lost if they try to do them with just a calculator.

Finish the learning episode

Ask students to give examples of the correct use of the terms energy, work and power. Then give them some incorrect examples (possibly taken from the media e.g., 'The most powerful person in the world') and ask them to explain why the examples are incorrect.

Students should complete the questions in the Student Book.

Technician's notes

Be sure to check the latest safety notes on these resources before proceeding.

The following resources are needed for the demonstrations in the introduction:

12 V d.c. electric laboratory motor; 12 V d.c. power supply
thread
mass hanger and two 100 g slotted masses
metre rule and stop clock
set of dumbbells

For the first starter suggestion the thread must be fastened to the motor shaft so that the mass rises as the motor turns.

The shaft of a large motor will turn slower, making it easier to time the rise of the mass.

The following resources are needed for the class practical P1f.4(1), per group

access to bathroom scales (N or kg)
metre rule and stop clock (or other timing device)
box or similar for students to step on to
20 g mass hanger and three 20 g slotted masses
thread to tie mass to finger

Do not use too large masses.

Learning episode P1f.5 Efficiency

Learning objectives

- Understand, qualitatively, the concept of efficiency of energy transfer
- Supplement Define efficiency as:
  a) (%) efficiency = (useful energy output)(total energy input) × (100%)

  b) (%) efficiency = (useful power output)(total power input) × (100%)
- Supplement Recall and use these equations

Common misconceptions

Efficiency is another term that has a different meaning in everyday use and in science. Some students will need a lot of convincing that when energy appears to 'go missing' in calculations, this is because some heating of the environment happens and energy is transferred into internal (thermal) stores.

Resources

Student Book Efficiency

Worksheet P1f.5(1) Heater efficiency

Supplement Worksheet P1f.5(2) Efficiency calculations

Resources for class practical (see Technician's notes, following)

Approach

Introduce the learning episode

Remind students what they know about energy transfers and conservation. Ask them what they understand by the word 'efficiency'. Give an example of the light bulb, where only some of the energy input is transferred as useful energy (light) and some is transferred as heat to the surroundings. Display an EU rating chart from an electrical appliance such as a lamp, washing machine or car. Ask questions to establish why it is good to increase energy efficiency [energy is expensive and limited].

Develop the learning episode

Demonstrate the practical on Worksheet P1f.5(1) to discuss energy transfers and efficiency (see Technician's notes). Note that the numerical calculations are not required for students studying the Core material only. For these students, discuss qualitatively what energy transfers are taking place, and the efficiency of the heater. Actively involve student in the demonstration, asking them to predict the findings and help them to interpret the results.

Supplement If you have enough equipment, students can work in small groups to carry out the practical on Worksheet P1f.5(1). They should also carry out the calculation using the results in the 'Analyse and interpret results' section. Some students may need support to do this.

SAFETY INFORMATION: Make sure that students DO NOT completely immerse the heater in the water, and that the cable connection must be above the water at all times. Remind students that the heater should not be operated when it is not in the water.

Supplement Introduce the equations for efficiency in the Student Book. Students should answer the questions on Worksheet P1f.5(2).

Finish the learning episode

Discuss the practical activity from this lesson. Did students encounter any difficulties? If so, how did they overcome them?

Technician's notes

Be sure to check the latest safety notes on these resources before proceeding.

The following resources are needed for the class practical P1f.5(1), per group:

12 V 50 W electrical heater (inspected for damage around the cable entry points to avoid any trapped water boiling and pressurising inside the heater)
ammeter (0 to 5 A); voltmeter (0 to 12 V at least); power pack (max 12 V)
three leads; 250 cm^3 glass beaker
thermometer
clamp, stand and boss
access to balance and stop clocks
insulation materials (bubble wrap, cotton wool, cloth, paper)
scissors and sticky tape

SAFETY INFORMATION: Students should not completely immerse the heater in the water. The cable connection must be above the water at all times. The heater should not be operated when it is not in the water.

Learning episode P1f.6 Consolidation and summary

Learning aims

- Review the learning points of the topic summarised in the End of topic checklist
- Test understanding of the topic content by answering the End of topic questions

Resources

Student Book End of topic checklist and End of topic questions

Approach

Introduce the learning episode

Ask students to work with a partner to make a list of key words from this topic. They could then work together to produce a spider diagram showing how the different concepts are linked. They could compare their list with the list of key terms given in the Student Book. Discuss the checklist in the Student Book and use questioning to see how much of the content students are comfortable with.

Students could make flashcards of the key content and then use the flashcards to quiz each other on the information.

Develop the learning episode

Ask students to work individually through the End of topic questions in the Student Book without looking at the text. As they work, walk around the classroom observing their answers and asking questions as necessary to find out which questions are causing difficulties.

Finish the learning episode

After a set period, ask the students to stop working and discuss any areas of difficulty you observed as you walked round the class. Students should complete any unanswered questions for homework, but you should stress that they should attempt the questions without looking at the text, so that they can see how much they have remembered.

P1g Pressure

Introduction

In this topic students are introduced to pressure as the ratio of force to area. They will use appropriate examples to relate pressure beneath a liquid surface to depth and density. They will learn more about the pressure in a gas in Topic P2a *Kinetic particle model of matter.*

Links to other topics

Sections	Essential background knowledge	Useful links
P1 Motion, forces and energy	P1a Physical quantities and measurement techniques P1b Motion P1c Mass and weight P1d Density P1e Forces	
P2 Thermal physics		P2a Kinetic particle model of matter

Topic overview

P1g.1	**Pressure, force and area** In this learning episode students explore the relationship between pressure, force and area.
P1g.2	**Consolidation and summary** This learning episode provides an opportunity for a quick recap on the ideas encountered in the topic as well as time for the students to answer the End of topic questions in the Student Book.

Career links

Examples of careers that requires an understanding of the concepts of pressure are plumbers who fit piping and heating systems in buildings.

Learning episode P1g.1 Pressure, force and area

Learning objectives

- Describe how pressure varies with force and area in the context of everyday examples
- Define pressure as force per unit area; recall and use the equation $p = FA$

Common misconceptions

The words pressure and force are used interchangeably in everyday life, so students will need to develop an understanding of the science meaning.

Resources

Student Book Measuring pressure

Resources for demonstration (see Technician's notes, following)

Approach

Introduce the learning episode

Ask students to look at the picture of the snowmobile in the Student Book (without reading the text around it). Ask them why the rider would probably sink into the snow if they stood on it, but the much heavier snowmobile does not sink. Draw out the fact that the snowmobile is covering a much larger area on the snow than its rider does, and therefore the pressure it exerts on the snow is less.

Develop the learning episode

Ask students to work in pairs to find a way of demonstrating how pressure depends on force applied and area over which the force acts. For example, they might want to think about how they could model the situation of the snowmobile and the rider above. What could they use to model the snow? How could they model the snowmobile? What about the rider? If you can provide materials such as those listed in the technician's notes, and if time allows, students could carry out their demonstration.

Allow time for the students to do the worked example in the Student Book. Ask them to try it first without the solution.

Ask students to read the Developing practical skills box in the Student Book and answer the questions with a partner. After a short period, bring the class back together and discuss students' responses and compare them with the sample responses. Discuss why the two sets differ (if they do!).

Finish the learning episode

Students should answer the questions in the Student Book to check understanding.

Technician's notes

Be sure to check the latest safety notes on these resources before proceeding.

The following resources are needed for the demonstration in the main part of the activity:

material such as wood chips that can be used to model snow
a mass with a wide base to model the snowmobile

Learning episode P1g.2 Consolidation and summary

Learning aims

- Review the learning points of the topic summarised in the End of topic checklist
- Test understanding of the topic content by answering the End of topic questions

Resources

Student Book End of topic checklist and End of topic questions

Approach

Introduce the learning episode

Ask students to work with a partner to make a list of key words from this topic. They could then work together to produce a spider diagram showing how the different concepts are linked. They could compare their list with the key terms given in the Student Book. Discuss the checklist in the Student Book and ask questions to see how much of the content students are comfortable with.

Students could make flashcards of the key content and then use the flashcards to quiz each other on the information.

Develop the learning episode

Ask students to work individually through the End of topic questions in the Student Book without looking at the text. As they work, walk around the classroom observing their answers and questioning them as necessary to find out which questions are causing difficulties.

Finish the learning episode

After a set period, ask students to stop working and discuss any areas of difficulty you observed as you walked round the class. Students should complete any unanswered questions for homework, but you should stress that they should attempt the questions without looking at the text, so that they can see how much they have remembered.

P2 Thermal physics

Contents

Overview of the section

This section begins by considering the three states of matter and the particle model for solids, liquids and gases. Students following the supplement curriculum also look at pressure changes in a gas at constant temperature.

Topic P2b considers the thermal expansion of solids, liquids and gases, before going on to consider the process of evaporation.

The third and final topic considers the transfer of thermal energy. Students look at conduction, convection and radiation in turn, and then learn about the everyday applications and consequences of thermal energy transfer.

Starting points

The Student Book section opener puts the ideas in the section into context and sets the scene. It also allows students to acknowledge and value their prior learning, and provides a benchmark against which future learning can be compared.

The questions provide a structure for introducing the section and can be used in a number of different ways:

- You could ask students to consider the questions as an introductory homework task.
- You could put students into groups to share their own ideas and understanding and then to report back to the whole class.
- Students could be given access to the Internet, preferably with a tight timescale, to find out the information required.

You could then use a spider chart or other form of wall chart to summarise everybody's ideas.

Recording these initial ideas allows you to retain them for reference as the individual topics are developed. In this way, your students' progress in learning can be readily acknowledged.

P2a Kinetic particle model of matter

Introduction

In this topic students are introduced to the three states of matter and the kinetic particle model for matter. Students following the supplementary curriculum also apply the kinetic particle model to explain what causes gas pressure, and learn how the pressure of a gas changes with volume if the temperature is constant and how the pressure changes with temperature if the volume is constant. Students first met pressure in Topic P1g, though the focus in that topic was on pressure of a solid on a surface.

Links to other topics

Topics	Essential background knowledge	Useful links
P1 Motion, forces and energy	P1g Pressure	
P2 Thermal physics		P2b Thermal properties and temperature P2c Transfer of thermal energy

Topic overview

P2a.1	**States of matter and the particle model** In this learning episode students recap simple ideas about states of matter and changes of state, and are introduced to the particle model. Supplement Students use the kinetic particle model to explain the properties of the three states of matter.
P2a.2	**Gas pressure and pressure changes** In this learning episode students learn what causes the pressure of a gas, and how changes in temperature or volume affect the pressure of a gas.
P2a.3	**Consolidation and summary** This learning episode provides an opportunity for a quick recap on the ideas encountered in the topic as well as giving time for the students to answer the End of topic questions in the Student Book.

Career links

One career which requires an understanding of the ideas in this section is a cylinder filler. These are the people who are responsible for maintaining and filling industrial gas cylinders so that they can then be delivered to the businesses and organisations that require gases such as oxygen and carbon dioxide.

Learning episode P2a.1 States of matter and the particle model

Learning objectives

- State the distinguishing properties of solids, liquids and gases
- Know the terms for the changes in state between solids, liquids and gases (gas to solid and solid to gas transfers are **not** required)
- Describe the structure of solids, liquids and gases in terms of the arrangement, separation and motion of the particles, and represent these states using simple particle diagrams
- Describe the relationship between the motion of particles and temperature
- Supplement Know that the forces and distances between particles and the motion of the particles affect the properties of solids, liquids and gases

Common misconceptions

Students often find it difficult to accept a model of a gas that consists of mostly empty space. They intuitively prefer a model in which the particles *expand* to fill the space available to a model in which the particles are free to *move apart* to fit the space available. Also, students often assume that particles in liquids and gases vibrate instead of moving around in random directions.

Resources

Student Book States of matter; Model of matter; Understanding gases

Resources for demonstration (see Technician's notes, following)

Approach

Introduce the topic

Ask students to name the three states of matter. Issue students with a large plastic bag and about six small polystyrene balls to place inside. Tell them to suck the air out of the bag using a hand pump or small vacuum cleaner, so that the balls are jammed together and cannot move, modelling a solid. Letting a little air in should allow the balls to stay close but slide past each other, modelling a liquid. Then get them to blow lots of air into the bag (using a balloon pump, if preferred), hold the end shut and shake it around so that the balls model a gas.

SAFETY INFORMATION: Students should not suck the air out of the bag using their mouths – the small polystyrene balls could be accidentally swallowed.

Develop the learning episode

Discuss the text and diagrams in the 'States of Matter' section in the Student Book. Take every opportunity at this stage to ensure that students have a clear mental image of particles in the three states of matter. Be particularly careful to ensure that their picture of a gas is correct (not simply empty space; particles do not expand).

Get the whole class to act out the effect of heating a solid until it becomes a gas. Use this role-play to develop the kinetic particle model, with an emphasis on how the arrangement of particles can account for the different large-scale properties of matter. Then act out the effect of cooling a gas until it becomes a solid.

As students are acting out the effects name the changes: melting, evaporation, condensation, solidification.

Use Table P2.1 in the Student Book to summarise the role-play.

Supplement Use the information in the Student Book to discuss how the arrangement of particles can explain the properties of solids, liquids and gases. It can be useful for students to work with a partner to discuss ideas.

Use the text 'Understanding gases' in the Student Book to discuss with students what the particles are that they have been drawing in their diagrams of solids, liquids and gases. Use the discussion to show how the kinetic particle model can explain the observations about gases that we can make. Table P2.2 summarises the link between the observations and the model.

Students could discuss the question in the Student Book.

Students need to realise that kinetic particle model is a scientific model. It therefore has to be able to account for **all** observed properties of matter, not just some. Their answers to the questions need to clearly distinguish between an observation and its explanation with the model.

Finish the learning episode

Discuss the questions in the Student Book. Show students flash cards of various drawings representing the arrangement of particles in solids, liquids and gases. Ask them to say which state of matter each card represents and to explain how the state can be changed.

Technician's notes

Be sure to check the latest safety notes on these resources before proceeding.

The following resources are needed for the practical activity in the introduction, per group:

large plastic bag
small polystyrene balls
pump to extract air from bags; balloon pump to put air back in (if required)
flashcards of drawings representing solids, liquids and gases

SAFETY INFORMATION: Students should not suck the air out of the bag using their mouths – the small polystyrene balls could be accidentally swallowed.

Learning episode P2a.2 Gas pressure and pressure changes

Learning objectives

- Supplement Describe the pressure of a gas in terms of the forces exerted by particles colliding with surfaces, creating a force per unit area
- Supplement Describe qualitatively, in terms of particles, the effect on the pressure of a fixed mass of gas of: (a) a change of temperature at constant volume and (b) a change of volume at constant temperature

Resources

Student Book Gas pressure; Pressure changes

Resources for demonstration (see Technician's notes, following)

Approach

Introduce the learning episode

Ask students to suggest why gas molecules exert pressure on a container. Introduce the idea of collisions causing the pressure

Develop the learning episode

Follow the introduction by discussing the section 'text and diagrams in the Student Book, which looks at how the pressure of a gas is caused by the movement of the gas particles.

If possible, show students a bicycle pump and operate its piston (see Technician's notes). Give students the opportunity to feel what happens to the piston if the air outlet is blocked. Ask students to attempt to explain the observation in terms of the molecules of the gas. After a short time, use the explanation given in the Student Book and discuss it.

Extend the discussion of pressure and gas particles by using the Student Book figure and ask students to predict the force exerted by the particles on the piston if its area was changed e.g., doubled, halved. They should be able to appreciate that the motion of the particles provides a force per unit area i.e., pressure.

In pairs, ask students to use the kinetic particle model to predict what will happen to the pressure of a gas if the temperature is increased. Take feedback from the groups and discuss their ideas. Ask them to check their explanations using the text on Pressure and temperature at constant volume in the Student Book.

Next ask students to predict what will happen to the pressure of a gas when the temperature is constant and the volume increases. Again, take feedback from the class and ask them to check their reasoning using the text on Pressure and volume at constant temperature in the Student Book.

Finish the learning episode

Students should work in pairs to construct a summary of the main points of the learning episode. These summaries should include the equation showing the relationship between pressure and volume.

Discuss any points that students found difficult in this learning episode.

Technician's notes

Be sure to check the latest safety notes on these resources before proceeding.

The following resources are needed for the demonstration on compressing a gas:

bicycle pump

Learning episode P2a.3 Consolidation and summary

Learning aims

- Review the learning points of the topic summarised in the End of topic checklist
- Test understanding of the topic content by answering the End of topic questions

Resources

Student Book End of topic checklist and End of topic questions

Approach

Introduce the learning episode

Ask students to work with a partner to make a list of key words from this topic. They could then work together to produce a spider diagram showing how the different concepts are linked. They could compare their list with the list of key terms given in the Student Book. Discuss the checklist in the Student Book and ask questions to see how much of the content students are comfortable with.

Students could make flashcards of the key content and then use the flashcards to quiz each other on the information.

Develop the learning episode

Ask students to work individually through the End of topic questions in the Student Book without looking at the text. As they work, walk around the classroom observing their answers and questioning them as necessary to find out which questions are causing difficulties.

Finish the learning episode

After a set period, ask the students to stop working and discuss any areas of difficulty you observed as you walked round the class. Students should complete any unanswered questions for homework, but you should stress that they should attempt the questions without looking at the text, so that they can see how much they have remembered.

P2b Thermal properties and temperature

Introduction

In this topic students learn about the thermal expansion of solids, liquids and gases and about evaporation in terms of the kinetic particle model. Students learned about the structure of the three states of matter in Topic P2a K*inetic particle model of matter.* The supplement material considers how various factors affect the rate of evaporation, and how to explain some of the everyday applications and consequences of thermal expansion.

Links to other topics

Sections	Essential background knowledge	Useful links
P1 Motion, forces, and energy		P1a Physical quantities and measurement techniques P1f Energy, work and power
P2 Thermal physics	P2a Kinetic particle model of matter	P2c Transfer of thermal energy

Topic overview

P2b.1	**Thermal expansion of solids, liquids and gases** In this learning episode students learn why solids, liquids and gases expand when heated. Supplement Students apply the kinetic particle model to explain some consequences and uses of thermal expansion.
P2b.2	**Evaporation** In this learning episode students learn about evaporation in terms of the kinetic particle model, and that evaporation causes cooling. Supplement Students describe how different factors affect evaporation.
P2b.3	**Consolidation and summary** This learning episode provides an opportunity for a quick recap on the ideas encountered in the topic as well as time for the students to answer the End of topic questions in the Student Book.

Career links

A refrigeration and air-conditioning installer needs to understand the topics in this section so that they can ensure that the systems they install are able to provide sufficient cooling and air conditioning for the needs of the customer.

Learning episode P2b.1 Thermal expansion of solids, liquids and gases

Learning objectives

- Describe, qualitatively, the thermal expansion of solids, liquids and gases at constant pressure
- Supplement Explain some of the everyday applications and consequences of thermal expansion

Resources

Student Book Thermal expansion of solids, liquids and gases

Resources for demonstration (see Technician's notes, following)

Approach

Introduce the learning episode

Ask students to work in pairs to draw diagrams of the arrangement of particles (atoms or molecules) in solids, liquids and gases. Then ask them to discuss with their partner what is going to happen to the particles when the substances are heated. Allow a few minutes for the discussion, then take feedback.

Develop the learning episode

Discuss the text in the Student Book, which explains that solids and liquids expand if they are heated. Ask students to work in pairs to make a summary of the content.

Now move on to discuss gases. You could use a bicycle pump and put a finger over the end. Ask students to work with a partner to explain what happens at the molecular level when a gas is compressed and then allowed to expand.

Supplement Ask students to use the text in the Student Book to make notes about everyday consequences of thermal expansion.

Finish the learning episode

Ask students to think of two things that they have learned in this learning episode and one thing that they found more difficult to grasp. Take feedback. Discuss the concepts that they identify as more difficult.

Technician's notes

Be sure to check the latest safety notes on these resources before proceeding.

The following resources are needed for the demonstration of compression and expansion:

piston or bicycle pump to demonstrate compression and expansion of gas

Learning episode P2b.2 Evaporation

Learning objectives

- Describe evaporation in terms of the escape of more energetic particles from the surface of a liquid.
- Know that evaporation causes cooling of a liquid
- Supplement Describe how temperature, surface area and air movement over a surface affect evaporation

Common misconceptions

Some students think that evaporation only occurs at higher temperatures. They may also think that evaporation only occurs from water.

Resources

Student Book Evaporation

Approach

Introduce the learning episode

Show students two cups – one with a wide surface area and one with a smaller surface area. Ask students which cup will allow coffee to cool more quickly.

Develop the learning episode

Discuss the section on evaporation in the Student Book with students. Highlight that evaporation causes cooling of a liquid and also that evaporation depends on the escape of more energetic particles from the surface of the liquid. Although not required by the syllabus, you could ask students to write their own explanation of how evaporation leads to cooling of a liquid.

Supplement Discuss with students the effect of temperature and air flow on the rate of evaporation. Link this to drying laundry or cooling drinks.

Students should complete the questions in the Student Book.

Finish the learning episode

Ask students to draw a simple flowchart to explain what happens in evaporation.

Learning episode P2b.3 Consolidation and summary

Learning aims

- Review the learning points of the topic summarised in the End of topic checklist
- Test understanding of the topic content by answering the End of topic questions

Resources

Student Book End of topic checklist and End of topic questions

Approach

Introduce the learning episode

Ask students to work with a partner to make a list of key words from this topic. They could then work together to produce a spider diagram showing how the different concepts are linked. They could compare their list with the list of key terms given in the Student Book. Discuss the checklist in the Student Book and ask questions to see how much of the content students are comfortable with.

Students could make flashcards of the key content and then use the flashcards to quiz each other on the information.

Develop the learning episode

Ask students to work individually through the End of topic questions in the Student Book without looking at the text. As they work, walk around the classroom observing their answers and questioning them as necessary to find out which questions are causing difficulties.

Finish the learning episode

After a set period, ask the students to stop working and discuss any areas of difficulty you observed as you walked round the class. Students should complete any unanswered questions for homework, but you should stress that they should attempt the questions without looking at the text, so that they can see how much they have remembered.

P2c Transfer of thermal energy

Introduction

In this topic students learn about thermal energy transfer by the processes of conduction, convection and radiation. Students following the supplementary curriculum apply their knowledge of density from Topic P1d *Density* to explain convection in liquids and gases. The process of radiation involves part of the electromagnetic spectrum known as infrared radiation, which students will meet again in Topic P3c *Electromagnetic spectrum*. The supplement material considers the Earth as an absorber and emitter of radiation, and how to distinguish good and bad emitters, and good and bad absorbers of radiation.

Links to other topics

Sections	Essential background knowledge	Useful links
P1 Motion, forces and energy	P1d Density	P1a Physical quantities and measurement techniques
P2 Thermal physics	P2a Kinetic particle model of matter P2b Thermal properties and temperature	
P3 Waves		P3a General properties of waves P3c Electromagnetic spectrum

Topic overview

P2c.1	**Conduction, convection and radiation** In this learning episode students learn about the three main methods of thermal energy transfer, and that unlike conduction and convection, radiation does not require a medium. Supplement Students learn about particle model explanations for thermal conduction in solids and convection in liquids and gases. They also learn about how to distinguish experimentally between good and bad emitters of radiation, and good and bad absorbers of radiation, and how the balance of radiation emitted and absorbed by the Earth affects its temperature.
P2c.2	**Examples of transfers of thermal energy** In this learning episode students apply their knowledge of thermal energy transfer to a range of everyday applications.
P2c.3	**Consolidation and summary** This learning episode provides an opportunity for a quick recap on the ideas encountered in the topic as well as time for the students to answer the End of topic questions in the Student Book.

Career links

Satellite payload system engineers have to have a good understanding of conduction, convection and radiation. In space, there are great extremes of temperature, and it is important that the electronic circuitry and equipment on the satellites are kept at suitable temperatures so they work efficiently.

Learning episode P2c.1 Conduction, convection and radiation

Learning objectives

- Identify and give examples of typically good thermal conductors and bad thermal conductors (thermal insulators)
- Supplement Describe thermal conduction in solids in terms of atomic or molecular lattice vibrations and also in terms of the movement of delocalised (mobile) electrons in metallic conductors
- Know that convection is an important method of energy transfer in liquids and gases
- Describe convection in liquids and gases
- Supplement Explain convection in liquids and gases in terms of density changes
- Know that thermal energy transfer by thermal radiation does not require a medium and is mainly due to infrared radiation
- Describe the effect of surface colour (black or white) and texture (dull or shiny) on the emission, absorption and reflection of thermal radiation
- Supplement Know that the temperature of the Earth is affected by the radiation absorbed by the Earth and the radiation emitted by the Earth
- Supplement Describe experiments to distinguish between good and bad emitters of thermal radiation
- Supplement Describe experiments to distinguish between good and bad absorbers of thermal radiation

Common misconceptions

Conduction and *convection* sound very similar, so students often confuse them. Many students have difficulty in accepting that a conductor and an insulator can both be at the same temperature, yet one feels colder than the other. Take time to explain this phenomenon in terms of energy transfer.

Many students also have difficulty with the idea that continually transferring energy into something can result in a steady temperature.

Resources

Student Book Conduction; Convection; Radiation

Worksheet P2c.1 Observing convection

Resources for demonstration and class practical (see Technician's notes, following)

Approach

This is a long learning episode with the opportunity for lots of practical work, and is likely to span two or more lessons.

Introduce the learning episode

Ask students what they understand by the terms *conduction* and *convection*. They may have ideas about conduction from previous learning about electricity. Ask them how the term applies to thermal energy transfer.

Set up a Bunsen burner on the front bench with a yellow flame. Introduce the idea of *convection currents*. Use a windmill to show that there is an upward movement of gas from the flame (see Technician's notes).

Develop the learning episode

Discuss conduction using the Student Book. If possible, demonstrate conduction using the equipment shown in the Student Book. In any case, make sure that students have noted this experiment and don't gloss over it. Students should answer the questions in of the Student Book with a partner to check understanding.

You could also ask students to touch conductors and insulators and explain whether they think they are thermal conductors or thermal insulators, and why.

Supplement Discuss the role of particle vibration in thermal conduction in solids (and electron motion, for conduction in metals) using the supplement text in the Student Book. Students should do the supplement question in the Student Book.

Next move on to convection, using the text in the Student Book. Worksheet P2c.1 can either be done as a demonstration or students can do it in small groups, depending on the availability of equipment (see Technician's notes). Condensation on the outside of the beaker can block the view of the convection currents. However, students should be discouraged from wiping it clear. For details, see Technician's notes below. Supplement Discuss how density changes cause convection.

SAFETY INFORMATION: Potassium manganate(VII), $KMnO_4$, is oxidising, harmful and dangerous to the environment. Wear eye protection.

Move on to discuss radiation using the Student Book text as a guide. If possible, set up a Leslie's cube and an infrared sensor with a large display meter. Show students that all four walls have the same thickness and so should transfer energy from hot water inside at the same rate. Then fill the cube with boiled water from the kettle and show that the black surface emits much more radiation (for the same surface area) than the white one. Students should try the questions in the Student Book to check their understanding.

Supplement Discuss the experiments to show how well different coloured cans emit and absorb infrared radiation using the diagrams shown in the Student Book. Extend the discussion to introduce the idea of the Earth as an absorber and emitter of radiation, as discussed in the Student Book. Link this to the average surface temperature of the Earth. Although not required by the syllabus you could discuss how greenhouse gases in the atmosphere absorb the longer wavelength infrared radiation emitted by the Earth's surface, causing an increase in the Earth's temperature.

You could discuss with students the Science in Context box in the Student Book to show the importance of radiation balance on the surface temperature of a planet (this is beyond the requirements of the syllabus but may allow students to explore the content more deeply).

Finish the learning episode

Ask students to write down their own definitions of conduction and convection. They could also draw diagrams to illustrate their definitions.

Technician's notes

Be sure to check the latest safety notes on these resources before proceeding.

The following resources are needed for the demonstration of convection, in the introduction:

aluminium foil windmill
Bunsen burner and heatproof mat

SAFETY INFORMATION: The windmill should not be made of paper or plastic.

A windmill can be constructed as follows:

- cut out a circle of aluminium foil
- make eight radial cuts from the edge of the circle about three-quarters of the way to the centre
- bend one edge of the foil 'blades' about 30° upwards
- mount the centre of the circle on a pin or similar
- clamp the pin in a clamp stand.

The following resources are needed for practical P2c.1, per group:

250 cm^3 glass beaker
Bunsen burner and heatproof mat
tripod and gauze
tweezers
tripod
hollow glass tube long and wide enough to deliver a potassium manganate(VII) crystal to base of beaker
potassium manganate(VII) crystals

SAFETY INFORMATION: Potassium manganate(VII) is oxidising, harmful and dangerous for the environment. Wear eye protection.

The following resources are needed for a demonstration of conduction:

- apparatus for demonstrating heat conduction shown in the Student Book:
  - Bunsen burner
  - rods of different metals arranged around central point with blob of wax on end of each rod
- blocks or sheets of, for example:
  - copper
  - iron
  - brass
  - aluminium
  - Perspex
  - glass
  - wood
  - nylon
  - rubber
  - polystyrene
  - expanded polystyrene
  - cardboard

The following resources are needed for a demonstration of radiation:

Leslie's cube
hot water from a kettle
large display meter (to connect to infrared sensor)
infrared sensor

Learning episode P2c.2 Examples of transfers of thermal energy

Learning objectives

- Identify and explain some of the basic everyday applications and consequences of conduction, convection and radiation

Common misconceptions

Many students think that radiators only transfer energy by radiation. In fact, ¾ of the energy transfer is by convection.

Resources

Student Book Examples of transfers of thermal energy

Resources for demonstration (see Technician's notes, following)

Approach

Introduce the learning episode

Use two blocks of ice (or ice cubes). Leave one block unwrapped. Wrap the second block in aluminium foil. Ask students to predict which cube will melt first, and then ask them to justify their prediction. Observe the blocks melting. The unwrapped block will melt much faster than the wrapped block. Use this demonstration to introduce consequences of thermal energy transfer (see Technician's notes).

Develop the learning episode

Discuss everyday consequences of thermal energy transfer using Student Book as a guide. Ask students to identify the importance of conduction, convection and radiation.

All students should work through the Developing Practical Skills box in the Student Book with a partner.

Finish the learning episode

Ask students to use their ideas of conduction, convection and energy transfer to explain these observations:

a) The walls of a refrigerator are mostly made of plastic.

b) Many cooking pans have a metal body and a plastic handle.

c) Cooking ovens have to use a fan to keep all of the shelves at the same temperature.

d) Many houses are built with a double outer layer of bricks with a narrow air gap between the two layers.

e) The cooling unit of a refrigerator is always at the top of the inside, above the food, never at the bottom.

Technician's notes

Be sure to check the latest safety notes on these resources before proceeding.

The following resources are needed for the demonstration in the introduction:

ice blocks / ice cubes
aluminium foil

Learning episode P2c.3 Consolidation and summary

Learning aims

- Review the learning points of the topic summarised in the End of topic checklist
- Test understanding of the topic content by answering the End of topic questions

Resources

Student Book End of topic checklist and End of topic questions

Approach

Introduce the learning episode

Ask students to work with a partner to make a list of key words from this topic. They could then work together to produce a spider diagram showing how the different concepts are linked. They could compare their list with the list of key terms given in the Student Book. Discuss the checklist in the Student Book and ask questions to see how much of the content students are comfortable with.

Students could make flashcards of the key content and then use the flashcards to quiz each other on the information.

Develop the learning episode

Ask students to work individually through the End of topic questions in the Student Book without looking at the text. As they work, walk around the classroom observing their answers and questioning them as necessary to find out which questions are causing difficulties.

Finish the learning episode

After a set period, ask the students to stop working and discuss any areas of difficulty you observed as you walked round the class. Students should complete any unanswered questions for homework, but you should stress that they should attempt to answer the questions without looking at to the text, so that they can see how much they have remembered.

P3 Waves

Contents

Overview of the section

The first topic in this section introduces students to the general properties of waves, and the vocabulary that they will need to use throughout the rest of the section. The second topic is about light. Students consider reflection, including the law of reflection, before going on to refraction. Refraction leads naturally to a consideration of the behaviour of a thin lens, and the conditions for the formation of a real image, before students learn about dispersion of light as shown by the action of light on a glass prism.

Students are then introduced to the electromagnetic spectrum and the applications of various electromagnetic radiations including in communication, microwave ovens, medicine and security.

The final topic in this section covers sound. Students learn the basic properties of sound waves and how to measure the speed of sound. They relate the loudness and pitch of sound waves to amplitude and frequency and describe how the reflection of sound may produce an echo.

Starting points

The Student Book section opener puts the ideas in the section into context and sets the scene. It also allows students to acknowledge and value their prior learning, and provides a benchmark against which future learning can be compared.

The questions provide a structure for introducing the section and can be used in a number of different ways:

- You could ask students to consider the questions as an introductory homework task.
- You could put students into groups to share their own ideas and understanding and then to report back to the whole class.
- Students could be given access to the Internet, preferably with a tight timescale, to find out the information required.

You could then use a spider chart or other form of wall chart to summarise everybody's ideas.

Recording these initial ideas allows you to retain them for reference as the individual topics are developed. In this way, your students' progress in learning can be readily acknowledged.

P3a General properties of waves

Introduction

In this topic students learn about general wave properties and the vocabulary that they will need in the rest of Section P3.

Supplement Students studying the supplement curriculum learn the difference between transverse and longitudinal waves, and examples of each. There are links for these students to Topic P5c *Electromagnetic spectrum* for calculations using the wave equation for electromagnetic waves and the speed of light. Allow time for this content.

Links to other topics

Sections	Essential background knowledge	Useful links
P1 Motion, forces and energy	P1a Physical quantities and measurement techniques P1b Motion	
P3 Waves		P3b Light P3c Electromagnetic spectrum P3d Sound

Topic overview

P3a.1	**General properties of waves** This learning episode lists examples of wave motion. It defines the terms amplitude, wavelength and frequency and explains the use of these terms to describe waves. Supplement Students describe transverse and longitudinal wave motion.
P3a.2	**The wave equation** This learning episode explains the use of the wave equation to link wavelength, frequency and wave speed.
P3a.3	**Water waves** In this learning episode students will explore the use of water waves to show reflection at a plane surface and refraction due to change of speed.
P3a.4	**Consolidation and summary** This learning episode provides an opportunity for a quick recap on the ideas encountered in the topic as well as time for the students to answer the End of topic questions in the Student Book.

Career links

An understanding of properties of waves is useful in many different careers, for example seismologists and some geologists study earthquakes and their effects, whilst there are structural engineers who specialise in designing and analysing buildings, bridges and other structures so that they are more resistant to damage from earthquakes. Environmental planners use their knowledge of water waves, and the energy transferred, to predict coastal erosion.

Learning episode P3a.1 General properties of waves

Learning objectives

- Know that waves transfer energy without transferring matter
- Describe what is meant by wave motion as illustrated by vibration (oscillation) in ropes and springs, and by experiments using water waves
- Describe the features of a wave in terms of wavelength, frequency, crest (peak), trough, amplitude and wave speed
- Supplement Know that for a transverse wave, the direction of vibration is at right angles to the direction of propagation and understand that electromagnetic radiation, water waves and seismic S-waves (secondary) are transverse
- Supplement Know that for a longitudinal wave, the direction of vibration is parallel to the direction of propagation and understand that sound waves and seismic P-waves (primary) are longitudinal

Resources

Student Book Understanding waves; Features of a wave

Worksheet P3a.1 Making waves

Resources for class demonstration and practical (see Technician's notes, following)

Approach

Begin the learning episode

Discuss examples that students can think of where the idea of 'waves' appears in everyday life. This could include 'scientific' examples that they already know, such as light, but draw out a wider range, such as 'Mexican wave' or 'brainwave'. Take all answers without comment at this stage.

Develop the learning episode

Demonstrate wave formation and pulses in the lab using a 'slinky' spring or a rope as shown in the Student Book (see Technician's notes). Introduce key terms – 'wavelength', 'speed', 'frequency', 'peak', 'trough', and 'amplitude'. Point out how each small portion of the spring or rope vibrates (oscillates) but ends up in the same position that it started in after the wave has passed through. Attaching a coloured ribbon to a coil of the spring will help show this.

Also demonstrate making waves on the surface of water, using a ripple tank if available, or a fish tank. Place a cork on the surface of the water and gently disturb the water surface nearby by dipping your finger in and out. Discuss how the cork (and the water surface) move up and down as the ripples spread. The ripples transfer energy but the water ends up in the same position that it started in.

Supplement Point out the key differences between transverse and longitudinal waves and give examples of each type.

As you discuss each wave, introduce examples of that wave: transverse – electromagnetic radiation, water waves, seismic S-waves (secondary); longitudinal – sound waves, seismic P-waves.

You should then demonstrate, or students can carry out, the practical on Worksheet P3a.1 (see Technician's notes). The method for the practical, on Worksheet P3a.1, is as follows:

1. Place the spring on the table.
2. Two students stretch it to a length of about 1 m.
3. Give one end of the spring a quick flick to one side.
4. The pulse will move down the spring and reflect off the hand of the further person.
5. Place the bottle on the table next to the spring about halfway along it.
6. Send pulses down the spring to knock the bottle over.

7. Shake the spring from side to side to make a transverse wave in the spring.
8. Change the size and speed of the sideways oscillation.
9. Shake the spring towards and away from you to make a longitudinal wave in the spring.
10. Change the size and speed of the backwards and forwards oscillation.

Discuss how the wave pulse transfers energy along the rope, and how the vibration of individual coils of the spring differs for longitudinal and transverse waves.

Finish the learning episode

Get the students to demonstrate a 'Mexican wave' as seen at sporting events. Discuss the energy transfer (kinetic) and, for supplementary students, the nature of the wave (longitudinal or transverse).

Technician's notes

Be sure to check the latest safety notes on these resources before proceeding.

The following resources are needed for the demonstration of wave formation:

slinky spring or rope
ripple tank, or fish tank or shallow tray filled with water
cork

The following resources are needed for the class practical P3a.1, per group:

slinky spring
small empty plastic drinks bottle
scissors

Learning episode P3a.2 The wave equation

Learning objective

- Recall and use the equation for wave speed $v = f\lambda$

Resources

Student Book Relationship between speed, frequency and wavelength

Resources for demonstration (see Technician's notes, following)

Approach

Introduce the learning episode

Give the students a number of cards, some with 'waves' words such as 'amplitude' and 'frequency', and some with 'non-waves' words such as 'mass'. Ask the students to group the words together, then extend the activity by calling for the correct definitions to go with them.

Develop the learning episode

Introduce the idea of 'wave speed' by sending a pulse along a rope (see Technician's notes). If you ask students to do this, ask them to try to send the pulse at different speeds – as long as the tension remains constant, they should find this very difficult to do. Alternatively, demonstrate or ask students to imagine making ripples on the surface of water and ask how the speed at which the ripples travel outwards could be calculated.

Introduce the connection between wave speed, frequency and wavelength.

Ask students to try the worked examples in the Student Book without the solutions. After a set period of time, compare their answers with the given solutions and discuss any differences. Give them further examples until they are confident in using the equation.

Supplement You may wish to introduce the speed of light at this point, and extend the calculations by using numbers relating to the electromagnetic spectrum. The frequencies for these waves can be very large (and the wavelengths very small), so students may need help to find efficient methods of handling them using standard form. These calculations will be revisited in Topic P3c *Electromagnetic spectrum* where the value for the speed of light is introduced.

Finish the learning episode

Remind students that speed can also be calculated from 'speed = distance travelled/time taken' and demonstrate that sometimes it is easier to find information using this method.

Technician's notes

Be sure to check the latest safety notes on these resources before proceeding.

The following resources are needed for the demonstrations in the main part of the learning episode:

rope
cards with wave words and cards with non-wave words (see above)
ripple tank, or fish tank or shallow tray filled with water (optional)

Learning episode P3a.3 Water waves

Learning objectives

- Describe how waves can undergo: (a) reflection at a plane surface (b) refraction due to a change of speed

Resources

Student Book Using water waves to show reflection and refraction

Resources for demonstration (see Technician's notes, following)

Approach

Introduce the learning episode

Ask students to work with a partner to write down what they understand by the words 'reflection' and 'refraction'. Take feedback.

Develop the learning episode

Explain that, in this lesson, you are going to demonstrate reflection and refraction using water waves, and that reflection and refraction will be studied more thoroughly in the next topic on light.

Begin by demonstrating reflection at a plane surface. Use the instructions in the Student Book to do this.

Now demonstrate refraction by using a clear glass block to change the depth of the water in a given region. Discuss the change of speed of the wave, using the Student Book to help you.

Finish the learning episode

Introduce a 'checklist' to check what should be counted as a wave. As a guide, use these six properties:

- waves are caused by a vibration/oscillation
- waves transfer energy without transferring matter
- waves reflect
- waves refract
- waves interfere/superpose*.

Give different examples and ask them to decide if they are waves or not waves.

*Note to teacher: Interference and superposition are not included in the syllabus, but are included in this checklist because students may have noticed the interference of the waves in the ripple tank demonstration.

Students should answer the questions in the Student Book in pairs.

Technician's notes

Be sure to check the latest safety notes on these resources before proceeding.

The following resources are needed for the demonstration of reflection and refraction:

ripple tank filled with water
plane surface
clear glass block

The ripple tank will need to be set up according to the manufacturer's instructions.

Learning episode P3a.4 Consolidation and summary

Learning aims

- Review the learning points of the topic summarised in the End of topic checklist
- Test understanding of the topic content by answering the End of topic questions

Resources

Student Book End of topic checklist and End of topic questions

Approach

Introduce the learning episode

Ask students to work with a partner to make a list of key words from this topic. They should also draw a diagram of a wave and label it with the key terms. They could then work together to produce a spider diagram showing how the different concepts are linked. They could compare their list with the list of key terms given in the Student Book. Discuss the checklist and use questioning to see how much of the content students are comfortable with.

Students could make flashcards of the key content and then use the flashcards to quiz each other on the information.

Develop the learning episode

Ask students to work individually through the End of topic questions in the Student Book without looking at the text. As they work, walk around the classroom observing their answers and questioning them as necessary to find out which questions are causing difficulties.

Finish the learning episode

After a set period, ask the students to stop working and discuss any areas of difficulty you observed as you walked round the class. Students should complete any unanswered questions for homework, but you should stress that they should try to answer the questions without looking at the text, so that they can see how much they have remembered.

P3b Light

Introduction

In this topic students learn about reflection and refraction. This continues the discussion from the previous topic, where students saw these effects using water waves. They consider the action of a thin converging lens and learn how to draw ray diagrams. The section finishes with a consideration of the dispersion of light.

Links to other topics

Sections	Essential background knowledge	Useful links
P3 Waves	P3a General properties of waves	P3c Electromagnetic spectrum P3d Sound

Topic overview

P3b.1	**Reflection of light** In this learning episode students learn about reflection of light at a plane surface. Through practical work they learn about the law of reflection. Supplement Students carry out further practical work to explore the formation of the image in a mirror and why it is virtual, using ray diagrams.
P3b.2	**Refraction of light** In this learning episode students learn about refraction of light.
P3b.3	**Thin converging lenses and dispersion of light** In this learning episode students learn about thin converging lenses and the images they form. They learn about dispersion of light through a prism.
P3b.4	**Consolidation and summary** This learning episode provides an opportunity for a quick recap on the ideas encountered in the topic as well as time for the students to answer the End of topic questions in the Student Book.

Career links

Optometrists and dispensing opticians are two careers that rely on an understanding of the refraction of light and lenses.

Learning episode P3b.1 Reflection of light

Learning objectives

- Use ray diagrams to define the terms normal, angle of incidence and angle of reflection
- Describe the formation of an optical image by a plane mirror and give its characteristics compared with the object, i.e., same size, same distance from mirror, laterally inverted
- State that for reflection, the angle of incidence is equal to the angle of reflection; recall and use this relationship
- Supplement Describe the formation of an optical image by a plane mirror and explain why it is virtual
- Supplement Use simple diagrams, measurements and calculations for reflection by plane mirrors

Common misconceptions

Students may have seen 'rays of light' (e.g., crepuscular rays through clouds at sunset). When ray diagrams are used some students assume that they are a direct representation of the world. They are not. Ray diagrams are models that physicists use. It is helpful for students to understand that we make simplifications to the real world to draw ray diagrams. For example, although beams of light do occur, the single ray from a point source that we often draw is simplified.

Resources

Student Book Reflection of light

Worksheet P3b.1(1) Reflection

Supplement Worksheet P3b.1(2) Images from rays

Supplement Worksheet P3b.1(3) Making a periscope

Resources for class demonstration and practicals (see Technician's notes, following)

Approach

Introduce the learning episode

Ask students to work with a partner to discuss where they see reflection in everyday life. Take feedback.

Develop the learning episode

Discuss the content in the Student Book. Students should carry out the practical on Worksheet P3b.1(1). If you do not have enough equipment, you could do this as a demonstration making sure that students are fully involved. The experiment works best in a darkened room (see Technician's notes).

Set up the ray box so that it sends a single beam of light across the white paper. Fasten the mirror to the wooden block with sticky putty and place on the paper so that it reflects the incident beam of light. Mark the beams on paper, draw in the normal lines and measure the angles of incidence and reflection. Students often find measuring angles with a protractor difficult. Encourage them to start off with 0 degrees lined up on the normal.

SAFETY INFORMATION: Remind students that ray boxes can become hot after prolonged use.

Supplement Consider the section on ray diagrams in the Student Book. Take time to make sure that students fully understand the concepts introduced. Ask them to carry out the practical on Worksheet P3b.1(2).

SAFETY INFORMATION: Remind students that ray boxes can become hot after prolonged use.

Supplement Consider the use of a periscope. Students should carry out the practical on worksheet P3b.1(3).

Students can then work in pairs to study the Developing Practical Skills box in the Student Book.

Finish the learning episode

Ask students to explain to each other how an image is formed by a mirror.

Supplement Students following the supplement curriculum could explain why an image in a plane mirror is virtual and how a periscope can be used to see round a corner.

Technician's notes

Be sure to check the latest safety notes on these resources before proceeding.

The following resources are needed for the class practical P3b.1(1), per group:

ray box; ray box comb
sticky putty
cylindrical lens
power supply and leads
wooden block
front silvered plastic mirror
protractor; white paper

This practical works best in a darkened room.

Take care. Some ray boxes become hot after prolonged use.

The following resources are needed for the class practical P3b.1(2), per group:

squared paper; protractor; 50 cm ruler
sticky putty
plastic surface coated mirror
power supply; ray box, comb and cylindrical lens
two 4 mm leads

This practical works best in a darkened room.

Take care. Some ray boxes become hot after prolonged use.

You will need the following resources for the class practical P3b.1(3), per group:

pair of front surface plastic mirrors
50 cm ruler
sticky putty

Learning episode P3b.2 Refraction of light

Learning objectives

- Define refraction as the change in direction of a light ray passing from one medium to another
- Define and use the terms normal, angle of incidence and angle of refraction using ray diagrams
- Describe the passage of light through a transparent material (limited to the boundaries between two media only)

Common misconceptions

Some students may find it difficult to transfer their understanding of refraction from a horizontal boundary (as shown in Fig. P3.16 in the Student Book) to a vertical boundary.

Resources

Student Book Refraction of light

Worksheet P3b.2 Refraction

Resources for class demonstration and practical (see Technician's notes, following)

Approach

Introduce the learning episode

Half fill a glass beaker with water and place a long pencil in it. Draw students' attention to the apparent break in the pencil where it enters the water and relate this to refraction.

Develop the learning episode

Students should carry out the practical on Worksheet P3b.2. Set up the ray box so that it sends a single beam of light across the white paper. Place the rectangular Perspex block in the path of the beam. Mark the beams on paper, draw in the normal lines and measure the angles. Repeat for the other blocks. The experiments work best in a room with a good blackout.

SAFETY INFORMATION: Remind students that some ray boxes can become hot after prolonged use.

Discuss the text on Refraction of light in the Student Book. Students can tackle the questions with a partner.

Finish the learning episode

Provide students with a diagram of refraction at boundaries set at different angles, including a vertical boundary. Ask students to draw the refraction of light across these boundaries. Ensure that they can transfer their understanding of refraction across a horizontal boundary to boundaries at different angles.

Ask students to tell you two things that they have learned in this learning episode and one thing that they found difficult. Discuss areas of difficulty and use the things that they have learned to produce a class summary of the learning episode.

Technician's notes

Be sure to check the latest safety notes on these resources before proceeding.

The following resources are needed for the demonstration in the introduction:

beaker half-filled with water
long pencil

You will need the following resources for the class practical P3b.2, per group:

ray box; cylindrical lens; ray box comb
power supply and leads
protractor; white paper
rectangular Perspex block; scalene triangle Perspex block; right-angled triangle Perspex block; semi-circular Perspex block

This practical works best in a darkened room.

SAFETY INFORMATION: Take care. Ray boxes can become hot after prolonged use.

Learning episode P3b.3 Thin converging lenses and dispersion of light

Learning objectives

- Describe the action of thin converging lenses on a parallel beam of light, and know that rays of light from an object at distance can be assumed to be parallel
- Define and use the terms principal axis, principal focus (focal point) and focal length
- Interpret ray diagrams for the formation of an image by a thin converging lens, limited to real images
- Describe the characteristics of an image using the terms enlarged/same size/diminished, and upright/inverted
- Describe the dispersion of light as illustrated by the refraction of white light by a glass prism
- Know the traditional seven colours (red, orange, yellow, green, blue, indigo, violet) of the visible spectrum in order of frequency and in order of wavelength

Common misconceptions

Many students find it difficult to interpret ray diagrams showing image formation by a lens. Ensure that you allow plenty of time. Trace each line to show that the light rays (the lines) are bent (refracted) by the lens, and that where they cross is where the image is formed (the focal point). Ensure students understand that light is emitted in *all* directions from each point on the source and hit the lens at different angles, but the diagrams in the Student Book show only rays from the top of the object as this makes it easy to see (from the position of the top of the image) how big the image is compared to the object.

Resources

Student Book Thin converging lenses; Dispersion of light

Worksheet P3b.3(1) The focal length of a lens

Worksheet P3b.3(2) Focusing on a non-distant object

Resources for class practicals (see Technician's notes, following)

Approach

Introduce the learning episode

Ask students to describe the characteristics of an image in a plane mirror.

Develop the learning episode

Demonstrate the effect of a converging lens using the Student Book.

At this point the students can carry out the practical on Worksheet P3b.3(1). They draw a dotted centre line on the paper, place the centre of the lens on the centre line and send light along five parallel paths, noting the path of the emergent rays. They should measure the centre line to ray intersection distance.

SAFETY INFORMATION: Remind students that some ray boxes can become hot after prolonged use.

Ask students to carry out the practical on Worksheet P3b.3(2). They arrange apparatus as shown on the worksheet. They should measure the diameter of the bulb before it is switched on. A clear image should be gained on the screen. To get students started, you may wish to suggest the initial bulb–lens distance; for a lens of power 5 D, a 1 m distance will result in an image being formed 25 cm beyond the lens. Students repeat for five separate bulb–lens distances.

SAFETY INFORMATION: Remind students that they should not touch the filament bulb once it has been switched on. It will become extremely hot. Also remind them that they should not look at the bulb directly.

They should also note the object size (bulb diameter – measure before switching the bulb on) and image size (measured on the screen).

Ask students to draw a graph of their results. This allows you to assess their graph drawing skills. Focus on appropriate scales, linear scales, labelled axes, accurate plotting, and a smooth curve of best fit. Select an appropriate result from a student to draw a ray diagram to represent this situation. (Note that students are not required to draw such ray diagrams, or to use ray diagrams to determine the position of an image.) Select an example for which the bulb–lens distance falls between *f* and 2*f* (where *f* is the lens focal length).

Point out that an image does not always form at the focal point. The closer an object is to a (convex) lens, the further the image forms away from the lens.

Make sure that students are able to interpret ray diagrams of the different examples. It is worth spending time here ensuring that they understand the terms principal axis, principal focus (focal point) and focal length, and understand that in this practical the object is close to the lens, and so the rays cannot be assumed to be parallel so they do cross at the focal point.

Students should answer the questions in the Student Book.

Move on to discuss dispersion of light using the section in the Student Book. You could demonstrate dispersion (see Technician's notes) or show a video of the effect. Relate dispersion to refraction.

Finish the learning episode

Ask students to define the terms 'enlarged/same size/diminished, upright/inverted and (supplement only) real/virtual in terms of images.

Technician's notes

Be sure to check the latest safety notes on these resources before proceeding.

The following resources are needed for the class practical P3b.4(1), per group:

acrylic demonstration lenses or cylindrical lenses, power approx. +5 – 10 dioptres (20–10 cm focal length); one lens per group
plain A4 paper and ruler
ray box and slits
power supply and leads

The following resources are needed for the class practical P3b.4(2), per group:

12 V filament bulb, bulb holder (that fits filament bulb) and leads
power pack
converging (convex) lens – approximate power +5 dioptres (20 cm focal length)
lens holder
metre rule
white A3 paper (to use as screen), sticky tape

The bulb can be a 5–10 W bulb.

Students arrange apparatus as shown on the practical sheet. A clear image should be gained on the screen (screens may be formed by attaching A3 paper to a block, an upright book or a stool laid upon a bench). To get students started, you may wish to suggest the initial bulb–lens distance; for a lens of power 5 D, a 1 m distance will result in an image being formed 25 cm beyond the lens. Students repeat for five separate bulb–lens distances.

Students should also note the object size (bulb diameter – measure before switching the bulb on) and image size (measured on the screen).

SAFETY INFORMATION: Warn students not to touch the filament bulb once switched on – it will become very hot.

The following resources are needed to demonstrate dispersion of light by a glass prism:

ray box, power supply and leads
converging (convex) lens
ray box comb
wooden block
sticky putty
white card (to use as screen)
triangular glass prism

Fasten the card onto the block with sticky putty to make a screen. Place the ray box about 1 m from the screen. Use the cylindrical lens to focus the lamp filament onto the screen. Place the prism on the bench as shown below. Move the screen round to obtain a clear spectrum of rainbow colours. The experiment requires a very dark room for the full effect.

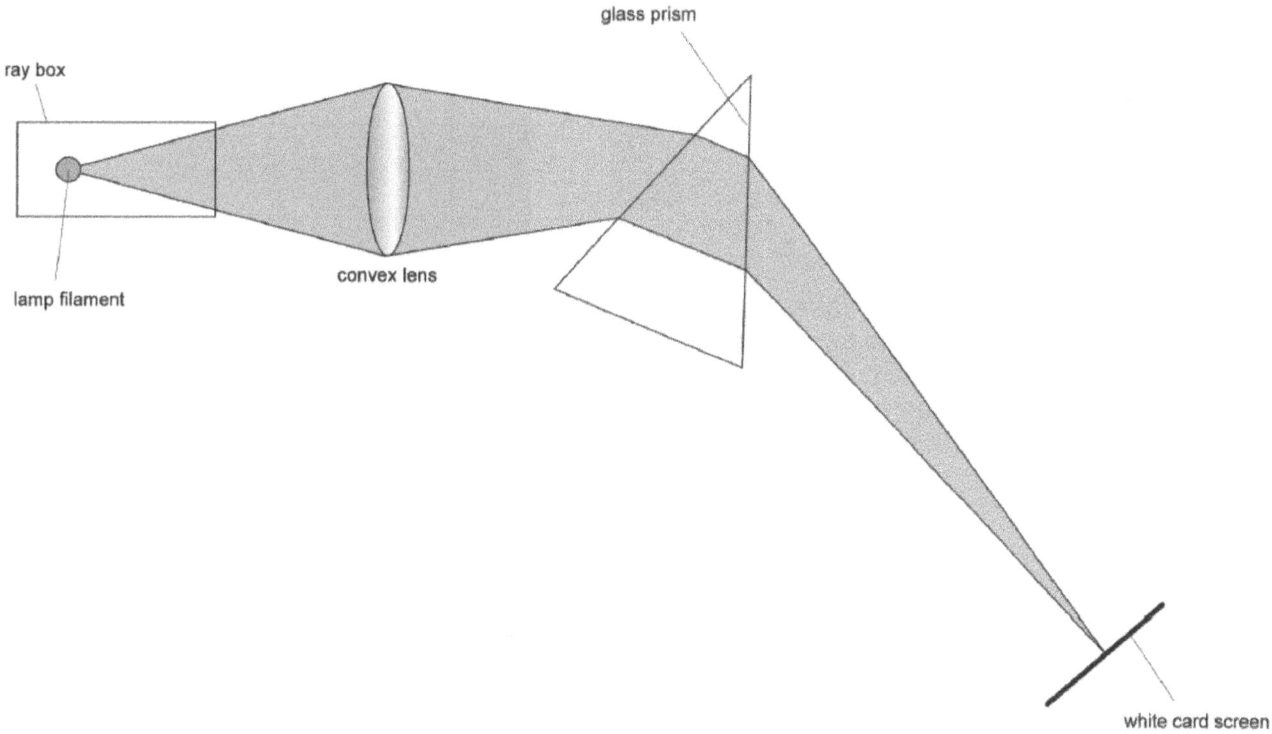

Learning episode P3b.4 Consolidation and summary

Learning aims

- Review the learning points of the topic summarised in the End of topic checklist
- Test understanding of the topic content by answering the End of topic questions

Resources

Student Book End of topic checklist and End of topic questions

Approach

Introduce the learning episode

Ask students to work with a partner to make a list of key words from this topic and then compare their list with the list of key terms given in the Student Book. Students could create a spider diagram to show the links between different key words. Discuss the checklist in the Student Book and use questioning to see how much of the content they are comfortable with.

Students could then create flashcards of the key information from the section. They can use the cards to quiz each other to assess their understanding.

Develop the learning episode

Ask students to work individually through the End of topic questions in the Student Book without looking at the text. As they work, walk around the classroom observing their answers and asking questions as necessary to find out which questions are causing difficulties.

Finish the learning episode

After a set period, ask the students to stop working and discuss any areas of difficulty you observed as you walked round the class. Students should complete any unanswered questions for homework, but you should stress that they should try to answer the questions without looking at the text, so that they can see how much they have remembered.

P3c Electromagnetic spectrum

Introduction

In this topic students learn about the electromagnetic spectrum.

Links to other topics

Sections	Essential background knowledge	Useful links
P3 Waves	P3a General properties of wave	P3d Sound

Topic overview

P3c.1	**Electromagnetic spectrum** In this learning episode students learn about the electromagnetic spectrum, the uses of electromagnetic waves and the harmless effects of some types of electromagnetic waves.
P3c.2	**Consolidation and summary** This learning episode provides an opportunity for a quick recap on the ideas encountered in the topic as well as time for the students to answer the End of topic questions in the Student Book.

Career links

Many careers require a knowledge of aspects of the electromagnetic spectrum. Broadcast and media systems technical operators need to understand communication systems to ensure that television, radio and online programmes are broadcast at high quality. Telecommunication technicians and engineers work on different communication systems such as satellites and fibre optic networks.

Learning episode P3c.1 Electromagnetic spectrum

Learning objectives

- Know the main regions of the electromagnetic spectrum (radio, microwave, infrared, visible, ultraviolet, x-ray, gamma) in order of frequency and in order of wavelength
- Know that all electromagnetic waves travel with the same high speed in a vacuum and at approximately the same speed in air
- Know some applications of the different regions of the electromagnetic spectrum including:
  (a) radio waves; radio and television transmissions, radar
  (b) microwaves; satellite television, mobile (cell) phones, microwave ovens
  (c) infrared; remote controllers for televisions, thermal imaging
  (d) visible light; vision, photography
  (e) ultraviolet; detecting fake bank notes
  (f) X-rays; medical scanning, security scanners
  (g) gamma rays; detection of cancer and its treatment
- Describe the harmful effects on people of excessive exposure to electromagnetic radiation, including:
  (a) ultraviolet; damage to surface cells and eyes, leading to skin cancer and eye conditions
  (b) X-rays and gamma rays; mutation or damage to cells in the body
- Supplement Know that the speed of electromagnetic waves in a vacuum is 3.0×10^8 m/s

Common misconceptions

The disturbance associated with electromagnetic waves is invisible, so students may have difficulty in grasping exactly what the amplitude of an electromagnetic wave is.

Students sometimes have difficulty in balancing risks and benefits, particularly when the benefits are immediate and the risks are only evident later.

Resources

Student Book Electromagnetic waves; Electromagnetic spectrum; Uses of electromagnetic waves; Harmful effects of electromagnetic waves

Resources for demonstrations (see below)

Approach

Introduce the learning episode

Show students pictures or other examples of objects that use electromagnetic radiation. Ask students what the objects have in common. Take answers and note them on the board.

Develop the learning episode

Discuss the text under the heading Electromagnetic waves in the Student Book, helping students to identify the properties that all electromagnetic waves share: they are transverse, can travel through a vacuum, all travel at the same speed.

Supplement Discuss with students the speed of light in vacuum. Some students may need support to understand standard form. Use the wave speed equation ($v = f\lambda$) to discuss the relationship between frequency and wavelength for electromagnetic radiation. If not carried out in Topic P3a, show how to carry out wave equation calculations using numbers relating to the electromagnetic spectrum.

Discuss the diagram of the electromagnetic spectrum in the Student Book. Students need to remember the waves in both frequency and wavelength order. They could work in pairs to create a mnemonic to help them remember the order. Then draw two arrows going in opposite directions on the board, one labelled wavelength and one labelled frequency. Give students the name of an electromagnetic wave and ask them to place a mark on the board to show where it is in order. Repeat until students are confident with the order of the waves.

Students should then answer the questions in the Student Book.

Using the section headed Uses of electromagnetic radiation in the Student Book, discuss the uses of electromagnetic waves. Return to the pictures and examples you used to introduce the learning episode as each use is discussed.

Students could create a table in their book with column headings 'electromagnetic wave' and 'uses'. You could assign one type of wave to pairs of students and ask them to identify the uses from the text. Students could then present their findings to the rest of the class who can complete the table in their book with the information.

Further information on the discovery of infrared radiation is given in the Science in Context box in the Student Book. If time allows, you could demonstrate the experiment described using the equipment given in the Technician's notes section. In pairs students can discuss the answer to the question at the end of the Science in Context box.

Now move on to discuss the harmful effects of electromagnetic waves using the text in the Student Book. Bring out a pair of sunglasses and a tube of sun cream. Ask students to suggest what these items protect against and how they work. Distinguish between burning skin (from infrared) and sunburn (from ultraviolet).

Demonstrate the safety features associated with handling and storing gamma ray sources (see Technician's notes):

- always handle them with tongs
- always store them in a lead-lined box
- never point them at people.

You could also discuss the use of gamma rays to treat cancer, perhaps considering risks and benefits of the treatment. The mention of cancer tumours needs sensitive handling as some of the class may have close relatives who have had to deal with it. Perhaps you could emphasise from the outset that a cancer diagnosis is often not fatal.

Show students a radiation badge or image of one and explain to them how it works. Introduce the term 'ionising radiation' and explain how the film works. The use of X-rays in diagnostics gives another opportunity to discuss risk and benefit trade-offs in the use of electromagnetic waves.

Students can answer the questions in the Student Book in pairs.

Finish the learning episode

You could ask students to work in pairs to make up quiz questions on the spectrum for other students (they must know the answers). For example, what type of electromagnetic radiation comes between microwaves and visible light? [infrared] What type of electromagnetic radiation is used for detecting broken bones? [X-rays]

Students could make notes on the Science in Context box in the Student Book. There is a question at the end of the section which could be set for homework to allow students to practice extended writing. Note this is beyond the requirements of the syllabus but may allow students to explore the content more deeply.

Technician's notes

Be sure to check the latest safety notes on these resources before proceeding.

The science department's radiation protection supervisor (schools) must provide training and information to all technicians and teachers who handle radioactive sources.

The following resources are needed for the demonstration of infrared radiation:

wood block
modelling putty and white card (to make a screen)
ray box
lens and comb
power supply and leads
glass prism and cylindrical lens
infrared sensor

The following resources are needed for the demonstration of the harmful effects of the electromagnetic spectrum:

sunglasses and sun cream
gamma ray source
handling tool
Geiger tube and counter
radiation monitoring badge

SAFETY INFORMATION: If a radiation monitoring badge is not available, show video or alternative online resource instead.

All gamma ray sources available in schools could be used, for example, cobalt-60 (although this is a little weak). Radioactive sources must be stored securely when not supervised by authorised staff. Standard operating procedures must always be followed when using radioactive materials.

Learning episode P3c.2 Consolidation and summary

Learning aims

Review the learning points of the topic summarised in the End of topic checklist
Test understanding of the topic content by answering the End of topic questions

Resources

Student Book End of topic checklist and Student Book End of topic questions

Approach

Introduce the learning episode

Ask students to work with a partner to make a list of key words from this topic. They could then work together to produce a spider diagram showing how the different concepts are linked. They could compare their list with the list of key terms given in the Student Book. Discuss the checklist in the Student Book and use questioning to see how much of the content they are comfortable with.

Students could make flashcards of the key content and then use the flashcards to quiz each other on the information.

Develop the learning episode

Ask students to work individually through the End of topic questions in the Student Book without looking at the text. As they work, walk around the classroom observing their answers and asking questions as necessary to find out which questions are causing difficulties.

Finish the learning episode

After a set period, ask the students to stop working and discuss any areas of difficulty you observed as you walked round the class. Students should complete any unanswered questions for homework, but you should stress that they should try to answer the questions without looking at the text, so that they can see how much they have remembered.

P3d Sound

Introduction

In this topic students learn about sound and its properties. They have met the general wave equation in Topic P3a *General properties of waves* and learned about speed in Topic P1b *Motion*. Students following the supplementary curriculum should also already be familiar from Topic P3a *General properties of waves* with the idea of sound as a longitudinal wave.

Links to other topics

Sections	Essential background knowledge	Useful links
P1 Motion, forces and energy	P1a Physical quantities and measurement techniques P1b Motion	
P2 Thermal physics		P2a Kinetic particle model of matter
P3 Waves	P3a General properties of waves	P3b Light P3c Electromagnetic spectrum

Topic overview

P3d.1	**Properties of sound** This learning episode applies wave ideas to the phenomenon of sound. Supplement Students describe sound waves in air as a longitudinal wave with a series of compressions and rarefactions.
P3d.2	**Measuring the speed of sound** This learning episode describes a method to allow the speed of sound to be measured.
P3d.3	**Consolidation and summary** This learning episode provides an opportunity for a quick recap on the ideas encountered in the topic as well as time for the students to answer the End of topic questions in the Student Book.

Career links

There are many careers which require an understanding of sound. For example, audio-visual technicians and live sound engineers are responsible for controlling and balancing the sound at events such as conferences and concerts. In healthcare, sonographers use ultrasound to 'look inside' the human body, and audiologists work with patients who have problems with their hearing.

Learning episode P3d.1 Properties of sound

Learning objectives

- Describe the production of sound by vibrating sources
- Supplement Describe the longitudinal nature of sound waves in air as a series of compressions and rarefactions
- Supplement Describe qualitatively compressions as regions of higher pressure due to particles being closer together and rarefactions as regions of lower pressure due to particles being spread further apart
- State the approximate range of frequencies audible to humans as 20 Hz to 20 000 Hz
- Know that a medium is needed to transmit sound waves
- Describe how changes in amplitude and frequency affect the loudness and pitch of sound waves
- Describe an echo as the reflection of a sound
- Define ultrasound as sound with a frequency higher than 20 kHz

Common misconceptions

Students have difficulty in relating the longitudinal pressure variations of the air to the apparently transverse wave often used to represent sounds. The Institute of Physics suggest the use of a 'Slink-o-scope' which uses a slinky spring and pivoted ruler to model the process of changing a longitudinal wave motion to a transverse trace.

Resources

Student Book The basics of sound; Pitch, frequency, amplitude and loudness; Echoes

Worksheet P3d.1 Sound sources

Resources for demonstrations and class practical (see Technician's notes, following)

Approach

Begin the learning episode

Use a loudspeaker, signal generator and candle flame to demonstrate low-frequency longitudinal waves (see Technician's notes). Ask students to feel the vibrations of their throats as they hum a low note. Then place a table tennis ball on a horizontal loudspeaker emitting a note at about 500 Hz to show that the cone is also vibrating.

Develop the learning episode

Discuss the text in the Student Book. Students should then carry out the first part of the practical on Worksheet P3d.1.

Then demonstrate the range of human hearing, introducing the ideas of amplitude and frequency at the same time. Explain that ultrasound is sound above 20 kHz that humans can't hear (see Technician's notes).

Show the effect of changing the amplitude and frequency of the sound from the signal generator. Use the section 'Pitch, frequency, amplitude and loudness' in the Student Book to show how these terms can be represented diagrammatically. Demonstrate the Slink-o-scope to show how the side-to-side motion of the longitudinal wave can be represented by the traces shown in Fig. P3.33 of the Student Book.

Supplement Introduce the terms 'compression' and 'rarefaction' to describe the longitudinal wave. Ask students to sketch a longitudinal wave and label areas of compression and rarefaction. Students should also label the wavelength of the wave as the distance between two points of rarefaction or compression.

Students should then carry out the second part of the practical on Worksheet P3d.1. They will need to ensure that they keep the string stretched so that vibrations can pass along it. Students should also try questions 1 to 3 in the Student Book.

Now move on to discuss echoes using the text on Echoes in the Student Book. If possible, take students to a location where there is a loud echo – perhaps in a large hall or some distance from a large wall. Clap your hands to produce a series of sharp clicks. Use the stop clock and tape measure/trundle wheel to take measurements which will allow the calculation of the speed of sound in Learning episode P3d.2.

Finish the learning episode

Ask students to work with a partner to produce a summary of the main points of the learning episode.

Technician's notes

Be sure to check the latest safety notes on these resources before proceeding.

The following resources are needed for the demonstrations in the learning episode:

signal generator set to about 10 Hz: signal generator range 20 Hz to 20 kHz
large loudspeaker with exposed cone
candle
clamp and stand to hold the lit candle about 20 cm in front of the loudspeaker
leads
table tennis ball
microphone
clamp and stand to hold microphone in front of loudspeaker
oscilloscope with large display
stop clock
tape measure or trundle wheel – your PE department may have one of these you can borrow.

The following resources are needed for the demonstration of the Slink-o-scope in the learning episode:

slinky spring
metre rule
rubber band
clamp stand
felt tip pen
sheets of A3 squared paper (stick two A4 pieces together)

For instructions on how to create the slink-o-scope can be found by searching on the Institute of Physics website (https://spark.iop.org/slink-o-scope)

The following resources are needed for the practical P3d.1, per group:

tuning fork
large rubber bung
small plastic beaker
30 cm plastic rulers; long length of thick string
two empty plastic pots with small holes in their bases
two used matchsticks

Learning episode P3d.2 Measuring the speed of sound

Learning objectives

- Determine the speed of sound in air using a method involving a measurement of distance and time
- Supplement Know that, in general, sound travels faster in solids than in liquids and faster in liquids that in gases

Resources

Student Book Determining the speed of sound in air; Determining the speed of sound by an echo method

Worksheet P3d.2 Measuring the speed of sound

Resources for demonstration and class practical (see Technician's notes, following)

Approach

Begin the learning episode

Recap the two equations that lead to a value for the speed of a wave (speed = frequency × wavelength and speed = distance travelled / time taken). Explain that in this lesson you are going to consider one or both of these equations to measure a value for the speed of sound.

Develop the learning episode

Introduce the method or methods you have chosen.

If you took data from an echo in Learning episode P3d.1, use that to demonstrate how to calculate the speed of sound (as long as the wall is at least 70 metres away). Study the worked example in the Student Book by setting students the problem to attempt in pairs (without the solution). After a given time, bring the class back together to compare students' solutions with the given answer.

Using a cathode ray oscilloscope with a microphone allows a sound pulse to be timed over a short distance. Use the section 'Determining the speed of sound in air' in the Student Book to help with this.

Worksheet P3d.2 gives instructions for carrying out a practical activity to measure the speed of sound. If you do not have enough apparatus, this could be carried out as a demonstration. (Alternatively, you could use the free software package 'Audacity' with separated stereo headphones used as microphones.)

SAFETY INFORMATION: Make sure that the sound source does not produce sound that is too loud, as this will be uncomfortable for ears.

Students can work in pairs to study the Developing Practical Skills activity in the Student Book.

Calculate a value based upon the measurements obtained and consider the sources of error in the method chosen. Use this opportunity to consider experimental skills and discuss aspects such as reliability, accuracy and precision.

Supplement Remind students about how sound waves propagate through the air with compressions and rarefactions. Ask students to predict the speed of sound through liquids and solids using their knowledge of the structure of solids, liquids and gases. Then use the two examples given in the supplement material in the Student Book to show that sound travels faster in solids than in liquids and gases.

Students should answer the questions in the Student Book.

Finish the learning episode

Set students the following problems.

1. A sound moving through the ground has a frequency of 500 Hz and a wavelength of 4 m. How fast is it moving?
2. The speed of sound in air is approximately 330–350 m/s. What is the frequency of a sound that has a wavelength of 0.34 m?
3. The speed of sound moving through a metal is 5000 m/s. What is the wavelength of a sound whose frequency is 10 kHz?
4. Supplement A sound of wavelength 2.0 m moves from air into a liquid. If the speed rises from 340 m/s to 3400 m/s, what happens to the frequency and the wavelength of the wave?

Technician's notes

Be sure to check the latest safety notes on these resources before proceeding.

The following resources are needed for the demonstration of echoes:

cathode ray oscilloscope and microphone OR stop clock and suitable area where echoes can be heard with a distance of at least 70 m between the wall and the observer

The following resources are needed for the class practical P3d.2, per group:

two microphones;
fast recording device such as digital storage oscilloscope (OR free software package Audacity and separated stereo headphones) and sound source
metre rule and stop clock

If you do not have access to a cathode ray oscilloscope, there are various oscilloscope programs available that you can use to display a sound signal on a computer screen.

Learning episode P3d.3 Consolidation and summary

Learning aims

- Review the learning points of the topic summarised in the End of topic checklist
- Test understanding of the topic content by answering the End of topic questions

Resources

Student Book End of topic checklist and End of topic questions

Approach

Introduce the learning episode

Ask students to work with a partner to make a list of key words from this topic. They could then work together to produce a spider diagram showing how the different concepts are linked. They could compare their list with the list of key terms given in the Student Book. Discuss the checklist in the Student Book and ask questions to see how much of the content students are comfortable with.

Students could make flashcards of the key content and then use the flashcards to quiz each other on the information.

Develop the learning episode

Ask students to work individually through the End of topic questions in the Student Book without looking at the text. As they work, walk around the classroom observing their answers and questioning them as necessary to find out which questions are causing difficulties.

Finish the learning episode

After a set period, ask students to stop working and discuss any areas of difficulty you observed as you walked round the class. Students should complete any unanswered questions for homework, but you should stress that they should try to answer the questions without looking at the text, so that they can see how much they have remembered.

P4 Electricity

Contents

Overview of the section

In this section students study electrical quantities: charge, current, electromotive force, potential difference, resistance and electrical energy. This is preparation for the following topic, which covers electrical circuits. Students are introduced to series and parallel circuits and various circuit components. A short topic on the hazards of electricity follows, and the final topic in the section summarises how energy is transferred from cells, batteries and generators to circuit components, and how energy is transferred by an electric motor.

Starting points

The Student Book section opener puts the ideas in the section into context and sets the scene. It also allows students to acknowledge and value their prior learning, and provides a benchmark against which future learning can be compared.

The questions provide a structure for introducing the section and can be used in a number of different ways:

- You could ask students to consider the questions as an introductory homework task.
- You could put students into groups to share their own ideas and understanding and then to report back to the whole class.
- Students could be given access to the Internet, preferably with a tight timescale, to find out the information required.

You could then use a spider chart or other form of wall chart to summarise everybody's ideas.

Recording these initial ideas allows you to retain them for reference as the individual topics are developed. In this way, your students' progress in learning can be readily acknowledged.

P4a Electrical quantities

Introduction

In this topic students learn about the electrical quantities: charge, current and resistance, and, for students studying the supplementary syllabus, electromotive force and potential difference. All students build on work in Topic P1f *Energy, work and power* to learn about electrical energy and electrical power, while students studying the supplementary syllabus also apply ideas about electrical work done by a source to the definition of electromotive force and electrical work done by a current to the definition of potential difference. This topic is essential knowledge for Topic P4c *Electrical circuits*.

Links to other topics

Topics	Essential background knowledge	Useful links
P1 Motion, forces and energy		P1f Energy, work and power
P4 Electricity and magnetism		P4c Electrical circuits P4c Electrical safety

Topic overview

P4a.1	**Electric charge** In this learning episode students see experiments that demonstrate simple electrostatic effects. They learn how to use the word 'charge' and why there is a need for two types. They learn an explanation for charge in terms of electrons.
P4a.2	**Electric current** In this learning episode students are introduced to current as the flow of charge including the flow of electrons in a metal. Supplement Students learn the definition of current and how it is calculated. They explore further the flow of electrons in a metal, which leads to the definition of conventional current.
P4a.3	**Voltage (electromotive force and potential difference)** In this learning episode students learn how to measure the voltage across a component. Supplement Students are also introduced to the definitions of electromotive force and potential difference.
P4a.4	**Resistance** In this learning episode students are introduced to resistance and how resistance is measured. Supplement Students learn about the effects of the length and cross-sectional area of a conductor on its resistance.
P4a.5	**Electrical energy and electrical power** This learning episode introduces energy transfers and calculations in electrical circuits, including the cost of using electrical appliances using the unit kilowatt-hour.

P4a.6	**Consolidation and summary** This learning episode provides an opportunity for a quick recap on the ideas encountered in the topic as well as time for the students to answer the End of topic questions in the Student Book.

Note on circuit construction

Students regularly struggle with constructing working circuits. There are a number of approaches that can be used to support students with this. Producing 'circuit mats' showing the circuit drawn out 'life size', that students place their components on before connecting can help. Having a correctly wired circuit which students can go and look at to check their circuits can also help. Another challenge is that the teacher may end up 'trouble shooting' electric circuits. Use 'test circuits' at the side of the laboratory to help avoid this. These 'test circuits' are simple bulb – battery circuits that are known to function correctly. A key component is then removed e.g., cell, bulb, wire. Students can then connect the component that they think is 'broken' into the appropriate circuit. If the bulb lights, their component is working correctly, if the bulb does not light, then the component is faulty and should be put to one side for your technician to deal with. Students can then sort out their circuits with much less recourse to the teacher.

Career links

An understanding of current and voltage (potential difference) is used in many careers in electrical generation and distribution. Possible careers which require use of ammeters and voltmeters (often as multimeters) include engineering maintenance technician and electronics engineering technician.

Learning episode P4a.1 Electric charge

Learning objectives

- State that there are positive and negative charges
- State that positive charges repel other positive charges, negative charges repel other negative charges, but positive charges attract negative charges
- Distinguish between electrical conductors and insulators, and give typical examples
- Supplement State that charge is measured in coulombs

Common misconceptions

Students need to appreciate the particular use of the word 'charge' to explain static electricity. For many, charging has to do with energy transfer, as in charging a mobile phone. The idea of 'charge' as an attribute of matter similar to mass that is responsible for forces is a difficult one to learn when the word already means something else.

Resources

Student Book Electric charge

Resources for demonstrations and class practical (see Technician's notes, following)

Approach

Introduce the learning episode

Inflate a balloon. Small 'water balloons' work well for this, but they need to be inflated using a balloon pump. Rub the balloon against your hair (or a student's hair – choose straight hair without gel) or a jumper. Notice the interaction between the balloon and the hair. The fact that charged objects attract neutral ones can be a barrier to understanding, particularly in practical work. It is best if students find this out early on, but there is no need at this stage for them to understand why it happens in terms of polarisation.

You could use a Van de Graaff generator to set up some large sparks. Explain that the sparks are caused by rapid heating of the air as electrons jump from one sphere to the other (see Technician's notes).

SAFETY INFORMATION: Standard safety precautions apply when using electrostatic generators. Please refer to the latest health and safety advice.

Develop the learning episode

Discuss the opening section headed Electric charge in the Student Book. Give students examples of different charges and ask if they will attract or repel.

Discuss the section headed Conductors and insulators in the Student Book. Use Table P4.1 to emphasise that most conductors are metals.

Demonstrate simple electrostatic experiments by charging balloons as follows.

Inflate each balloon. Take the first one and, holding it at the tied end, briskly rub a cloth over the other end. Suspend the balloon by a thread from a clamp stand. Repeat with the second balloon so that you have two charged balloons, each hanging from a separate clamp stand. (Keep the balloons the same distance from each other on the bench at this point.)

Now bring the two balloons towards each other by moving the stands together. If all is well, they should repel each other, suggesting a rule: **similarly charged objects repel**.

Charge the cellulose acetate strip and polythene strip in turn and hold close to one of the balloons.

The attraction to one of the strips suggests that there is more than one type of charge, leading to another rule: **differently charged objects attract**.

The demonstration works best if the equipment is dry. Liberal use of a hair dryer may help.

Supplement Introduce the unit of charge as coulombs (C) using the Student Book.

Students should answer the questions in the Student Book in pairs. Question 3 is a supplement question.

Students should work through the Developing Practical Skills box in the Student Book with a partner.

Finish the learning episode

Use the Van de Graaff generator to charge a metallised plastic ball on a nylon thread (see Technician's notes). Note the reaction of the ball to strips of charged cellulose acetate and charged polythene. Ask the class to deduce the charge on the dome of the generator.

SAFETY INFORMATION: Standard safety precautions apply when using electrostatic generators. Please refer to the latest health and safety advice.

Students could make notes on the Science in Context box in the Student Book as part of their homework (this is beyond the requirements of the syllabus but may allow students to explore the content more deeply).

Technician's notes

Be sure to check the latest safety notes on these resources before proceeding.

The following resources are needed for the demonstrations in the introduction and at the end of the activity:

rubber balloon (Small 'water balloons' work particularly well, but must be inflated using a balloon pump)
Van de Graaff generator
metal sphere on insulating handle
lead to connect sphere to base of generator
strips of cellulose acetate and strips of polythene

SAFETY INFORMATION: General safety precautions apply when using electrostatic generators. It is important for the safety of the demonstrator that the base of the generator is properly earthed. No student or member of staff should have electrical items in their pockets, metal in their body or a pacemaker.

The following resources are needed for the class practical P4b.2a.1, per group:

woollen cloth
two strips of cellulose acetate, two strips of polythene and one strip of wood
hair dryer
small evaporating dish

The experiment works best if the materials are dry.

The following resources are needed for the demonstration on charging balloons in the main part of the activity:

pair of rubber balloons and thread to suspend the balloons
pair of clamp stands from which to suspend balloons
cellulose acetate strip and polythene strip
woollen cloth and hair dryer

The demonstration works best if the equipment is dry. Liberal use of a hair dryer may help.

Learning episode P4a.2 Electric current

Learning objectives

- Know that electric current is related to the flow of charge
- Know that electric current in metals is related to the flow of electrons
- Describe the use of ammeters (analogue and digital) with different ranges
- Know the difference between direct current (d.c.) and alternating current (a.c.)
- Supplement Describe electrical conduction in metals in terms of the movement of delocalised (mobile) electrons
- Supplement Define electric current as the charge passing a point per unit time
  Supplement Recall and use the equation $I = Qt$
- Supplement State that conventional current is from positive to negative and the flow of electrons is from negative to positive

Common misconceptions

Students find it difficult to grasp that batteries and Van de Graaff generators do essentially the same job. They may also be confused about the difference between the direction of conventional current (arbitrarily chosen by Benjamin Franklin before electrons were discovered) and that of electron flow in a metallic conductor.

Often students think of electricity as a 'thing' that moves around in wires. Discourage them from using the word 'electricity'. Talk instead about electric current. This will help to remove the confusion as you build up the electron model during the topic.

Resources

Student Book Electric current; Delocalised (mobile) electrons; Equation for current; Alternating current and direct current; Measuring current

Resources for class demonstrations (see Technician's notes, following)

Approach

Introduce the learning episode

Place a wig of long hair on the dome of a Van de Graaff generator. Charge up the dome and use class questioning or encourage the students to apply their learning from the previous lesson to explain why the hairs stand on end (see Technician's notes).

SAFETY INFORMATION: Standard safety precautions apply when using electrostatic generators. Please refer to the latest health and safety advice.

Develop the learning episode

Discuss the section headed 'Electric Current' in the Student Book.

Carry out the following demonstration to show that charge can be made to flow when a circuit is completed. Connect the dome of the Van de Graaff generator in series with a neon lamp, a microammeter and the base of the generator. Switch on the generator. Note that the lamp glows (perhaps faintly) and the microammeter indicates a current.

Make sure that students understand what the role of the microammeter is in the circuit.

Set up a simple circuit with a bulb. Demonstrate how to use digital and analogue ammeters to measure the current in the circuit. When using an analogue ammeter, make sure students understand the purpose of the backing mirror and using parallax to increase the accuracy of the reading. Students may not have met electric circuits in their previous study, so reassure them that they will study circuits and build their own in the next section of the course.

Supplement Point out that electrons flow in the direction of negative current – early scientists assumed that positive charge flowed through metals! Then replace the generator with a high voltage supply and repeat the experiment. Make sure that students understand the difference between the flow of electrons and the direction of conventional current.

Discuss alternating current (a.c.) and direct current (d.c.) using the section headed Alternating current and direct current in the Student Book. Students could sketch the nature of each current (as shown in Fig. P4.9 and Fig. P4.10) and then annotate their diagram with information from the Student Book.

Supplement Introduce the definition of electric current as the charge passing a point in a circuit per unit time. You may need to remind students about charge from the previous learning episode. Ask students to work through the example given in the Student Book. You may wish to give them other questions to try to ensure that they can correctly use the equation $I = Qt$.

Supplement Use Fig. P4.8 to discuss the difference between conventional current and electron flow.

Supplement Students could then answer the questions in the Student Book in pairs.

Finish the learning episode

Ask students to write down two things that they have learned in the learning episode and one thing that surprised them. Take feedback and discuss responses.

Technician's notes

Be sure to check the latest safety notes on these resources before proceeding.

The following resources are needed for the static wig demonstration:

Van de Graaff generator
wig of long hair to place on dome

It is important for the safety of the operator that the base of the Van de Graaff generator is properly earthed.

A wig made from strips of paper is effective.

The following resources are needed for the static circuits demonstration:

Van de Graaff generator
neon lamp
microammeter (1 μA resolution)
high voltage supply
leads (preferably with transparent insulation so that the metal core is visible)

Subdued lighting makes it easier for students to see the glow of the neon lamp.

A scalamp or light-spot galvanometer works best, but you have to be careful to work down through the ranges from the least sensitive to the most sensitive in order to avoid overloading it.

Be careful when dealing with high voltage supplies – always use the output with the built-in safety resistor.

The following resources are needed for the ammeter demonstration:

simple circuit consisting of a cell and a lamp.
digital ammeter
analogue ammeter

Learning episode P4a.3 Voltage (electromotive force and potential difference)

Learning objectives

- Describe the voltage of a source as the cause of current in a circuit
- Know that the voltage of the source is shared between the components in a series circuit
- Describe the use of voltmeters (analogue and digital) with different ranges
- Supplement Define electromotive force (e.m.f.) as the electrical work done by a source in moving a unit charge around a complete circuit
- Supplement Know that e.m.f. is measured in volts (V)
- Supplement Define potential difference (p.d.) as the work done by a unit charge passing through a component
- Supplement Know that the p.d. between two points is measured in volts

Common misconceptions

Students often confuse the terms cell and battery (two or more cells) because battery is used commonly to describe what appears to be a single component. Students also struggle to distinguish between electromotive force and potential difference, especially as both are measured in volts, and potential difference has the symbol *V*. Try to always use the correct terms for e.m.f. and p.d. when discussing electric circuits e.g., ask, 'What is the e.m.f. of the cell?' rather than, 'How many volts can the cell produce?'.

Resources

Student Book Voltage (electromotive force and potential difference); Measuring voltage

Resources for class demonstration (see Technician's notes, following)

Approach

Introduce the learning episode

Ask students to work in pairs to write down what a cell does. Take feedback. Explain that in this learning episode they are going to learn about the function of a cell in an electrical circuit.

Develop the learning episode

Supplement Discuss the meaning of voltage and the term 'electromotive force' in the Student Book. Emphasise that the e.m.f. can be thought of as the energy transferred into a circuit.

Supplement Discuss the section on Potential difference in the Student Book using the concept of energy transfer to explain the difference between e.m.f. and p.d.

Show the students a voltmeter and set up the circuits shown in the Student Book. Note that they have not yet met circuit symbols in this course, and they will have the opportunity to build circuits for themselves in the next topic, P4c *Electrical circuits.*

Stress that the voltmeter goes in parallel with the component, whereas the ammeter goes in series. (Again, students will meet the terms 'parallel' and 'series' in more detail in the next topic.) Show that the sum of the potential differences across each identical lamp is the same as the potential difference across the cell in the first circuit.

Finish the learning episode

Discuss the main points of the learning episode and check that students have achieved the learning objectives.

Technician's notes

Be sure to check the latest safety notes on these resources before proceeding.

The following resources are needed for the potential difference demonstration:

three voltmeters and one ammeter
two lamps
cell and connecting leads

Learning episode P4a.4 Resistance

Learning objectives

- Recall and use the equation for resistance, $R = VI$
- Supplement Recall and use the following relationship for a metallic electrical conductor:(a) resistance is directly proportional to length (b) resistance is inversely proportional to cross-sectional area

Common misconceptions

As soon as physics becomes quantitative it becomes much more difficult for many students. Voltmeters and ammeters look very similar, so students often confuse current with potential difference.

Some uncertainty at this point is perfectly normal. As long as the correct language is used and misconceptions are corrected every time they come up, students should get things straight in their heads by the end of the topic.

Resources

Student Book Resistance

Approach

Introduce the learning episode

Introduce the idea that all circuit components will provide a resistance to current moving through them. Some components provide a lot of resistance, and others less resistance.

Develop the learning episode

Introduce the quantitative aspects of resistance at this point. Make sure that students can rearrange the equation to make R the subject.

Students can do the worked examples in the Student Book without the solutions at first. After a given time, compare their answers to the given solutions and discuss any differences. They should then answer the questions.

Discuss the section headed Determining resistance in the Student Book.

Students can work through the Developing Practical Skills activity in the Student Book in pairs and explain what it shows.

Supplement Students should read the text on Resistance of a metallic wire in the Student Book and make notes to answer the following two questions: (a) how does the length of a wire affect its resistance? (b) how does the cross-sectional area of a wire affect its resistance?

Supplement Study the worked example in the Student Book by setting students the problem to attempt in pairs (without the solution). After a given time, bring the class back together to compare students' solutions with the given answer. Students should answer the questions in the Student Book.

Finish the learning episode

You could discuss the Science in Context box in the Student Book (this is beyond the requirements of the syllabus but may allow students to explore the content more deeply).

Learning episode P4a.5 Electrical energy and electrical power

Learning objectives

- Understand that electric circuits transfer energy from a source of electrical energy, such as an electrical cell or mains supply, to the circuit components and then into the surroundings
- Recall and use the equation for electrical power $P = IV$
- Recall and use the equation for electrical energy $E = IVt$
- Define the kilowatt-hour (kW h) and calculate the cost of using electrical appliances where the energy unit is the kW h

Common misconceptions

Students will not yet be used to dealing with mixed units in science, so will need to be made aware of the need to express values in basic units (W, J or s) before putting them into a formula.

Resources

Student Book Electrical energy and electrical power; The kilowatt-hour

Worksheet P4a.5 Kettle power

Resources for demonstrations (see Technician's notes, following)

Approach

Introduce the learning episode

Arrange a circus (display) of different mains electrical appliances around the room. Give students ten minutes to find out the power rating (W or kW) of each appliance from its base plate. If the appliances are not available, then provide photographs of base plates for students to use instead.

SAFETY INFORMATION: It is safer to turn off the mains supply to the room while students are handling the appliances.

Develop the learning episode

Using the text on Electrical energy and electrical power in the Student Book as a guide, explain the significance of the power rating, and do an energy transfer calculation, involving a power rating in kW and a time in minutes or hours, to emphasise the need to get quantities into base units before doing calculations.

Study the worked examples in the Student Book by setting students the problem to attempt in pairs (without the solution). After a given time, bring the class back together to compare students' solutions with the given answer. Some students may find these questions challenging because they will not only need to be able to transpose the formula $E = IVt = Pt$ to do these questions but also deal with conversions of power and time to basic units. You may need to provide additional support to students. Students should then answer the questions in the Student Book.

Discuss the cost of electricity in your region, and how electricity bills are calculated. Using the text in the Student Book on The kilowatt-hour, show students how the kilowatt-hour is derived. Study the worked example using the process described previously.

Students should complete the questions in the Student Book.

Finish the learning episode

Carry out the kettle demonstration, which provides the measurements needed to do Worksheet P4a.5. You use a kettle to boil a known mass of water.

SAFETY INFORMATION: Boiling water is scalding, and appropriate care should be taken.

Encourage students to help with taking measurements, but do not let them take measurements of the temperature of the boiling water: this is too dangerous. The measurements you need to calculate the heating power of the kettle are the: mass of empty kettle, mass of kettle + water, mass of water and the temperature rise of the water (note that students do not need to recall the details of this calculation). From these measurements students can calculate the power rating of the kettle. They can use Worksheet P4a.5 to carry out the calculation.

Ask students to write down two things that they did well in the learning episode and one thing that they feel they could improve. Discuss responses and deal with any issues that arise.

Technician's notes

Be sure to check the latest safety notes on these resources before proceeding.

The following resources are needed for the demonstrations in the introduction:

A range of mains electrical appliances with base plates that show their power ratings (could include kettle, TV, kitchen blender, hair dryer, desk lamp, computer)

SAFETY INFORMATION: All mains appliances should be school property and have been inspected and tested according to the normal employer's routine.

It is safer to turn off the mains supply to the room while students are handling the appliances.

The appliances should be set out as a circus – at different places around the room. Students go to each appliance in turn.

Place a potato in the oven to cook and use a stop clock to measure how long it takes.

The following resources are needed for the kettle demonstration (Worksheet P4a.5):

plastic kettle (white if possible); stop clock; scales; thermometer

SAFETY INFORMATION: Boiling water is scalding, and appropriate care should be taken.

Students should be encouraged to help with measurement taking, but no measurements of the temperature of the boiling water should be made, as this is too hazardous.

Learning episode P4a.6 Consolidation and summary

Learning aims

- Review the learning points of the topic summarised in the End of topic checklist
- Test understanding of the topic content by answering the End of topic questions

Resources

Student Book End of topic checklist and End of topic questions

Approach

Introduce the learning episode

Ask students to work with a partner to make a list of key words from this topic. They could then work together to produce a spider diagram showing how the different concepts are linked. They could compare their list with the list of key terms given in the Student Book. Discuss the checklist in the Student Book and ask questions to see how much of the content students are comfortable with. Students could make flashcards of the key content and then use the flashcards to quiz each other on the information.

Develop the learning episode

Ask students to work individually through the End of topic questions in the Student Book without looking at the text. As they work, walk around the classroom observing their answers and questioning them as necessary to find out which questions are causing difficulties.

Finish the learning episode

After a set period, ask the students to stop working and discuss any areas of difficulty you observed as you walked round the class. Students should complete any unanswered questions for homework, but you should stress that they should attempt the questions without looking at the text, so that they can see how much they have remembered.

P4b Electrical circuits

Introduction

In this topic students learn about circuit diagrams, series and parallel circuits, and the action and use of circuit components. This continues the learning from Topic P4a *Electrical quantities*. The action of fuses will be covered in more detail in Topic P4c *Electrical safety*, and the action of generators and motors will be covered again in Topic P4d *Cells, batteries, generators and motors*.

Links to other topics

Section	Essential background knowledge	Useful links
P1 Motion, forces, and energy		P1f Energy, work and power
P4 Electricity	P4a Electrical quantities	P4c Electrical safety P4d Cells, batteries, generators and motors

Topic overview

P4b.1	**Circuit diagrams and circuit components** In this short learning episode students recall the symbols used to draw circuit diagrams. Supplement Students following the supplement curriculum will additionally learn the symbols for generators, light-emitting diodes (LEDs), transformers and magnetising coils.
P4b.2	**Series and parallel circuits** In this learning episode students make use of the circuit symbols to construct series and parallel circuits and describe what happens to the current and voltage in series circuits. They will also consider the use of lamps and resistors in parallel circuits.
P4b.3	**Behaviour of some components** In this learning episode students will learn about motors and variable resistors. Supplement Students following the supplement curriculum will also learn about the action of a generator and LEDs in a circuit.
P4b.4	**Consolidation and summary** This learning episode provides an opportunity for a recap on the ideas encountered in the topic as well as time for the students to answer the End of topic questions in the Student Book.

Note on circuit construction

Students regularly struggle with constructing working circuits. There are a number of approaches that can be used to support students with this. Producing ‘circuit mats’ which show the circuit drawn out ‘life size’, that students place their components on before connecting can help. Having a correctly wired circuit which students can go and look at to check their circuits can also help. Another challenge is that the teacher may end up ‘trouble shooting’ electric circuits. Use ‘test circuits’ at the side of the laboratory to help avoid this. These circuits are simple bulb – battery circuits that are known to function correctly. A key component is then removed e.g., cell, bulb, wire. Students can then connect the component that they think is ‘broken’ into the appropriate circuit.

If the bulb lights, their component is working correctly, if the bulb does not light, then the component is faulty and should be put to one side for your technician to deal with. Students can then sort out their circuits with much less recourse to the teacher.

Career links

An understanding of circuit construction and circuit components is important in many careers. Electricians must be able to read circuit diagrams and use them to safely install electrical products. Robotics engineers use electrical circuits and components to design and build machines to do automated jobs in industries.

Learning episode P4b.1 Circuit diagrams and circuit components

Learning objectives

- Draw and interpret circuit diagrams containing cells, batteries, power supplies, switches, resistors (fixed and variable), heaters, lamps, motors, ammeters, voltmeters and fuses, and know how these components behave in the circuit
- Supplement Draw and interpret circuit diagrams containing generators and light emitting diodes (LEDs), and know how these components behave in the circuit
- Know that the current at every point in a series circuit is the same
- Supplement Recall and use in calculations, the fact that: a) the sum of the currents entering a junction in a parallel circuit is equal to the sum of the currents that leave the junction b) the total p.d. across the components in a series circuit is equal to the sum of the individual p.ds across each component (c) the p.d. across each branch of a parallel arrangement of components is the p.d. across the whole arrangement

Common misconceptions

Students expect something to be used up as it goes around a circuit. This gets in the way of their acceptance of the idea of current conservation in a series circuit. It helps to emphasise from the start that the current carries energy, which is transferred on its way round the circuit. There are a number of models that can be used to help explain electric current, and it can be helpful to identify how the model is good at explaining the physical phenomena, but also where it is less good.

Resources

Student Book Circuit diagrams and circuit components

Worksheet P4b.1 Circuit symbol dominoes

Resources for demonstration (see Technician's notes, following)

Approach

Introduce the learning episode

Display the circuit shown below on the board. Assemble the circuit methodically, talking about the function of each component as you put it into place. Explain that the blobs in the diagram show where leads must be stacked to obtain the parallel connection.

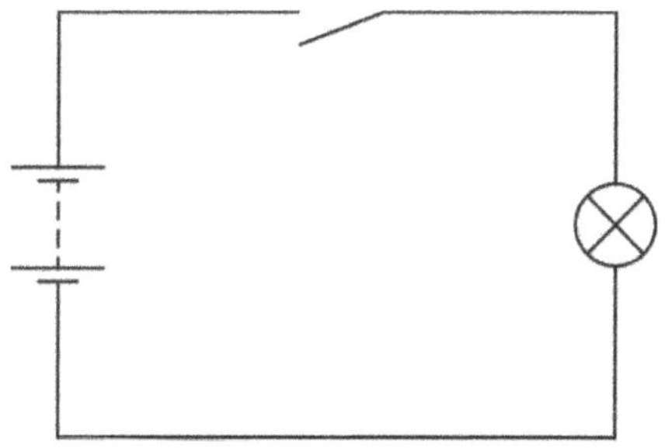

Discuss the difference between a cell and a battery.

Explain that in this activity students will learn what the symbols in the circuit diagrams represent.

Develop the learning episode

Students should play 'symbol dominoes' in pairs using dominoes made using Worksheet P4c.1. Students place all the dominoes face down on the table and take turns to pick them up at random until all have gone.

Students should keep their dominoes hidden from each other. They then take it in turn to place a domino on the table (if they can), starting with the battery symbol, matching each symbol with appropriate words.

Other players can challenge an incorrect placement, requiring the player to miss a turn. The first one to lay down all their dominoes wins.

Note that there are two devices that emit light (LED (supplement) and lamp), two that impede electrons (resistor and variable resistor) and one that stops them (open switch).

Supplement Introduce the circuit symbols for generators and light emitting diodes (LEDs). Students are very likely to already know what LEDs are, and will have met the idea of generating electricity when studying energy resources in Topic P1f *Energy, work and power*.

Finish the learning episode

Students should quiz each other on the circuit symbols using the domino cards.

Technician's notes

Be sure to check the latest safety notes on these resources before proceeding.

The following resources are needed for the demonstration in the introduction:

two 1.5 V batteries
ammeter (0–20 mA); voltmeter (0–2 V)
lamp (1.25 V) and bulb holder
4 mm leads

Learning episode P4b.2 Series and parallel circuits

Learning objectives

- Draw and interpret circuit diagrams containing cells, batteries, power supplies, switches, resistors (fixed and variable), ... lamps, ... ammeters, voltmeters ..., and know how these components behave in the circuit
- Know that the current at every point in a series circuit is the same
- Know how to construct and use series and parallel circuits
- Calculate the combined resistance of two or more resistors in series
- Know that, for a parallel circuit, the current from the source is larger than the current in each branch
- Know that the combined resistance of two resistors in parallel is less than that of either resistor by itself
- Know the advantages of connecting lamps in parallel in a lighting circuit
- Supplement Recall and use in calculations, the fact that: (a) the sum of the currents entering a junction in a parallel circuit is equal to the sum of the currents that leave the junction (b) the total p.d. across components in a series circuit is equal to the sum of the individual p.d.s across each component (c) the p.d. across each branch of a parallel arrangement of components is the p.d across the whole arrangement
- Supplement Calculate the combined resistance of two resistors in parallel

Common misconceptions

Many students find it difficult to assemble parallel circuits from a circuit diagram. The trick is to assemble the series parts first and then put the parallel ones in second.

Students expect something to be used up as it goes around a circuit. This gets in the way of them accepting the idea of conservation of current in a series circuit. It helps to emphasise from the start that the current carries energy, which is transferred on its way round the circuit.

Resources

Student Book Series and parallel circuits

Worksheet P4b.2(1) Simple series circuits

Worksheet P4b.2(2) Parallel circuits

Resources for demonstrations and class practicals (see Technician's notes, following)

Approach

Introduce the learning episode

Ask students to write down as many circuit symbols and what they represent as they can. Some students may find it helpful to have the dominoes used in the previous lesson to help them.

Develop the learning episode

Students should carry out the practical on Worksheet P4b.2(1) Simple series circuits. They will follow the circuit diagrams to assemble four different circuits. Here is a list of what each circuit does:

Circuit 1: The lamp comes on when the switch is pressed.

Circuit 2: The lamp goes off when the switch is pressed.

Circuit 3: The lamp changes brightness as the variable resistor is changed. The voltmeter reading increases as the brightness increases.

Circuit 4: The lamp gets brighter as the switch is pressed. The ammeter reading increases as the brightness of the lamp increases.

Series circuits

Once students have completed the worksheet, move on to discuss current in a series circuit.

Use a rope loop to model the movement of current in an electric circuit. The movement of the rope represents the flow of charge (current) in a simple series circuit (see Technician's notes). The teacher acts as the cell/battery and a student acts as the bulb. The teacher and student hold the rope loop and stand far enough apart so that the rope is not loose. The student should hold the rope very lightly in one hand only. You should then make the rope circulate. Point out that the movement of the rope (current) is the same throughout the circuit, and that it takes no time for the energy to travel from the source (the battery) to the bulb. Ask the student volunteer what is happening at their hand. It will be getting warmer as energy is transferred from the rope to their hand. This represents the resistance of the bulb.

SAFETY NOTE: make sure that the student does not grip the rope too hard to avoid potential friction burns.

Supplement It is worthwhile emphasising at this stage that the potential difference across a component is a measure of how much energy is transferred by the current, so it can get transferred from the electrons to the components on its way round the circuit.

Add another 'bulb' to the rope circuit by asking another student to lightly hold the rope in one hand. As the rope is moved, point out that the rope moves at the same rate throughout the circuit – each bulb is experiencing the same current.

Relate the demonstration to the text and diagrams in the Student Book on Lamps in series.

Continue the rope model demonstration. Ask one student 'bulb' to increase slightly how tightly they are holding the rope (increase the resistance). Make the rope move and ask the two students what is happening at their hands. The student whose hand is tighter will feel their hand getting warmer. Point out that the 'resistance' of the bulbs is the sum of the individual resistances.

Relate the demonstration to the text in the Student Book on Resistors in series.

Supplement Ask students to look at Fig. P4.27 and the data table given in the Student Book. Ask them how the data supports the statement about p.d. at the bottom of the section.

Students should answer the questions in the Student Book. Question 3 is Supplement.

Parallel circuits

Connect a single lamp to a mains power pack so that it glows with normal brightness. Then connect a second lamp in parallel. Use class questioning to establish that because both lamps have the same brightness as the original one they must have the same current and potential difference.

Then establish that each electron that leaves the negative terminal of the power supply must pass through one of the lamps but not through both. Challenge the class to decide how the current in the battery is related to the current in each lamp.

At this stage, the students can carry out the practical on Worksheet P4b.2(2). Students connect the resistors in parallel with the battery. Then they place the ammeter in different places. Some students tend to build parallel circuits as though they were series ones, so it is advisable to keep an eye on the circuits that they assemble. It may be helpful to demonstrate using a simple series circuit containing one resistor and show students how to 'add on' the parallel sections one at a time so that they can see them added on 'across' the original circuit (see Technician's notes).

Supplement With students following the supplementary curriculum, establish that the current in and out of the battery is the sum of the currents in each resistor.

Next ask students to decide if house lights are wired in a series circuit or a parallel circuit. They should then read the text in the Student Book on Lamps in parallel to find out if their answer was correct.

Now move on to discuss combinations of resistors in circuits using the text in the Student Book on Resistors in parallel. It is important for students to realise that when resistors are combined in parallel, the total resistance is lower than the value of any of the individual resistors; and when resistors are combined in series, the total resistance is higher than any of the individual resistors. This is a useful way to check calculations. Work through the example data given in the Student Book.

Supplement Work through the analysis in the Student Book with the students to show them how to approach this type of question. They may need support to carry out the calculation. You could provide further questions for them to practice using the equation to calculate the total resistance of resistors in parallel.

Students should answer the questions in the Student Book. Questions 2, 3 and 4 are Supplement.

Finish the learning episode

Provide the class with one power supply, two lamps and three switches. Challenge them to guide you through assembling a circuit in which each lamp can be turned on and off independently, but only when the master switch is closed. Here is the circuit diagram.

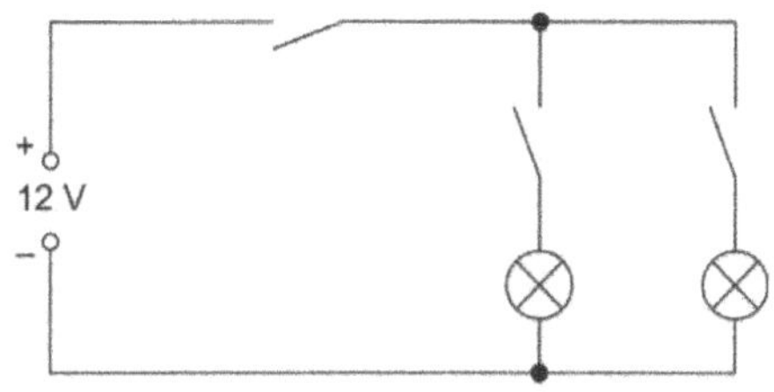

Technician's notes

Be sure to check the latest safety notes on these resources before proceeding.

The following resources are needed for the current demonstration:

rope loop – about a 3 m length of rope tied into a loop. If a nylon rope is used, the ends can be melted together.

The following resources are needed for the for the class practical P4b.2(1), per group:

battery
lamp
normally open switch
resistor
ammeter (0–1 A) and voltmeter (0–10 V)
variable resistor

The lamp rating needs to match the battery voltage.

3.5 V MES lamps are easy to match with a 3 V battery of two 1.5 V cells.

The resistor needs to be about the same as the working resistance of the lamp – 12 Ω is fine for a 3.5 V lamp which draws 0.3 A.

The variable resistor needs to go from 0 to about 10 times the working resistance of the lamp – 100 Ω is fine for a 3.5 V lamp.

The following resources are needed for the parallel circuit demonstration:

two 12 V ray box lamps
mains 12 V d.c. power supply
four 4 mm leads

A low output resistance power supply is essential, so that the p.d. across the supply terminals does not drop appreciably when the second lamp is connected.

The following resources are needed for the demonstration at the end of the activity:

12 V power supply
two 12 V ray box lamps
three press-to-make switches
nine 4 mm leads

The following resources are needed for the class practical P4b.2(2), per group:

3 V battery
ammeter (0–1 A)
22 Ω resistor and 10 Ω resistor
seven 4 mm leads

A 2 W power rating for the resistors should be enough to stop them getting too hot.

Warn students not to touch them until they have had a chance to cool down a little.

Learning episode P4b.3 Behaviour of some components

Learning objectives

- Draw and interpret circuit diagrams containing cells, batteries, power supplies, switches, resistors (fixed and variable), heaters, lamps, motors, ammeters, voltmeters and fuses, and know how these components behave in the circuit
- Supplement Draw and interpret circuit diagrams containing generators and light emitting diodes (LEDs), and know how these components behave in the circuit

Resources

Student Book Behaviour of some components

Resources for demonstrations (see Technician's notes, following)

Approach

Introduce the learning episode

Ask students to draw the circuit symbols for a selection of the following components: cells, batteries, power supplies, switches, resistors (fixed and variable), heaters, lamps, motors, ammeters, voltmeters and fuses. Students could reuse the domino cards from Learning episode P4b.1 to support their recall.

Supplement Also ask students to draw the circuit symbols for a generator and an LED.

Develop the learning episode

Introduce variable resistors using the text and diagrams in the Student Book under Variable resistor. Demonstrate the motor circuit to show the behaviour of a variable resistor and the change in motor as the resistance of the resistor is altered. Discuss possible uses for these two circuit components.

Supplement Revisit lamps in parallel by asking students to draw a circuit diagram for a circuit containing one power supply, two LEDs and some switches, in which each LED can be turned on and off independently.

Supplement If possible, demonstrate a simple generator being turned by hand to light a lamp.

Finish the learning episode

Students can work with a partner to make a summary of the main points of this learning episode. Bring the class back together after a given period and discuss their summaries. Allow students to add to their summaries during the discussion.

Technician's notes

Be sure to check the latest safety notes on these resources before proceeding.

The following resources are needed for the motor circuit demonstration:

voltmeter
motor (12 V)
power supply (12 V)
variable resistor
connecting leads
switch

The following resources are needed for the generator demonstration:

motor kit and power pack and connecting leads
or generator
1.5V lamp

Learning episode P4b.4 Consolidation and summary

Learning aims

- Review the learning points of the topic summarised in the End of topic checklist
- Test understanding of the topic content by answering the End of topic questions

Resources

Student Book End of topic checklist and End of topic questions

Approach

Introduce the learning episode

Ask students to work with a partner to make a list of key words from this topic. They could then work together to produce a spider diagram showing how the different concepts are linked. They could compare their list with the list of key terms given in the Student Book. Discuss the checklist in the Student Book and ask questions to see how much of the content students are comfortable with. Students could make flashcards of the key content and then use the flashcards to quiz each other on the information.

Develop the learning episode

Ask students to work individually through the End of topic questions in the Student Book without looking at the text. As they work, walk around the classroom observing their answers and questioning them as necessary to find out which questions are causing difficulties.

Finish the learning episode

After a set period, ask the students to stop working and discuss any areas of difficulty you observed as you walked round the class. Students should complete any unanswered questions for homework, but you should stress that they should try to answer the questions without looking at the text, so that they can see how much they have remembered.

P4c Electrical safety

Introduction

In this topic students learn about the hazards associated with electricity and some of the safety features that help to protect consumers.

Links to other topics

Sections	Essential background knowledge	Useful links
P4 Electricity	P4a Electrical quantities P4c Electrical circuits	P4d Cells, batteries, generators and motors

Topic overview

P4c.1	**Electrical hazards and domestic appliances** This learning episode identifies common electrical hazards, including the reason electrical appliances using mains supplies can be dangerous to humans. Students learn about electrical safety features in a domestic context.
P4c.2	**Consolidation and summary** This learning episode provides an opportunity for a quick recap on the ideas encountered in the topic as well as time for the students to answer the End of topic questions in the Student Book.

Career links

All careers which involve the use of electrical circuits require an understanding of electrical hazards and electrical safety features.

Learning episode P4c.1 Electrical hazards and domestic appliances

Learning objectives

- Understand the heating effect of current
- State the hazards of (a) damaged insulation, (b) overheating cables, (c) damp conditions, (d) excess current from overloading of plugs, extension leads, single and multiple sockets when using a mains supply
- Explain the use and operation of trip switches and fuses and choose appropriate fuse ratings and trip switch settings (knowledge of RCDs (Residual Current Devices) is **not** required)
- Explain why the outer casing of an electrical appliance must be either non-conducting (double-insulated) or earthed

Common misconceptions

Often students think of electricity as a 'thing' that moves around in wires. Discourage them from using the word 'electricity'. Talk instead about electric current. This will help to remove the confusion as you build up the electron model during the topic.

Students often fail to realise that a neutral wire is actually connected to the ground at the electricity substation. Current that flows from the live wire to ground through the earth wire gets back into the neutral wire at the substation, thus completing the circuit.

Resources

Student Book Electrical hazards

Worksheet P4c.1 Fuse blowing

Resources for demonstrations and class practical (see Technician's notes, following)

Approach

Introduce the learning episode

You could ask the class to tell you what they know about electrical safety features in the home, and generate a mind map on the whiteboard. Don't correct any errors – leave that until the plenary.

Develop the learning episode

Gather information about our everyday use of electricity. This could be through discussion in small groups, describing electrical appliances from home, or through a survey of the room or building.

Show students pictures of typical hazards: frayed cables, over-long cables, damaged plugs, damaged sockets, water near sockets, incorrect objects inserted into sockets, metal objects put into electrical appliances such as toasters and overloaded sockets (through use of multiplugs). Use Table P4.2 in the Student Book at this point. Explain why faults in electrical appliances can be dangerous. Explain how to avoid these hazards and what action to take to correct them.

Now move on to discuss the text in the Student Book on Domestic appliances, drawing attention to the fuse in the plug of appliances. Students carry out the practical on Worksheet P4c.1 Fuse blowing. Check that they know what the different components in the circuit are for and how they are connected.

Students will connect the fuse, meter and lamp in series. They will slowly increase the voltage from the power supply and observe what happens to the lamp, meter and fuse. Take care as the fuse is very hot. The lamp needs to be rated at 12 W or 24 W, so that the fuse blows before the lamp does! Low lighting makes it easier for the students to observe the glow of the fuse just before it melts. Students often rush the practical and fail to observe key details. You might want to have two fuses for each group.

SAFETY INFORMATION: Remind students to take care as the fuse is very hot.

Earth safety demonstration

Students could observe the earth safety demonstration described below (see Technician's notes).

Assemble the circuit, explaining the role of each cable as it is connected. The loose wire can be left out at this stage.

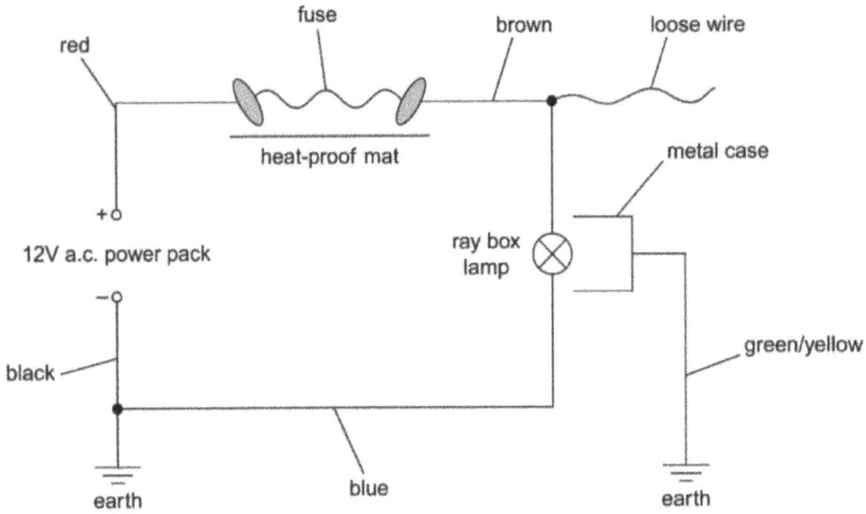

Coil the fuse wire on the heatproof mat so that it is not illuminated by the lamp of the ray box.

The earthing of the ray box casing could be done through a second power pack.

Turn on the a.c. supply so that the lamp shines brightly. The fuse should get slightly warm.

Now add a loose lead to the circuit somewhere between the fuse and the lamp. Explain that this is a potentially deadly situation.

Touch the other end of the loose lead to the casing and watch the fuse blow and cut the lamp off from the supply. Take care as the fuse will be very hot.

You might want to repeat with the casing not earthed.

Students should answer the questions in the Student Book in pairs. Then you could ask them to discuss the Science in Context box in the Student Book with a partner (this is beyond the requirements of the syllabus but may allow students to explore the content more deeply).

Finish the learning episode

Use class questioning to correct/improve the mind map generated in the starter activity.

Issue whiteboards and marker pens or sheets of paper. Set students some general questions about connecting appliance to the mains, then finish off with a fuse rating calculation or two.

Technician's notes

Be sure to check the latest safety notes on these resources before proceeding.

The following resources are needed for the demonstration in the introduction:

pictures of typical electrical hazards, for example, frayed cables, over-long cables, damaged plugs, damaged sockets, water near sockets, incorrect objects inserted into sockets, metal object put into electrical appliances such as toasters or overloaded sockets (through use of multiplugs)

The following resources are needed for the earth safety demonstration:

12 V a.c. supply with earthing point
ray box lamp with metal outer casing
coil of 5 A fuse wire
two red 4 mm leads; black 4 mm lead, green or yellow 4 mm lead, brown 4 mm lead, blue 4 mm lead
heatproof mat
two crocodile clips

The power pack needs to supply at least 6 A to ensure that the fuse blows.

Check that the voltage used is sufficient to blow the fuse before the activity.

The following resources are needed for the class practical P4c.1, per group:

20 mm glass bodied fuse rated at 500 mA, part number RS415654 or similar
12 V ray box lamp or similar
ammeter (0–1 A)
four 4 mm leads
two crocodile clips
variable d.c. power supply for ray box

The lamp needs to be rated at 12 V and 24 W, so that the fuse blows before the lamp does! The fuse will get very hot. It should not be touched during the activity or picked up immediately afterwards. Subdued lighting makes it easier for the students to observe the glow of the fuse just before it melts.

SAFETY INFORMATION: Take care as the fuse is very hot.

Learning episode P4c.2 Consolidation and summary

Learning aims

- Review the learning points of the topic summarised in the End of topic checklist
- Test understanding of the topic content by answering the End of topic questions

Resources

Student Book End of topic checklist and End of topic questions

Approach

Introduce the learning episode

Ask students to work with a partner to make a list of key words from this topic. They could then work together to produce a spider diagram showing how the different concepts are linked. They could compare their list with the list of key terms given in the Student Book. Discuss the checklist in the Student Book and ask questions to see how much of the content students are comfortable with. Students could make flashcards of the key content and then use the flashcards to quiz each other on the information.

Develop the learning episode

Ask students to work individually through the End of topic questions in the Student Book without looking at the text. As they work, walk around the classroom observing their answers and questioning them as necessary to find out which questions are causing difficulties.

Finish the learning episode

After a set period, ask the students to stop working and discuss any areas of difficulty you observed as you walked round the class. Students should complete any unanswered questions for homework, but you should stress that they should try to answer the questions without looking at the text, so that they can see how much they have remembered.

P4d Cells, batteries, generators and motors

Introduction

In this topic students learn how cells and batteries transfer chemical energy into electrical energy and how generators transfer kinetic energy into electrical energy. This topic provides opportunities to revise earlier work on energy stores and transfers from Topic *P1f Energy, work and power*, and the uses of circuit components from Topic *P4b Electrical circuits*.

Links to other topics

Topics	Essential background knowledge	Useful links
P1 Motion, forces and energy		P1f Energy, work and power
P4 Electricity		P4a Electrical quantities P4b Electrical circuits

Topic overview

P4d.1	**Cells, batteries, generators and motors** In this learning episode students will learn about the action and use of cells, batteries, generators and motors in terms of energy transfer.
P4d.2	**Consolidation and summary** This learning episode provides an opportunity for a quick recap on the ideas encountered in the topic as well as time for the students to answer the End of topic questions in the Student Book.

Career links

Careers in the electricity generation sector require an understanding of the concepts in this topic. There are jobs in both renewable and non-renewable energy sectors.

Learning episode P4d.1 Cells, batteries, generators and motors

Learning objectives

- Know that cells and batteries transfer chemical energy into electrical energy
- Know that generators transfer kinetic energy into electrical energy
- Know that electric motors transfer electrical energy into kinetic energy

Common misconceptions

As noted in Section 1, students may have developed an understanding of energy from the common uses of the term, which does not match to how the term is used in physics. There is no easy definition of energy which can be used with students, so you will need to develop the concept carefully. Emphasise that energy does not explain 'why' things happen. Rather, energy is a method of calculating if activities are possible.

The Institute of Physics has produced teacher resources which explore misconceptions in energy in more detail.

Resources

Student Book Cells and batteries; Generators; Electric motors

Resources for class demonstration (see Technician's notes, following)

Approach

Introduce the learning episode

Ask students to work with a partner to make a list of as many devices as possible that use electric motors. After a short period of time, bring the class back together to discuss ideas.

Develop the learning episode

Show that connecting a battery to a motor produces rotation (see Technician's notes).

Explain that what has just been demonstrated is electrical energy being transferred from a source (the battery) to movement of the motor (kinetic energy).

Remind students (from Topic P1f *Energy, work and power*) of the idea of energy transfers as a way for energy to move between different stores. Point out the mechanism that is causing the change – electric current, electrical working. You may wish to show a simple flow diagram to represent the energy transfer.

Point out to students that some energy is transferred to internal (thermal) stores due to heating (caused by the resistance of the wires, or by friction between the moving parts in the motor). The amount of energy input equals the overall amount of energy output – energy is conserved, but the motor is not 100% efficient so not all the electrical energy is transferred usefully as kinetic energy.

Now ask students if they can recall an example of the reverse – how movement can be used to generate electricity. Students may have built a model wind turbine in Topic P1f *Energy, work and power*, and may know that some bicycles have dynamo lights that operate when the wheels are turning. If possible, demonstrate a simple generator being turned by hand to light a lamp. Note that the operation of a generator, in terms of electromagnetic induction, is not required in the Combined Science syllabus.

Finish the learning episode

Ask students to work with a partner to make a summary of the main points covered in this learning episode. After a period of time, bring the class together to discuss the summaries and ensure that the learning objectives have been noted.

Technician's notes

Be sure to check the latest safety notes on these resources before proceeding.

The following resources are needed for the motor demonstration:

motor (12 V)
12 V battery
connecting leads
switch

The shaft of a large motor will turn slower, making it easier to see that the shaft rotates.

The following resources are needed for the generator demonstration:

motor kit and power pack and connecting leads
or generator
1.5V lamp

Learning episode P4d.2 Consolidation and summary

Learning aims

- Review the learning points of the topic summarised in the End of topic checklist
- Test understanding of the topic content by answering the End of topic questions

Resources

Student Book End of topic checklist and End of topic questions

Approach

Introduce the learning episode

Ask students to work with a partner to make a list of key words from this topic. They could then work together to produce a spider diagram showing how the different concepts are linked. They could compare their list with the list of key terms given in the Student Book. Discuss the checklist in the Student Book and ask questions to see how much of the content students are comfortable with. Students could make flashcards of the key content and then use the flashcards to quiz each other on the information.

Develop the learning episode

Ask students to work individually through the End of topic questions in the Student Book without looking at the text. As they work, walk around the classroom observing their answers and questioning them as necessary to find out which questions are causing difficulties.

Finish the learning episode

After a set period, ask the students to stop working and discuss any areas of difficulty you observed as you walked round the class. Students should complete any unanswered questions for homework, but you should stress that they should try to answer the questions without looking at the text, so that they can see how much they have remembered.

P5 Space physics

Contents

Overview of the section

There are just two topics in this section. The first topic introduces students to the Solar System: the Sun, the eight named planets, minor planets, including Pluto, and asteroids in the asteroid belt. Students also learn about moons that orbit planets.

In the second topic, students learn about the general characteristics of the Sun as the star closest to Earth, are introduced to the light year as a unit of distance, and carry out calculations using this unit. Students learn how gravity is the force that drives the orbital motion of planets and other objects in orbit around the Sun. Students studying the supplementary material are also introduced to the equation for average orbital speed, and how stars are powered by nuclear reactions involving the fusion of hydrogen.

Students also find out about the lifecycle of stars, and supplementary students learn about the Big Bang theory.

Starting points

The Student Book section opener puts the ideas in the section into context and sets the scene. It also allows students to acknowledge and value their prior learning, and provides a benchmark against which future learning can be compared.

The questions provide a structure for introducing the section and can be used in a number of different ways:

- You could ask students to consider the questions as an introductory homework task.
- You could put students into groups to share their own ideas and understanding and then to report back to the whole class.
- Students could be given access to the Internet, preferably with a tight timescale, to find out the information required.

You could then use a spider chart or other form of wall chart to summarise everybody's ideas.

Recording these initial ideas allows you to retain them for reference as the individual topics are developed. In this way, your students' progress in learning can be readily acknowledged.

P5a The Solar System

Introduction

In this topic students learn about the Solar System. They learn about the Sun as the closest star to the Earth.

Links to other topics

Sections	Essential background knowledge	Useful links
P1 Motion, forces and energy	P1a Physical quantities and measurement techniques	
P3 Waves		P3c Electromagnetic spectrum

Topic overview

P5a.1	**The Solar System** In this learning episode students will learn about the present-day structure and scale of the Solar System.
P5a.2	**Consolidation and summary** This learning episode provides an opportunity for a quick recap on the ideas encountered in the topic as well as time for the students to answer the End of topic questions in the Student Book.

Career links

There are many different careers that are linked to the Solar System, including those linked to the design and launching of satellites such as propulsion engineers responsible for the launch vehicles which put satellites in space, and digital electronic hardware technicians who work to assemble the computer systems used on the satellites, and those linked to the information and data produced by satellites, including disaster response coordinators who use satellite data to support on-the-ground rescue responses, meteorologists who study the Earth's atmosphere to predict weather and climate, or even space lawyers.

Learning episode P5a.1 The Solar System

Learning objectives

- Describe the Solar System as containing:
  (a) one star, the Sun
  (b) the eight named planets and know their order from the Sun
  (c) minor planets that orbit the Sun, including dwarf planets such as Pluto and asteroids in the asteroid belt
  (d) moons, that orbit the planets

Common misconceptions

Many students still think that Pluto is still a planet. An explanation during the introduction would be a good idea.

Many textbooks and online prints show planetary orbits as being circular. They are all elliptical, but with differing degrees of ellipticity (Venus' orbit is closest to a circle, Mercury's is the most elliptical).

Resources

Student Book The Solar System

Worksheet P5a.1 The Solar System

Resources for demonstration (see Technician's notes, following)

Approach

Introduce the learning episode

Ask students to work in pairs and list all the objects that they think make up the Solar System. They could then be asked to group the objects. You could then discuss their collective groups and objects.

Develop the learning episode

Ask students to use the first part of Worksheet P5a.1 to complete a schematic diagram of the Solar System, using the information in the Student Book.

Students should then complete the questions in the Student Book.

You could read the Science in Context box: 'Life beyond our Solar System' with students and discuss the meaning of the term 'exoplanet' (this is beyond the requirements of the syllabus but may allow students to explore the content more deeply). Discuss the question posed in the Science in Context box and follow up by asking the students to suggest if there is 'life' in other parts of the Universe.

Finish the learning episode

You should finish by getting the students to create a scale model of the Solar System in the classroom (planets only). They can use an online Solar System scale calculator to work out rough dimensions and then get students to stand in the appropriate positions (using a tape measure) holding a piece of paper with the name of the planet and a scale diagram of the planet (most planets will only be dots, unless you have a large scale outside on a playground or a field). There are many online scale calculators that could be used in real-time in the classroom, or you could pre-prepare one for use that fits your classroom. If you exclude Pluto, you could finish by asking the students to estimate where the position of Pluto might be.

Technician's notes

The following resources are needed for demonstration purposes:

tape measure (to suit dimension of classroom or playground/open space)

Learning episode P5a.2 Consolidation and summary

Learning aims

- Review the learning points of the topic summarised in the End of topic checklist
- Test understanding of the topic content by answering the End of topic questions

Resources

Student Book End of topic checklist and End of topic question

Approach

Introduce the learning episode

Ask students to work with a partner to make a list of key words from this topic. They could then work together to produce a spider diagram showing how the different concepts are linked. They could compare their list with the list of key terms given in the Student Book. Discuss the checklist in the Student Book and ask questions to see how much of the content students are comfortable with.

Develop the learning episode

Ask students to work individually through the End of topic questions in the Student Book without looking at the text. As they work, walk around the classroom observing their answers and questioning them as necessary to find out which questions are causing difficulties.

Finish the learning episode together

After a set period, ask the students to stop working and discuss any areas of difficulty you observed as you walked round the class. Students should complete any unanswered questions for homework, but you should stress that they should try to answer the questions without looking at the text, so that they can see how much they have remembered.

P5b Stars and the Universe

Introduction

In this topic students learn about the general characteristics of the Sun and its place in the Milky Way, including how astronomical distances are measured. They learn why Solar System objects orbit the Sun and students following the supplementary syllabus also consider how orbital speed depends on the radius of the orbit. They also find out about the life cycle of stars and about the Big Bang, and the age of the Universe.

Links to other topics

Sections	Essential background knowledge	Useful links
P1 Motion, forces and energy	P1a Physical quantities and measurement techniques P1b Motion	
P3 Waves	P3b Light P3c Electromagnetic spectrum	

Topic overview

P5b.1	**The Sun as a star** In this learning episode students will learn about the characteristics of the Sun and the electromagnetic radiation it emits, and about light-years. Supplement Students will learn how the Sun releases energy by the process of nuclear fusion.
P5b.2	**Gravity and orbits** In this learning episode students will learn that most of the mass of the Solar System is contained in the Sun and how this means that the other bodies in the Solar System orbit the Sun. Supplement Students will also learn that the mean orbital speed decreases with distance from the Sun.
P5b.3	**Stars, galaxies and the Universe** In this learning episode students will learn about the Milky Way, galaxies and the life cycle of stars. They revisit the idea of light-years for the measurement of astronomical distances. Supplement Students will also learn about the Big Bang Theory.
P5b.4	**Consolidation and summary** This learning episode provides an opportunity for a quick recap on the ideas encountered in the topic as well as time for the students to answer the End of topic questions in the Student's Book.

Career links

An understanding of the fusion processes within stars is used by solar physicists, some of whom use the information to predict space weather. Fusion physicists are also working to recreate fusion on Earth. Astro-chemists study the elements and chemical reactions which occur in space and use the knowledge to predict conditions elsewhere in the universe. Astro-biologists extend our understanding of the conditions required for life on Earth to exoplanets to consider what signs alien life might leave. Astronomers and astrophysicists study objects in the universe to develop our understanding of how stars and other objects formed.

Learning episode P5b.1 The Sun as a star

Learning objectives

- Know that the Sun is a star of medium size, consisting mostly of hydrogen and helium, and that it radiates most of its energy in the infrared, visible and ultraviolet regions of the electromagnetic spectrum
- Know that: (a) the Sun is the closest star to the Earth (b) astronomical distances can be measured in light-years, where one light-year is the distance travelled in (the vacuum of) space by light in one year
- Calculate the time it takes light to travel a significant distance such as between objects in the Solar System
- Supplement Know that stars are powered by nuclear reactions that release energy and that in stable stars the nuclear reactions involve the fusion of hydrogen into helium

Resources

Student Book The Sun as a star; Light-years; Time taken by light to travel across the Solar System; Nuclear fusion

Worksheet P5b.1(1) The Sun as a star

Supplement Worksheet P5b.1(2) Comparing nuclear fusion and combustion

Approach

Introduce the learning episode

Start the learning episode by showing the students a series of images of the Sun at different wavelengths as it is now (or nearly now). Images can be taken from NASA's Solar Dynamics Observatory (search 'Solar Dynamics Observatory)

You can either show the website live or download/copy images taken at different wavelengths in preparation for the activity. Ask students to work in pairs making observations on the images such as: sunspots; solar plumes; flares; prominences; the corona; coronal holes. There are a great many images online that show diagrams of these features

Point out to students that the Sun is a dynamic entity and that it looks very different at different observational wavelengths.

Alternatively, students should read the Student Book text on The Sun as a star.

Develop the learning episode

Students should then work in pairs on Worksheet P5.b(1). Students should use the copy of the Hertzsprung-Russell diagram to plot the position of the Sun and to label the four quadrants of the diagram (this is beyond the requirements of the syllabus but may allow students to explore the content more deeply). For some students this is a good opportunity to compare power scales to linear scales. Point out that the Sun is a stable star presently, but that other stars on the diagram are at different stages in their life cycle and are no longer stable. This idea will be revisited in Learning episode P5b.3.

Relate the diagram showing the spectrum of the Sun back to the pictures of the Sun at different wavelengths from the NASA SDO website. Point out that the spectrum of the Sun extends beyond the traditional seven colours of the visible spectrum. Revisit work from Topic P3b *Light* and Topic P3c *Electromagnetic spectrum* to link the ultraviolet and infrared regions to the regions of the electromagnetic spectrum in order of wavelength.

Students should work through the text about the time taken for light to travel across the Solar System. They should do the worked example without the solution and then compare their solution with that given in the Student Book.

With all students, discuss that the Sun is a medium mass star consisting mostly of hydrogen and helium.

Supplement Summarise the process of nuclear fusion in the Sun. Note: the process shown in the Student Book is a simplification of the actual process (called the proton-proton chain). Keen students, and particularly those interested in studying Physics at a higher level should be encouraged to research this process. Students should attempt the questions in the Student Book.

Supplement Students then work through worksheet P5b.1(2).

Finish the learning episode

Students should produce a small mind-map of the key points of this learning episode.

Learning episode P5b.2 Gravity and orbits

Learning objectives

- Know that the Sun contains most of the mass of the Solar System and this explains why the planets orbit the Sun
- Know that the force that keeps an object in orbit around the Sun is due to the gravitational attraction of the Sun
- Supplement Define orbital speed from the equation:

  $v=2\pi rT$

  where r is the radius of the orbit and T is the orbital period; recall and use this equation
- Supplement Know that the strength of the Sun's gravitational field decreases and that the orbital speeds of the planets decrease as the distance from the Sun increases

Common misconceptions

Many students find it difficult to understand the concept of mutual gravitational attraction. Gravitational attraction occurs between two masses, and the same force is exerted on each object, however, if one object is substantially more massive than the other, the object with the smaller mass will appear to move more.

Resources

Student Book The role of gravity

Worksheet P5b.2.1 Gravity and the Solar System

Supplement Worksheet P5b.2.2 Orbits 1

Supplement Worksheet P5b.2.3 Orbits 2

Resources for (optional) demonstration (see Technician's notes, following)

Approach

Introduce the learning episode

There are many excellent orrery simulations online.

You should use this simulation to illustrate the relative motions of some of the objects in the Solar System. If you choose the 'Solar System with dwarf planets' scenario, you will be able to show the students the elliptical nature of some of the orbits, and how the orbital speed varies with distance away from the Sun. You should ask students to work in pairs jotting down as many key points about the model and motions as possible. Alternatively, you could show students a mechanical orrery model if you have one.

Develop the learning episode

You initially need to discuss with the students the role of gravity as an attractive force between masses and ask the students to suggest the factors that affect the size of the gravitational force between two objects. Once the students have correctly identified mass as a factor you should then get the students to read 'The role of gravity' in the Student Book. Summarise that the force required to keep an object in orbit around the Sun is provided by the gravitational force acting on the object. Explain that while the Earth and the Sun are both attracted to each other, the Sun is much more massive (only about 0.14% of the Solar System's mass is found in the planets) and this is why planets move around the Sun and not the Sun around a planet. With some students you may wish to ask them to work through Worksheet P5b.2(1) Gravity and the Solar System. Note however that the relationship between surface gravitational field strength and mass, and the variation of gravitational field strength with distance from Earth's surface both extend beyond the syllabus requirements.

Supplement You should then run through the orbital speed equation in the Student Book, reminding students that they need to learn and recall the equation. Get students to go through the worked example and then attempt the question in the Student Book. You need to reinforce the fact that the equation only applies for orbits that are approximately circular. Students should then complete Worksheet P5b.2(2) Orbits 1. Note that the geometry of elliptical orbits extends beyond the syllabus requirements.

Supplement Ask the students to then read the 'Why orbital speed varies' section in the Student Book, then work in pairs to complete the questions that follow, followed by Worksheet P5b.2(3) Orbits 2.

Finish the learning episode

You should finish by asking students to make a mind-map of the key points from the learning objectives for this learning episode.

Technician's notes

The following resources may be needed:

(Optional) Orrery e.g., Geosafari Motorized Solar System	

Learning episode P5b.3 Stars, galaxies and the Universe

Learning objectives

- Know that stable stars are formed as protostars from interstellar clouds of gas and dust due to gravitational attraction
- Know that the next stages of the life cycle of a star depend on its mass, limited to:
  (a) a small mass star (about the same mass as the Sun): red giant → white dwarf + planetary nebula
  (b) a large mass star: red supergiant → supernova → neutron star
  (c) a very large mass star: red supergiant → supernova → black hole
- Supplement Know that the nebula from a supernova may form new stars with orbiting planets
- Know that:
  (a) galaxies are each made up of many billions of stars
  (b) the Sun is a star in the galaxy known as the Milky Way
  (c) other stars that make up the Milky Way are much further away from the Earth than the Sun is from the Earth
- Know that the Milky Way is one of many billions of galaxies making up the Universe and that the diameter of the Milky Way is approximately 100 000 light-years
- Supplement Know that the Big Bang Theory explains many astronomical observations and states that
  (a) the Universe expanded from a single point of high density and temperature
  (b) the Universe is still expanding
  (c) the Universe is approximately 13.8 billion years old

Resources

Student Book Life cycle of a star; Galaxies; The Universe

Worksheet P5b.3(1) Stellar lifecycles

Worksheet P5b.3(2) Stars, galaxies and distances

Approach

Introduce the learning episode

Start the learning episode by asking students to write down their address and getting them to start to think about larger scales, such as continents and the planet and so on. How many steps to their address can they add? You could alternatively show students various animations showing the scale of the Universe – a good place to start is: https://scaleofuniverse.com/, but there are many alternatives. Once any of the animations/videos are shown, ask the students to revisit their address – is there anything they could add?

Develop the learning episode

Students should then read the section on Life cycle of stars in the Student Book.

Students should then work in pairs to complete Worksheet P5b.3(1). Note that one of the cards relates only to the supplement content, and could be omitted for some students. Students should finish this activity by completing the questions in the Student Book.

Remind students of the scale of the Universe from the introduction, and ask them to recall the unit for measuring astronomical distances such as the distance between stars. Use the text and images in the Galaxies section of the Student Book to discuss what is meant by a galaxy, and the size of the Milky Way.

Students should then work in pairs to complete Worksheet P5b.3(2). Some students may need support using standard form in the calculation in question 5.

Students should finish this activity by completing questions in the Student Book.

Supplement Students should then read and discuss the material about the Big Bang Theory in the Student Book.

Finish the learning episode

Students could read Science in Context: White dwarfs in the Student Book and think about the question given at the end of the section (this is beyond the requirements of the syllabus but may allow students to explore the content more deeply). (This is similar to Olbers' Paradox 'Why is the night sky dark?') There are many references to Olbers' Paradox on the internet.

Technician's Notes

The following resources are needed for P5b.3(1) Stellar life cycles

scissors	
glue	
sheets of plain paper	

Learning episode P5b.4 Consolidation and summary

Learning aims

- Review the learning points of the topic summarised in the End of topic checklist
- Test understanding of the topic content by answering the End of topic questions

Resources

Student Book End of topic checklist and End of topic questions

Approach

Introduce the learning episode

Ask students to work with a partner to make a list of key words from this topic. They could then work together to produce a spider diagram showing how the different concepts are linked. They could compare their list with the list of key terms given in the Student Book. Discuss the checklist in the Student Book and ask questions to see how much of the content students are comfortable with.

Develop the learning episode

Ask students to work individually through the End of topic questions in the Student Book without looking at the text. As they work, walk around the classroom observing their answers and

questioning them as necessary to find out which questions are causing difficulties.

Finish the learning episode

After a set period, ask the students to stop working and discuss any areas of difficulty you observed as you walked round the class. Students should complete any unanswered questions for homework, but you should stress that they should try to answer the questions without looking at the text, so that they can see how much they have remembered.

Physics Teaching sequence

The course can be taught in syllabus order and it will work well to do so. If you want to change the teaching sequence for the course, you should keep in mind the considerations below. An example alternative teaching sequence has been provided.

- There should be a logical progression in the development of concepts and knowledge in order to aid students' understanding.
- The course should have a 'storyline' so that wherever possible the sequence can be easily explained to students.
- Consider carefully how to introduce the more difficult concepts. For example, ensure that they are not introduced too early in the course, which may demotivate students, or put in a sequence where they all just follow one after another.
- Maintain the emphasis on practical and investigative work throughout the course.
- You will need to allocate different amounts of teaching time to different topics, depending on levels of difficulty or extent of the content.
- It is important to retain some flexibility so that you can revisit difficult ideas, or apply new ideas in different contexts.
- Some topics may need consideration of geographical location of teaching, for example if students are encouraged to view the night sky.
- You will need time for revision at the end of the course. It is also recommended that you encourage students to revisit material for revision throughout the course. This could be done using the flashcards that they create at the end of each topic.

Alternative teaching sequence

This example alternative teaching sequence is based on a two-year course taught in six terms. It is mapped over five terms on the assumption that much of the sixth term is spent on revision. For different numbers of terms or semesters, where the course is started early, or where the time allocated each week is either generous or limited, you will need to adjust this sequence.

First year, 1st term

P1a Physical quantities and measurement techniques
P1b Motion
P1c Mass and weight
P1f Energy, work and power

First year, 2nd term

P3a General properties of waves
P3b Light
P3c Electromagnetic spectrum
P1d Density
P1g Pressure
P4a Electrical quantities

First year, 3rd term

P1e Forces
P5a The Solar System

Second year, 1st term

P2a Kinetic particle model of matter
P2b Thermal properties and temperature
P2c Transfer of thermal energy
P3d Sound

Second year, 2nd term

P4b Electrical circuits
P4c Electrical safety
P4d Cells, batteries, generators and motors
P5b Stars and the Universe

Physics Scheme of work

Learning episode	**Learning objectives and learning aims** Learning objectives are given which link to the syllabus. Learning aims reflect the intended outcomes of the consolidation and summary learning episodes.	**Syllabus reference**	**Student's Book heading/ subheading**
P1a.1 Length and time	Learning objectives • Describe the use of rulers and measuring cylinders to find a length or a volume. • Describe how to measure a variety of time intervals using clocks and digital timers. • Determine an average value for a small distance and for a short interval of time by measuring multiples (including the period of oscillation of a pendulum).	P1.1	Making measurements
P1a.2 Consolidation and summary	Learning aims • Review the learning points of the topic summarised in the End of topic checklist. • Test understanding of the topic content by answering the End of topic questions.	P1.1	End of topic checklist and End of topic questions
P1b.1 Speed and graphs of motion	Learning objectives • Define speed as distance travelled per unit time; recall and use the equation $v = st$. • Recall and use the equation average speed = total distance travelledtotal time taken. • Sketch, plot and interpret distance-time and speed-time graphs. • Know that an object moving with increasing speed is accelerating, and that an object moving with decreasing speed is decelerating. • Determine, qualitatively from the shape of a distance−time graph when an object is: (a) at rest (b) moving with constant speed. • Determine, qualitatively from the shape of a speed−time graph when an object is: at rest, moving with constant speed, accelerating, decelerating. • Calculate speed from the gradient of a straight-line section of a distance–time graph. • Supplement Calculate the area under a speed−time graph to work out the distance travelled for motion with (a) constant speed (b) constant acceleration.	P1.2	Calculating speed; Using graphs to study motion; Speed–time graph

Learning episode	**Learning objectives and learning aims** Learning objectives are given which link to the syllabus. Learning aims reflect the intended outcomes of the consolidation and summary learning episodes.	**Syllabus reference**	**Student's Book heading/ subheading**
P1b.2 Acceleration	Learning objectives • Supplement Define acceleration for an object moving in a straight line as change in speed per unit time; recall and use the equation $a = \Delta v \Delta t$. • Supplement Determine from given data or the shape of a speed–time graph when an object is moving with constant acceleration. • Supplement Calculate acceleration from the gradient of a straight-line section of a speed–time graph. • Supplement Know that deceleration is a negative acceleration and use this in calculations. • Supplement Know that the acceleration of free fall g for an object near to the surface of the Earth is approximately constant and is approximately 9.8 m/s^2.	P1.2	Acceleration; Falling objects
P1b.3 Consolidation and summary	Learning aims • Review the learning points of the topic summarised in the End of topic checklist. • Test understanding of the topic content by answering the End of topic questions.	P1.2	End of topic checklist and End of topic questions
P1c.1 Mass and weight	Learning objectives • Know that mass is a measure of the quantity of matter in an object. • Know that weight is the gravitational force on an object that has mass. • Define gravitational field strength, g, as the gravitational force per unit mass; recall and use the equation $g = Wm$ and know that near the surface of the Earth g is approximately 9.8 N/kg. • Supplement Describe, and use the concept of, weight as the effect of a gravitational field on a mass. • Supplement Know gravitational field strength is equivalent to the acceleration of free fall.	P1.3	Mass and weight; The Earth's gravitational field
P1d.1 Density	Learning objectives • Define density as mass per unit volume; recall and use the equation $\rho = mV$. • Describe how to determine the density of a liquid, of a regularly shaped solid and of an irregularly shaped solid which sinks in a liquid (volume by displacement), including appropriate calculations. • Determine whether an object floats or sinks based on density data.	P1.4	Density
P1d.2 Consolidation and summary	Learning aims • Review the learning points of the topic summarised in the End of topic checklist. • Test understanding of the topic content by answering the End of topic questions.	P1.4	End of topic checklist and End of topic questions

Learning episode	Learning objectives and learning aims Learning objectives are given which link to the syllabus. Learning aims reflect the intended outcomes of the consolidation and summary learning episodes.	Syllabus reference	Student's Book heading/ subheading
P1e.1 Introducing forces	Learning objectives • Know that forces may produce changes in the size, shape, and motion of an object. • Describe friction as the force between two surfaces that may impede motion and produce heating. • Know that friction (drag) acts on an object moving through a liquid. • Know that friction (drag) acts on an object moving through a gas (e.g., air resistance). • Know that an object either remains at rest or continues in a straight line at constant speed unless there is a resultant force on the object.	P1.5	What are forces?; What is friction?
P1e.2 Resultant forces	Learning objectives • Determine the resultant of two or more forces acting along the same straight line. • Know that an object either remains at rest or continues in a straight line at constant speed unless there is a resultant force on the object.	P1.5	Adding forces
P1e.3 Forces and acceleration	Learning objectives • Supplement Recall and use the equation $F = ma$ and know that the resultant force and the acceleration are in the same direction.	P1.5	How are mass, force and acceleration related?
P1e.4 Consolidation and summary	Learning aims • Review the learning points of the topic summarised in the End of topic checklist. • Test understanding of the topic content by answering the End of topic questions.	P1.5	End of topic checklist and End of topic questions
P1f.1 Energy stores and energy transfers	Learning objectives • State that energy may be stored as kinetic, gravitational potential, chemical, elastic (strain), nuclear, electrostatic, and internal (thermal). • Describe how energy is transferred between stores during events and processes, including examples of transfer by forces (mechanical work done), electrical currents (electrical work done), heating, and by electromagnetic, sound, and other waves. • Know the principle of the conservation of energy and apply this principle to simple examples including the interpretation of simple flow diagrams (Sankey diagrams are **not** required). • Supplement Recall and use the equation for kinetic energy $E_k = \frac{1}{2}mv^2$. • Supplement Recall and use the equation for the change in gravitational potential energy $\Delta E_p = mg\Delta h$.	P1.6.1	Energy stores; Energy transfers

Learning episode	**Learning objectives and learning aims** Learning objectives are given which link to the syllabus. Learning aims reflect the intended outcomes of the consolidation and summary learning episodes.	**Syllabus reference**	**Student's Book heading/ subheading**
P1f.2 Energy resources	Learning objectives • Describe how useful energy may be obtained, or electrical power generated, from: (a) fossil fuels; (b) biofuels; (c) water, including waves, tides, and hydroelectric dams; (d) geothermal resources; (e) nuclear fission; (f) light from the Sun (solar cells); (g) infrared and other electromagnetic waves from the Sun to heat water (solar thermal collectors); (h) wind (wind turbines) including references to a boiler, turbine and generator where they are used. • Supplement Know that radiation from the Sun in the main source of energy for all our energy resources except geothermal, nuclear, and tidal. • Supplement Know that energy is released by nuclear fusion in the Sun (detailed knowledge of the process of fusion is **not** required). • Supplement Know that energy is released by nuclear fission in nuclear reactors (detailed knowledge of the process of fission is **not** required).	P1.6.3	Energy resources
P1f.3 Work	Learning objectives • Understand that mechanical or electrical work done is equal to the energy transferred. • Recall and use the equation for mechanical working $W = Fd = \Delta E$.	P1.6.2	Work
P1f.4 Power	Learning objectives • Define power as work done per unit time and also as energy transferred per unit time. • Recall and use the equations: (a) $P = Wt$ (b) $P = E t$.	P1.6.4	Power
P1f.5 Efficiency	Learning objectives • Understand, qualitatively, the concept of efficiency of energy transfer. • Supplement Define efficiency as: (a) (%) efficiency = (useful energy output)(total energy input) × (100%) (b) (%) efficiency = (useful power output)(total power input) × (100%) • Supplement Recall and use these equations.	P1.6.3	Efficiency
P1f.6 Consolidation and summary	Learning aims • Review the learning points of the topic summarised in the End of topic checklist. • Test understanding of the topic content by answering the End of topic questions.	P1.6	End of topic checklist and End of topic questions

Learning episode	**Learning objectives and learning aims** Learning objectives are given which link to the syllabus. Learning aims reflect the intended outcomes of the consolidation and summary learning episodes.	**Syllabus reference**	**Student's Book heading/ subheading**
P1g.1 Pressure, force and area	Learning objectives • Describe how pressure varies with force and area in the context of everyday examples. • Define pressure as force per unit area; recall and use the equation $p = FA$.	P1.7	Measuring pressure
P1g.2 Consolidation and summary	Learning aims • Review the learning points of the topic summarised in the End of topic checklist. • Test understanding of the topic content by answering the End of topic questions.	P1.7	End of topic checklist and End of topic questions
P2a.1 States of matter and the particle model	Learning objectives • State the distinguishing properties of solids, liquids, and gases. • Know the terms for the changes in state between solids, liquids, and gases (gas to solid and solid to gas transfers are **not** required). • Describe the structure of solids, liquids, and gases in terms of the arrangement, separation, and motion of the particles, and represent these states using simple particle diagrams. • Describe the relationship between the motion of particles and temperature. • Supplement Know that the forces and distances between particles and the motion of the particles affect the properties of solids, liquids, and gases.	P2.1.1, P2.1.2	States of matter; Model of matter; Understanding gases
P2a.2 Gas pressure and pressure changes	Learning objectives • Supplement Describe the pressure of a gas in terms of the forces exerted by particles colliding with surfaces, creating a force per unit area. • Supplement Describe qualitatively, in terms of particles, the effect on the pressure of a fixed mass of gas of: (a) a change of temperature at constant volume and (b) a change of volume at constant temperature.	P2.1.2, P2.1.3	Understanding gases; Pressure changes
P2a.3 Consolidation and summary	Learning aims • Review the learning points of the topic summarised in the End of topic checklist. • Test understanding of the topic content by answering the End of topic questions.	P2.1	End of topic checklist and End of topic questions
P2b.1 Thermal expansion of solids, liquids and gases	Learning objectives • Describe, qualitatively, the thermal expansion of solids, liquids and gases at constant pressure. • Supplement Explain some of the everyday applications and consequences of thermal expansion.	P2.2.1	Thermal expansion of solids, liquids and gases

Learning episode	**Learning objectives and learning aims** Learning objectives are given which link to the syllabus. Learning aims reflect the intended outcomes of the consolidation and summary learning episodes.	**Syllabus reference**	**Student's Book heading/ subheading**
P2b.2 Evaporation	Learning objectives • Describe evaporation in terms of the escape of more energetic particles from the surface of a liquid. • Know that evaporation causes cooling of a liquid. • Supplement Describe how temperature, surface area and air movement over a surface affect evaporation.	P2.2.2	Evaporation
P2b.3 Consolidation and summary	Learning aims • Review the learning points of the topic summarised in the End of topic checklist. • Test understanding of the topic content by answering the End of topic questions.	P2.2	End of topic checklist and End of topic questions
P2c.1 Conduction, convection and radiation	Learning objectives • Identify and give examples of typically good thermal conductors and bad thermal conductors (thermal insulators). • Supplement Describe thermal conduction in solids in terms of atomic or molecular lattice vibrations and also in terms of the movement of delocalised (mobile) electrons in metallic conductors. • Know that convection is an important method of energy transfer in liquids and gases. • Describe convection in liquids and gases. • Supplement Explain convection in liquids and gases in terms of density changes. • Know that thermal energy transfer by thermal radiation does not require a medium and is mainly due to infrared radiation. • Describe the effect of surface colour (black or white) and texture (dull or shiny) on the emission, absorption and reflection of thermal radiation. • Supplement Know that the temperature of the Earth is affected by the radiation absorbed by the Earth and the radiation emitted by the Earth. • Supplement Describe experiments to distinguish between good and bad emitters of thermal radiation. • Supplement Describe experiments to distinguish between good and bad absorbers of thermal radiation.	P2.3.1, P2.3.2, P2.3.3	Conduction; Convection; Radiation
P2c.2 Examples of transfers of thermal energy	Learning objectives • Identify and explain some of the basic everyday applications and consequences of conduction, convection and radiation.	P2.3.4	Examples of transfers of thermal energy

Learning episode	**Learning objectives and learning aims** Learning objectives are given which link to the syllabus. Learning aims reflect the intended outcomes of the consolidation and summary learning episodes.	**Syllabus reference**	**Student's Book heading/ subheading**
P2c.3 Consolidation and summary	Learning aims • Review the learning points of the topic summarised in the End of topic checklist. • Test understanding of the topic content by answering the End of topic questions.	P2.3	End of topic checklist and End of topic questions
P3a.1 General properties of waves	Learning objectives • Know that waves transfer energy without transferring matter. • Describe what is meant by wave motion as illustrated by vibration (oscillation) in ropes and springs, and by experiments using water waves. • Describe the features of a wave in terms of wavelength, frequency, crest (peak), trough, amplitude and wave speed. • Supplement Know that for a transverse wave, the direction of vibration is at right angles to the direction of propagation and understand that electromagnetic radiation, water waves and seismic S-waves (secondary) are transverse. • Supplement Know that for a longitudinal wave, the direction of vibration is parallel to the direction of propagation and understand that sound waves and seismic P-waves (primary) are longitudinal.	P3.1	Understanding waves; Features of a wave
P3a.2 The wave equation	Learning objectives • Recall and use the equation for wave speed $v = f\lambda$.	P3.1	Relationship between speed, frequency and wavelength
P3a.3 Water waves	Learning objectives • Describe how waves can undergo: reflection at a plane surface; refraction due to a change of speed.	P3.1	Using water waves to show reflection and refraction
P3a.4 Consolidation and summary	Learning aims • Review the learning points of the topic summarised in the End of topic checklist. • Test understanding of the topic content by answering the End of topic questions.	P3.1	End of topic checklist and End of topic questions

Learning episode	**Learning objectives and learning aims** Learning objectives are given which link to the syllabus. Learning aims reflect the intended outcomes of the consolidation and summary learning episodes.	**Syllabus reference**	**Student's Book heading/ subheading**
P3.b1 Reflection of light	Learning objectives • Use ray diagrams to define the terms normal, angle of incidence and angle of reflection. • Describe the formation of an optical image by a plane mirror and give its characteristics compared with the object, i.e., same size, same distance from mirror, laterally inverted. • State that for reflection, the angle of incidence is equal to the angle of reflection; recall and use this relationship. • Supplement Describe the formation of an optical image by a plane mirror and explain why it is virtual. • Supplement Use simple diagrams, measurements, and calculations for reflection by plane mirrors.	P3.2.1	Reflection of light
P3b.2 Refraction of light	Learning objectives • Define refraction as the change in direction of a light ray passing from one medium to another. • Define and use the terms normal, angle of incidence and angle of refraction using ray diagrams. • Describe the passage of light through a transparent material (limited to the boundaries between two media only).	P3.2.2	Refraction of light
P3b.3 Thin converging lenses and dispersion of light	Learning objectives • Describe the action of thin converging lenses on a parallel beam of light, and know that rays of light from an object at distance can be assumed to be parallel. • Define and use the terms principal axis, principal focus (focal point) and focal length. • Interpret ray diagrams for the formation of an image by a thin converging lens, limited to real images. • Describe the characteristics of an image using the terms enlarged/same size/diminished, and upright/inverted. • Describe the dispersion of light as illustrated by the refraction of white light by a glass prism. • Know the traditional seven colours (red, orange, yellow, green, blue, indigo, violet) of the visible spectrum in order of frequency and in order of wavelength.	P3.2.3, P3.2.4	Thin converging lenses; Dispersion of light
P3b.4 Consolidation and summary	Learning aims • Review the learning points of the topic summarised in the End of topic checklist. • Test understanding of the topic content by answering the End of topic questions.	P3.2	End of topic checklist and End of topic questions

Learning episode	**Learning objectives and learning aims** Learning objectives are given which link to the syllabus. Learning aims reflect the intended outcomes of the consolidation and summary learning episodes.	**Syllabus reference**	**Student's Book heading/ subheading**
P3c.1 Electromagnetic spectrum	Learning objectives • Know the main regions of the electromagnetic spectrum (radio, microwave, infrared, visible, ultraviolet, x-ray, gamma) in order of frequency and in order of wavelength. • Know that all electromagnetic waves travel with the same high speed in a vacuum and at approximately the same speed in air. • Know some applications of the different regions of the electromagnetic spectrum including: (a) radio waves; radio and television transmissions, radar (b) microwaves; satellite television, mobile (cell) phone, microwave ovens (c) infrared; remote controllers for televisions, thermal imaging (d) visible light; vision, photography (e) ultraviolet; detecting fake bank notes (f) X-rays; medical scanning, security scanners (g) gamma rays; detection of cancer and its treatment. • Describe the harmful effects on people of excessive exposure to electromagnetic radiation, including: (a) ultraviolet; damage to surface cells and eyes, leading to skin cancer and eye conditions (b) X-rays and gamma rays; mutation or damage to cells in the body. • Supplement Know that the speed of electromagnetic waves in a vacuum is 3.0×10^8 m/s.	P3.3	Electromagnetic waves; Uses of electromagnetic waves; Harmful effects of electromagnetic waves
P3c.2 Consolidation and summary	Learning aims • Review the learning points of the topic summarised in the End of topic checklist. • Test understanding of the topic content by answering the End of topic questions.	P3.3	End of topic checklist and End of topic questions
P3d.1 Properties of sound	Learning objectives • Describe the production of sound by vibrating sources. • Supplement Describe the longitudinal nature of sound waves in air as a series of compressions and rarefactions. • Supplement Describe qualitatively compressions as regions of higher pressure due to particles being closer together and rarefactions as regions of lower pressure due to particles being spread further apart. • State the approximate range of frequencies audible to humans as 20 Hz to 20 000 Hz. • Know that a medium is needed to transmit sound waves. • Describe how changes in amplitude and frequency affect the loudness and pitch of sound waves. • Describe an echo as the reflection of sound waves • Define ultrasound as sound with a frequency higher than 20 kHz.	P3.4	The basics of sound; Pitch, frequency, amplitude and loudness; Echoes

Learning episode	**Learning objectives and learning aims** Learning objectives are given which link to the syllabus. Learning aims reflect the intended outcomes of the consolidation and summary learning episodes.	**Syllabus reference**	**Student's Book heading/ subheading**
P3d.2 Measuring the speed of sound	Learning objectives • Determine the speed of sound in air using a method involving a measurement of distance and time. • Supplement Know that, in general, sound travels faster in solids than in liquids and faster in liquids than in gases.	P3.4	Examples of transfers of thermal energy
P3d.3 Consolidation and summary	Learning aims • Review the learning points of the topic summarised in the End of topic checklist. • Test understanding of the topic content by answering the End of topic questions.	P3.4	End of topic checklist and End of topic questions
P4a.1 Electric charge	Learning objectives • State that there are positive and negative charges. • State that positive charges repel other positive charges, negative charges repel other negative charges, but positive charges attract negative charges. • Distinguish between electrical conductors and insulators and give typical examples. • Supplement State that charge is measured in coulombs.	P4.1.1	Electric charge
P4a.2 Electric current	Learning objectives • Know that electric current is related to the flow of charge. • Know that electric current in metals is related to the flow of electrons. • Describe the use of ammeters (analogue and digital) with different ranges. • Know the difference between direct current (d.c.) and alternating current (a.c.). • Supplement Describe electrical conduction in metals in terms of the movement of delocalised (mobile) electrons. • Supplement Define electric current as the charge passing a point per unit time. • Supplement Recall and use the equation $I = Qt$. • Supplement State that conventional current is from positive to negative and the flow of electrons is from negative to positive.	P4.1.2	Delocalised (mobile) electrons; Equation for current; Alternating current and direct current; Measuring current

Learning episode	**Learning objectives and learning aims** Learning objectives are given which link to the syllabus. Learning aims reflect the intended outcomes of the consolidation and summary learning episodes.	**Syllabus reference**	**Student's Book heading/ subheading**
P4a.3 Voltage (electromotive force and potential difference)	Learning objectives • Describe the voltage of a source as the cause of current in a circuit. • Know that the voltage of the source is shared between the components in a series circuit. • Describe the use of voltmeters (analogue and digital) with different ranges. • Supplement Define electromotive force (e.m.f.) as the electrical work done by a source in moving a unit charge around a complete circuit. • Supplement Know that e.m.f. is measured in volts (V). • Supplement Define potential difference (p.d.) as the work done by a unit charge passing through a component. • Supplement Know that the p.d. between two points is measured in volts.	P4.1.3	Voltage (electromotive force and potential difference); Measuring voltage
P4a.4 Resistance	Learning objectives • Recall and use the equation for resistance, $R = VI$. • Supplement Recall and use the following relationship for a metallic electrical conductor: resistance is directly proportional to length; resistance is inversely proportional to cross-sectional area.	P4.1.4	Resistance
P4a.5 Electrical energy and electrical power	Learning objectives • Understand that electric circuits transfer energy from a source of electrical energy, such as an electrical cell or mains supply, to the circuit components and then into the surroundings. • Recall and use the equation for electrical power $P = IV$. • Recall and use the equation for electrical energy $E = IVt$. • Define the kilowatt-hour (kW h) and calculate the cost of using electrical appliances where the energy unit is the kW h.	P4.1.5	Electrical energy and electrical power; The kilowatt-hour
P4a.6 Consolidation and summary	Learning aims • Review the learning points of the topic summarised in the End of topic checklist. • Test understanding of the topic content by answering the End of topic questions.	P4.1	End of topic checklist and End of topic questions

Learning episode	**Learning objectives and learning aims** Learning objectives are given which link to the syllabus. Learning aims reflect the intended outcomes of the consolidation and summary learning episodes.	**Syllabus reference**	**Student's Book heading/ subheading**
P4b.1 Circuit diagrams and circuit components	Learning objectives • Draw and interpret circuit diagrams containing cells, batteries, power supplies, switches, resistors (fixed and variable), heaters, lamps, motors, ammeters, voltmeters and fuses, and know how these components behave in the circuit. • Supplement Draw and interpret circuit diagrams containing generators and light emitting diodes (LEDs), and know how these components behave in the circuit. • Know that the current at every point in a series circuit is the same. • Supplement Recall and use in calculations, the fact that: a) the sum of the currents entering a junction in a parallel circuit is equal to the sum of the currents that leave the junction b) the total p.d. across the components in a series circuit is equal to the sum of the individual p.d.s across each component (c) the p.d. across each branch of a parallel arrangement of components is the p.d. across the whole arrangement.	P4.2.1, P4.2.2	Circuit diagrams and circuit components
P4b.2 Series and parallel circuits	Learning objectives • Draw and interpret circuit diagrams containing cells, batteries, power supplies, switches, resistors (fixed and variable), lamps, ammeters, voltmeters, and know how these components behave in the circuit. • Know that the current at every point in a series circuit is the same. • Know how to construct and use series and parallel circuits. • Calculate the combined resistance of two or more resistors in series. • Know that, for a parallel circuit, the current from the source is larger than the current in each branch • Know that the combined resistance of two resistors in parallel is less than that of either resistor by itself. • Know the advantages of connecting lamps in parallel in a lighting circuit. • Supplement Recall and use in calculations, the fact that: (a) the sum of the currents entering a junction in a parallel circuit is equal to the sum of the currents that leave the junction (b) the total p.d. across components in a series circuit is equal to the sum of the individual p.d.s across each component (c) the p.d. across each branch of a parallel arrangement of components is the p.d across the whole arrangement. • Supplement Calculate the combined resistance of two resistors in parallel.	P4.2.2	Series and parallel circuits

Learning episode	**Learning objectives and learning aims** Learning objectives are given which link to the syllabus. Learning aims reflect the intended outcomes of the consolidation and summary learning episodes.	**Syllabus reference**	**Student's Book heading/ subheading**
P4b.3 Behaviour of some components	Learning objectives • Draw and interpret circuit diagrams containing cells, batteries, power supplies, switches, resistors (fixed and variable), heaters, lamps, motors, ammeters, voltmeters and fuses, and know how these components behave in the circuit. • Supplement Draw and interpret circuit diagrams containing generators and light emitting diodes (LEDs), and know how these components behave in the circuit.	P4.2.1	Characteristics and use of some circuit components
P4b.4 Consolidation and summary	Learning aims • Review the learning points of the topic summarised in the End of topic checklist. • Test understanding of the topic content by answering the End of topic questions.	P4.2	End of topic checklist and End of topic questions
P4c.1 Electrical hazards and domestic appliances	Learning objectives • Understand the heating effect of current. • State the hazards of (a) damaged insulation, (b) overheating cables, (c) damp conditions, (d) excess current from overloading of plugs, extension leads, single and multiple sockets when using a mains supply. • Explain the use and operation of trip switches and fuses and choose appropriate fuse ratings and trip switch settings (knowledge of RCDs (Residual Current Devices) is **not** required). • Explain why the outer casing of an electrical appliance must be either non-conducting (double-insulated) or earthed.	P4.3	Electrical hazards
P4c.2 Consolidation and summary	Learning aims • Review the learning points of the topic summarised in the End of topic checklist. • Test understanding of the topic content by answering the End of topic questions.	P4.3	End of topic checklist and End of topic questions
P4d.1 Cells, batteries, generators and motors	Learning objectives • Know that cells and batteries transfer chemical energy into electrical energy. • Know that generators transfer kinetic energy into electrical energy. • Know that electric motors transfer electrical energy into kinetic energy.	P4.4	Cells and batteries; Generators; Electric motors
P4d.2 Consolidation and summary	Learning aims • Review the learning points of the topic summarised in the End of topic checklist. • Test understanding of the topic content by answering the End of topic questions.	P4.4	End of topic checklist and End of topic questions

Learning episode	**Learning objectives and learning aims** Learning objectives are given which link to the syllabus. Learning aims reflect the intended outcomes of the consolidation and summary learning episodes.	**Syllabus reference**	**Student's Book heading/ subheading**
P5a.1 The Solar System	Learning objectives • Describe the Solar System as containing: (a) one star, the Sun (b) the eight named planets and know their order from the Sun (c) minor planets that orbit the Sun, including dwarf planets such as Pluto and asteroids in the asteroid belt (d) moons, that orbit the planets.	P5.1.1	The Solar System
P5a.2 Consolidation and summary	Learning aims • Review the learning points of the topic summarised in the End of topic checklist. • Test understanding of the topic content by answering the End of topic questions.	P5.1.1	End of topic checklist and End of topic questions
P5b.1 The Sun as a star	Learning objectives • Know that the Sun is a star of medium size, consisting mostly of hydrogen and helium, and that it radiates most of its energy in the infrared, visible and ultraviolet regions of the electromagnetic spectrum. • Know that: (a) the Sun is the closest star to the Earth (b) astronomical distances can be measured in light-years, where one light-year is the distance travelled in (the vacuum of) space by light in one year. • Calculate the time it takes light to travel a significant distance such as between objects in the Solar System. • Supplement Know that stars are powered by nuclear reactions that release energy and that in stable stars the nuclear reactions involve the fusion of hydrogen into helium.	P5.2.1	The Sun as a star; Light-years; Time taken by light to travel across the Solar System; Nuclear fusion
P5b.2 Gravity and orbits	Learning objectives • Know that the Sun contains most of the mass of the Solar System and this explains why the planets orbit the Sun. • Know that the force that keeps an object in orbit around the Sun is due to the gravitational attraction of the Sun. • Supplement Define orbital speed from the equation: $v=2\pi rT$ where r is the radius of the orbit and T is the orbital period; recall and use this equation. • Supplement Know that the strength of the Sun's gravitational field decreases and that the orbital speeds of the planets decrease as the distance from the Sun increases.	P5.2.1	The role of gravity

Learning episode	**Learning objectives and learning aims** Learning objectives are given which link to the syllabus. Learning aims reflect the intended outcomes of the consolidation and summary learning episodes.	**Syllabus reference**	**Student's Book heading/ subheading**
P5b.3 Stars, galaxies and the Universe	Learning objectives • Know that stable stars are formed as protostars from interstellar clouds of gas and dust due to gravitational attraction. • Know that the next stages of the life cycle of a star depend on its mass, limited to: (a) a small mass star (about the same mass as the Sun): red giant → white dwarf + planetary nebula (b) a large mass star: red supergiant → supernova → neutron star (c) a very large mass star: red supergiant → supernova → black hole. • Supplement Know that the nebula from a supernova may form new stars with orbiting planets. • Know that: (a) galaxies are each made up of many billions of stars (b) the Sun is a star in the galaxy known as the Milky Way (c) other stars that make up the Milky Way are much further away from the Earth than the Sun is from the Earth. • Know that the Milky Way is one of many billions of galaxies making up the Universe and that the diameter of the Milky Way is approximately 100 000 light-years. • Supplement Know that the Big Bang Theory explains many astronomical observations and states that: (a) the Universe expanded from a single point of high density and temperature (b) the Universe is still expanding (c) the Universe is approximately 13.8 billion years old.	P5.2.2, P5.2.3	Life cycle of a star; Galaxies; The Universe
P5b.4 Consolidation and summary	Learning aims • Review the learning points of the topic summarised in the End of topic checklist. • Test understanding of the topic content by answering the End of topic questions.	P5.2	End of topic checklist and End of topic questions

